MUNICIPAL WASTE
DISPOSAL
in the 1990s

MUNICIPAL WASTE DISPOSAL
in the 1990s

Béla G. Lipták

CHILTON BOOK COMPANY

Radnor, Pennsylvania

Library of Congress Cataloging in Publication Data

Lipták, Béla.
    Municipal waste disposal in the 1990s / Béla Lipták.
        p.   cm.
    Includes bibliographical references and index.
    ISBN 0-8019-7867-X
    1. Refuse and refuse disposal.   2. Recycling (Waste, etc.)
    I. Title.
    TD791.L57   1991
    628.4—dc20   90-55321
                                                     CIP

1 2 3 4 5 6 7 8 9 0   0 9 8 7 6 5 4 3 2 1

To my beloved children
Ágnes and *Ádam*, and to Ádam's wife, *Jennifer*
And also to my nephews, nieces and god-children
Aron, Caitlin, Noémi, Tamás,
Tibor, Tímea and *Yvonne*

We do not inherit the earth from our parents.
We borrow it from our children.
—Native American saying

Contents

Chapter 3   **Hazardous Waste Disposal**   **46**

Chapter 4   **Sewage Sludge Disposal**   **57**

Chapter 5   **The Incineration Process**   **87**

Preface

■ ■

THERE ARE only three ways to handle solid waste: throw it away, reuse it, or do not generate it in the first place. Municipal solid waste, or MSW, can be landfilled or deep-welled, dumped in the ocean, or incinerated into the atmosphere.

Recycling, a newer option, will require cooperation among government, industry, and the average citizen. Ways must be found to include disposal and environmental impact costs in the prices of goods. Industrial products should be labeled if they are "reuseable" or "recyclable." Stable markets for recycled materials should be established by requiring that newsprint, plastics and other products contain a certain percentage of recycled materials. Society as a whole will have to change its buying habits, separate its household wastes, and become more aware of the overall environmental impact of its lifestyle. This will have to come through education.

In addition to recycling and reuse, it is also time to get serious about precycling. Precycling suggests that waste should not be produced unnecessarily in the first place. Examples of such unnecessary wastes are poorly made goods that last only one season and the multiplicity of telephone books, junk mail, and excessive packaging. Government should provide incentives not only to reduce the amount of packaging but also to make it safer and less toxic. In the broader sense governments will have to pay more attention to the longer-range considerations of the sustainability of the life-support systems and to the depletion of resources than to the short-range considerations of conventionally measured economic growth.

While the above goals are highly desirable, the purpose of this book is not to editorialize, but to present information and provide practical answers. This is a practical book that summarizes the worldwide experience of MSW disposal and

xiii

recycling. The book concentrates on what works and what does not, so that the reader can learn from the mistakes of others and also benefit from their successes.

This book is directed toward both the scientifically trained and the nontechnical reader. The technical and scientific information is presented in an easily understandable form, and mathematical derivations, undefined abbrevations, and engineering jargon have been eliminated. Economic, social, biological, and legal aspects are included to give a complete treatment of the problems and potential solutions of MSW disposal.

The book consists of ten chapters. Chapter 1 describes the nature of MSW and the trends affecting its composition and quantity. The next three chapters deal with the more traditional disposal methods: landfills (Chapter 2), hazardous waste disposal (Chapter 3), and sludge disposal, including ocean dumping (Chapter 4). The next four chapters discuss the various aspects of incineration: Chapter 5 describes the incineration process, its controls and accessories; Chapter 6 is devoted to the pollution and health effects of incineration; Chapter 7 outlines the operation and performance capabilities of the different pollution-control devices; and Chapter 8 provides information on the operation and performance of heat-recovery incinerators (HRIs).

The last two chapters deal with waste material reuse and recovery. Chapter 9 covers the broad topic of MSW recycling and discusses the laws, regulations, and market conditions that affect recycling. In addition, it outlines the various strategies for separation at the source and the various approaches to the recycling and reuse of waste paper, plastics, glass, metals, organics, and incinerator residues. Chapter 10 describes two kinds of resource-recovery plants: the traditional refuse-derived fuel (RDF) plants and the more complex materials-recovery facilities (MRFs). The Postscript focuses on the larger environmental issues of which the MSW disposal problem is only one.

A generation ago, nuclear reactors were being built before the costs, risks, and nuclear waste disposal problems were fully understood. The well-deserved consequence was a black eye for the whole nuclear industry. Today, paradoxically, incinerators are being built before their costs, risks, and residue disposal problems have been fully understood or resolved. The purpose of this book is to help prevent the repetition of the nuclear experience. This book provides the information needed to evaluate the MSW disposal options available and to select solutions that will not be regretted by the next generation.

Béla G. Lipták
Stamford, Connecticut

MUNICIPAL WASTE DISPOSAL
DISPOSAL
in the 1990s

Chapter 1
Municipal Solid Waste (MSW)

THIS CHAPTER provides information on the trends affecting solid-waste compositions, quantities, disposal methods, and costs and describes the various solid-waste classifications. Before discussing specific topics, the underlying causes of the solid-waste problem need to be determined.

The vast proliferation of municipal solid waste is a symptom of an overindustrialized, urbanized, and centralized society and of a culture that places a high value on comfort and economic growth. The deeply rooted belief in continuous growth assumes that the planet is inexhaustible and implies an increasing consumption of raw materials and energy. This belief in continuous material growth also assumes both a continuously expanding economy and a technology that serves that economy by conquering, regulating, and controlling nature. Such cultural attitudes result in exploiting nature as a commodity and using the oceans, atmosphere, and land as free dumps. Fortunately, these attitudes are changing. The notion of an inexhaustible planet is beginning to be discarded as we gradually come to understand and accept the fixed dimensions of the biosphere. This process, however, has barely gotten started.

To illustrate this point, one might look at the present meaning of the gross national product (GNP), which is still considered to be the indicator of the health of the economy. When the GNP rises, as the *quantity* of goods produced increases, the economy is considered healthy. Yet the GNP can also be seen as an indicator of the rate at which the natural resources are being converted into wastes. Only a culture that believes in an inexhaustible planet can accept an economy that operates as an open pipeline, taking in natural resources at one end and spilling out products

1

AN ESSENTIALLY "LINEAR" OR OPEN MATERIALS ECONOMY. THE OBJECTIVE IS TO INCREASE ANNUAL PRODUCTION (GNP) BY MAXIMIZING THE FLOW OF MATERIALS. THE NATURAL PRESSURE, THEREFORE, IS TO DECREASE THE LIFE OR QUALITY OF THE ITEMS PRODUCED.

A CIRCULAR OR CLOSED MATERIALS ECONOMY. LIMITS ON THE TOTAL AMOUNT OF MATERIALS OR WEALTH WILL DEPEND UPON THE AVAILABILITY OF RESOURCES AND ENERGY AND THE EARTH'S ECOLOGICAL, BIOLOGICAL AND PHYSICAL SYSTEM. WITHIN THESE LIMITS, THE LOWER THE RATE OF MATERIAL FLOW, THE GREATER THE WEALTH OF THE POPULATION. THE OBJECTIVE WOULD BE TO MAXIMIZE THE LIFE EXPECTANCY AND, HENCE, QUALITY OF ITEMS PRODUCED.

Fig. 1.1 The "open" and "closed" material-flow economies. (Courtesy of Robert Bordeaux, Damascus, Maryland [125].)

2

destined for the garbage dump at the other (Fig. 1.1). Maximizing the GNP only speeds the flow of materials through this pipeline, thereby minimizing both the lifespan of the products and the time left before raw materials are exhausted.

If environmental and disposal costs were reflected in the total costs of products, a new economic model would evolve. In this model the profitability of industries would increase when the flow of materials in the "closed" pipeline of Figure 1.1 is reduced and when the *quality and durability* of the products is increased. These and other, broader issues that contribute to the solid-waste problem are discussed in the Postscript. Here the discussion is limited to the recycling and disposal of municipal solid waste.

The Composition of MSW

The quantities and compositions of solid waste generated by a society reflect on the prevailing culture (Table 1.1). Farming communities in centuries past had practically no waste at all. Leftover food and bones were eaten by the household animals, paper and wood waste was used for kindling, glass jars were reused, and even the ash ended up on the manure pile to be used as fertilizer. By the end of the nineteenth century, as cities began to grow and population density increased, MSW had to be collected. The bulk of this early MSW, as shown in Table 1.2, was ash.

By the middle of the twentieth century a drastic change had occurred in both the lifestyle of the population and in the composition of its refuse. The ash content of MSW has decreased, while the amount of paper and food wastes has increased. Tables 1.3 and 1.4 show the domestic waste compositions in the United States in the late 1960s and early 1970s. These numbers also reflect the differences in lifestyles, where, for example, in Santa Clara County, California, garden waste accounts for a great proportion of MSW, while in Madison, Wisconsin, construction-related dirt and glass accounts for a greater proportion.

By the 1980s the composition of MSW had changed again. The most noticeable change was an increase in the amount of plastics discarded. In the 1960s and 1970s the concentration of plastics was between 1% and 3%. In New York in the 1980s about 5% of the MSW consisted of plastics (Table 1.5); in Baltimore County, Mary-

Table 1.1
MSW QUANTITIES, COMPOSITIONS AND
RECYCLING IN THE U.S., 1988

	Amount Generated	Amount Recovered	Percentage Recovered
Paper and paperboard	71.8	18.4	25.6%
Glass	12.5	1.5	12.0
Metals	15.3	2.2	14.6
Plastics	14.4	0.2	1.1
Rubber and leather	4.6	0.1	2.3
Yard waste	31.6	0.5	1.6

Source: Environmental Protection Agency

Table 1.2
CONTENT OF REFUSE
COLLECTED IN LONDON,
1892 (31)

Ash and cinders	83%
Vegetable, putrescibles, bones	8%
Metal	1%
Paper and rugs	5%
Glass	2%
Miscellaneous	1%

Table 1.3
COMPOSITIONS OF DOMESTIC WASTES

TYPICAL URBAN WASTES (13)

Component	Average Weight, %
Food Wastes	18.2
Garden Wastes	7.9
Paper Products	43.8
Metals	9.1
Glass and Ceramics	9.0
Plastics, Leather and Rubber	3.0
Textiles	2.7
Wood	2.5
Rock and Ash	3.7
Ultimate Residue Analysis	
Moisture	15–35
Carbon	15–30
Oxygen	12–24
Hydrogen	2–5
Nitrogen	0.2–1.0
Sulfur	0.02–0.1
Others	15–25

Note: Overall calorific value is 3000–6000 Btu/lb.

SANTA CLARA COUNTY[a] REFUSE IN 1967 (15)

Classification	Percentage of Total	Waste Production Rate	
		lb/day per capita	*tons/yr*
Garbage	12	0.42	75,240
Rubbish			
paper	50	1.75	313,500
wood	2	0.07	12,540
cloth	2	0.07	12,540
rubber	1	0.04	6,270
leather	1	0.04	6,270
garden wastes	9	0.31	56,430
metals	8	0.28	50,160
plastics	1	0.04	6,270
ceramics & glass	7	0.24	43,890
nonclassified	7	0.24	43,890
Total	100	3.50	627,000

[a] There were 980,000 persons in this county in 1967.

Table 1.4
CHARACTERIZATION OF
SOLID WASTE COLLECTED
IN MADISON, WISCONSIN
(83)

Component	Proportion, %
Metal:	
ferrous	5.7
nonferrous	0.3
Moisture	12.4
Nonmetal:	
paper	41.0
cardboard	15.3
cloth and leather	4.4
plastic	2.9
wood and plants	1.5
rubber	0.6
dirt and glass	15.9

land, in 1983, the concentration of plastics varied from 3% to 8% (Table 1.6); and by 1986 the concentration of plastics in the MSW of the United States reached 6.5% (Fig. 1.2). Table 1.7 provides a breakdown of the composition of per capita MSW generated in the United States in 1985. According to the New York City Sanitation Department, the city's garbage in 1990 consisted of 37% organics and yard wastes, 31% paper, 11% white goods (appliances, etc.), 8.5% plastics, 5% glass, 5% metals (including 1% aluminum), 2% inorganics (dirt, concrete) and 0.5% hazardous wastes.

A similar trend is noticeable overseas. In Rome, for example, between 1962 and 1985 the concentration of kitchen waste, ash, and dirt decreased, while the amount of paper and glass nearly doubled, and the concentration of plastics has risen from 0% to 3% (Table 1.8). Table 1.9 compares the MSW composition of Rome in 1985 with rural and urban MSW compositions in the United States. The concentration of kitchen wastes and sweepings are much higher in the MSW of Rome, while paper, glass, metals, and plastics are more prevalent in the urban wastes of the United States. When comparing urban and rural MSW compositions in the United States, there tends to be more moisture and sweepings in rural MSW and more kitchen waste, glass, metals, and plastics in urban MSW.

Table 1.5
COMPOSITION OF NEW YORK AREA REFUSE
IN THE 1980s (12)

Paper	30%	Plastics	5%
Yard and garden	19%	Wood	3.5%
Food	17%	Textile	2.5
Glass and ceramics	10%	Rubber and leather	2%
Metals	9%	Rock, ash & miscellaneous	2%

Table 1.6

COMPOSITION OF MSW OBTAINED JANUARY 10–21, 1983, FROM THE BALTIMORE COUNTY RESOURCE AND RECOVERY FACILITY (65)

	Mass Percent (Dry)			
	Day 4	6	7	9
Initial Moisture	27.6	34.3	25.1	22.0
Metal (Magnetic)	5.0	6.0	2.6	2.2
Metal (Nonmagnetic)	1.3	1.5	1.3	0.6
Wood, Vegetable	0.8	1.3	0.4	2.5
Textiles	1.6	3.2	5.2	13.8
Plastics	3.3	6.2	8.2	6.6
Paper	66.2	55.8	67.7	64.3
Glass, Ceramics	2.6	2.4	0.8	0.3
"Fines"	19.2	23.6	13.8	9.7
Metals	6.3	7.5	3.9	2.8
Combustible Content	71.9	66.5	81.5	87.2
Noncombustible Content	21.8	26.0	14.6	10.0
Samples as Burned				
Combustible Content	76.7	71.9	84.8	89.7
Noncombustible Content	23.3	28.1	15.2	10.3

Fig. 1.2 Composition of refuse, U.S. household, commercial, and industrial sources in 1986, about 158 million tons. By 1988 the total reached 180 trillion tons, composed of 40% paper, 25% food and yard waste, 8.5% metals, 8% plastics, 7% glass, and 11.5% miscellaneous including durable goods (199). (Franklin Associates, Ltd., INFORM, Inc. [156].)

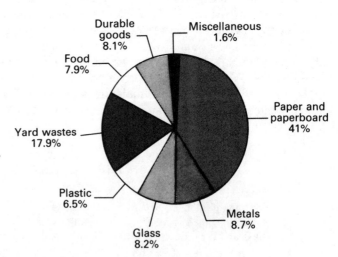

Table 1.7
COMPOSITION OF TRASH
IN 1985 PER CAPITA IN
POUNDS (156)

Newspaper	101.5
Books and magazines	66.2
Office paper	47.3
Corrugated	136
Mixed paper	130.1
Beer and soft drink cans	4.7
Food cans	18.5
Other nonfood cans	6.7
Barrels, drums, etc.	2.1
Aluminum cans	8.6
Aluminum foil	2.8
Beer and soft drink bottles	61.7
Wine and liquor bottles	21.8
Food and other glass	40.2
Plastic	36.5
Yard waste	244.4
Food waste	214.1
Durable goods	159.3
Miscellaneous	79.5
Total	1382.9

Source: Massachusetts Department of Environmental Management

Table 1.8
CHANGES IN THE MSW COMPOSITION OF
ROME, ITALY (94)

Fraction	Percent 1962	Percent 1985
Organic matter (kitchen waste)	45	33
Paper and cardboard	12	22
Glass	3	7
Ferrous metals	4	4
Fines (less than 1 in.)	20	12
Plastic, film	—	1
Plastic, hard	—	2
Textiles	2	2
Miscellaneous (unclassified)	14	17
Total	100%	100%

Table 1.9
COMPARISON OF ROME AND U.S. WASTE
(WEIGHT PERCENT ASSUMED) (64)

Rome			U.S.	Rural	Urban
Organic matter (kitchen waste)	33		Wood/vegetable	1	6
Paper and cardboard	22		Paper	44	40
Glass	7		Glass/ceramics	4	15
Ferrous metals	4		Metals	5	8
Fines (less than 1 in.)	12		Fines (sweepings)	11	3
Plastic, film	1		Plastics, soft	4	3
Plastic, hard	2		Plastics, hard	2	7
Textiles	2		Textiles	3	2
Miscellaneous	17		Moisture	26	16
	Total	100		100	100

The trend to replace paper products with plastics seems to be accelerating. At the end of the 1980s the plastic concentration in the MSW of the United States had reached 8% by weight and 25% by volume. This amounts to some 14 million tons a year, or 0.35 pounds of discarded plastics per day for every person in the country. These statistics have resulted in some misconceptions. While plastics are replacing some of the paper products, the overall use of paper is still increasing. The view that paper is preferable to plastics because it is degradable is not valid; in dry, compacted landfills neither will degrade. While plastics at the source amount to 25% by volume, by the time they are compacted in the garbage truck and are further compressed in the landfill, the weight and volume percentages of plastics are much closer. In existing landfills the weight and volume percentages of paper are between 40 and 50%, while plastics are less than 5% by weight and around 12% by volume (179). The concentration of plastics is also on the rise in Europe, but the actual percentages are smaller (Table 1.9). As will be seen later, plastics and heavy metals account for some of the most serious health problems related to waste disposal.

Quantities of MSW Generated

As social priorities shift toward comfort and convenience and as commercial considerations place more emphasis on packaging, the generation of solid waste tends to increase. The average amount of MSW generated in the United States was 2.9 pounds per person per day in 1969 (Table 1.10). Today it approaches 4 pounds (Fig. 1.3, Tables 1.11 and 1.12) and projections range from 3.5 to 6 pounds by the turn of the century (110). The solid waste generated by offices, hospitals, hotels, and schools is also rising (Table 1.13).

The total MSW generated in the United States has been rising for several decades (Fig. 1.4) and today we are generating more than 180 million tons of MSW per year. This is an 80% rise since 1960. By the year 2000 the yearly MSW production

Table 1.10
SOLID WASTES COLLECTED PER CAPITA IN THE
UNITED STATES (1969) (13)

Type	Quantity, lb/Capita-Calendar Day
Residential (domestic)	1.5–5.0
Commercial (restaurants, businesses or stores)	1.0–3.0
Bulky solid wastes (furniture, fixtures and construction wastes)	0.3–2.5

is expected to rise to 216 million tons. In comparison, Germany generates 23 million tons of MSW a year (62) and Canada 16 million (48). Larger cities like New York and Los Angeles, with their respective MSW generation of 10 and 15 million tons per year, create gigantic MSW disposal problems. Among different cities of the world, Los Angeles is at the top of the list in per capita generation of MSW (6.2 pounds/day), while large, overpopulated cities in underdeveloped countries, such as Calcutta (1.1 pounds/day), are at the bottom (Fig. 1.5).

To put these numbers in perspective, we can say that it takes several hundred 20-ton 40-foot-long trucks a day to haul the MSW from Long Island alone. If we loaded the MSW produced in the United States this year on these trucks, they would

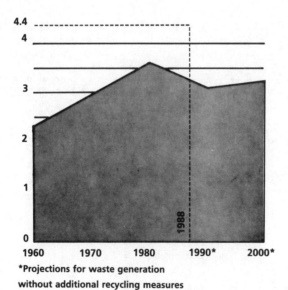

*Projections for waste generation
without additional recycling measures

Fig. 1.3 Pounds of household, commercial, and industrial waste generated per person in the United States per day. (Environmental Protection Agency [156].)
Note: Actual increase exceeded projections and reached 4.4 pounds by 1988 (199).

Table 1.11
MUNICIPAL SOLID WASTE GENERATION, RECOVERY, AND DISPOSAL, 1960 TO 1986 (178)

Item and Material	1960	1965	1970	1975	1980	1981	1982	1983	1984	1985	1986
Gross waste generated	87.5	102.3	120.5	125.3	142.6	144.8	142.0	148.3	153.6	152.5	157.7
per person per day (lb.)	2.65	2.88	3.22	3.18	3.43	3.45	3.35	3.47	3.56	3.49	3.58
Materials recovered	5.8	6.2	8.0	9.1	13.4	13.2	12.9	13.9	15.3	15.3	16.9
per person per day (lb.)	.18	.17	.21	.23	.32	.31	.30	.32	.35	.35	.39
Processed for energy recovery	(na)	.2	.4	.7	2.7	2.3	3.5	5.0	6.5	7.6	9.6
per person per day (lb.)	(na)	.01	.01	.02	.06	.05	.08	.12	.15	.17	.22
Net waste disposed of	81.7	95.9	112.1	115.5	126.5	129.3	125.6	129.5	131.8	129.7	131.2
per person per day (lb.)	2.48	2.70	3.00	2.93	3.04	3.08	2.96	3.03	3.05	2.97	2.98
Percent distribution of net discards:[a]											
paper and paperboard	30.0	33.5	32.4	29.6	32.5	33.1	32.1	34.1	35.7	35.5	35.6
glass	7.8	8.8	11.1	11.4	11.0	10.9	10.7	9.9	9.3	8.9	8.4
metals	12.8	11.1	12.0	11.5	10.1	9.8	9.7	9.6	9.3	9.0	8.9
plastics	.5	1.5	2.7	3.8	5.9	5.9	6.5	6.8	6.9	7.1	7.3
rubber and leather	2.1	2.3	2.7	3.2	3.2	3.1	2.9	2.5	2.4	2.5	2.8
textiles	2.1	2.0	1.8	1.9	2.0	2.6	2.2	2.1	2.0	2.0	2.0
wood	3.7	3.6	3.6	3.8	3.8	3.3	3.9	3.9	3.7	3.9	4.1
food wastes	14.9	12.9	11.4	11.5	9.2	9.2	9.3	8.9	8.8	9.0	8.9
yard wastes	24.5	22.5	20.6	21.7	20.5	20.3	20.9	20.4	20.1	20.4	20.1
other wastes	1.6	1.7	1.6	1.7	1.7	1.7	1.8	1.8	1.7	1.8	1.8

Source: Franklin Associates, Ltd., Prairie Village, KS, *Characterization of Municipal Solid Waste in the United States, 1960 to 2000*, 1988. Prepared for the U.S. Environmental Protection Agency.
Note: In millions of tons, except as indicated. Covers post-consumer residential and commercial solid wastes which comprise the major portion of typical municipal collections. Excludes mining, agricultural and industrial processing, demolition and construction wastes, sewage sludge, and junked autos and obsolete equipment wastes. Based on material-flows estimating procedure and wet weight as generated.
[a] Net discards after materials recovery and before energy recovery.

Table 1.12

ESTIMATED QUANTITIES, COMPOSITION AND RECYCLING OF MSW GENERATED
IN THE U.S. IN 1990

	Total Quantity Generated (millions of tons/year)	Weight Percentage of This Fraction in the Total MSW	Percentage of this Fraction That Is Being Recycled or Reused in 1990	Average Value of the Recycled Material (units of $/ton)
All paper products	72	40%	26% overall 33% newsprint	+40 to −15
Organic food and yard waste	45	25%	1–2% composted	25
Appliances, durable goods, steel cans and other ferrous metal items	25	14%	15%	40–80
Glass	13	7%	12%	50
Plastics	14	8%	1–2%	100–200 (PET)
Inorganics and miscellaneous	8.5	4.5%	—	—
Aluminum	2.5	1.5%	63%	500–1000
Total	180	100%	13%	NA

Table 1.13
TYPICAL DAILY SOLID WASTE
GENERATION (13)

Building Types	Estimated Waste
Private homes	5 lb basic + 1 lb per bedroom/day
Apartments	4 lb per sleeping room/day
Warehouses	2 lb per 100 sq ft/day
Office buildings	1½ lb per 100 sq ft/day
Department stores	4 lb per 100 sq ft/day
Restaurants	2 lb per meal/day
Grade schools	10 lb per room + ¼ lb per pupil/day
High schools	8 lb per room + ¼ lb per pupil/day
Hospitals	15 to 18 lb per bed/day
Nursing homes	3 lb per person/day
Hotels, class I	3 lb per room + 2 lb per meal/day
Hotels, medium	1½ lb per room + 1 lb per meal/day
Motels	2 lb per room/day
Trailer camps	6 to 10 lb per trailer/day

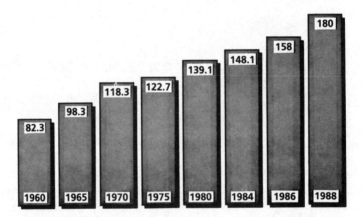

Fig. 1.4 MSW, excluding industrial and agricultural waste, generated in the United States (in millions of tons). (Statistical Abstract
of the United States, 1988 [118 and 199].)

12

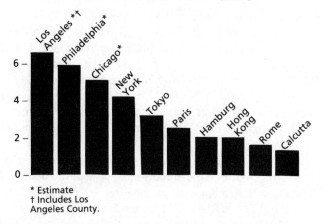

* Estimate
† Includes Los
Angeles County.

Fig. 1.5 Pounds of garbage generated per capita per day in se-
lected cities throughout the world in 1987. (National Solid Wastes
Management Association, city agencies, and Gershman, Brickner
& Bratton.)

stand bumper to bumper more than three times around the earth. Indeed the 3.6
billion cubic feet of the Great Wall of China is being challenged by the Fresh Kills
landfill on Staten Island, now nearing a volume of 3 billion cubic feet. If it grows at
its present rate, it will reach a height of over 500 feet by 2005 and will become the
tallest mountain on the Atlantic seashore south of Maine (184).

In the United States today, each population block of one million requires a
2,000-tons/day solid waste disposal plant. This means that New York requires 14
such plants and Los Angeles 20. While in the United States each individual generated
over 4 pounds of MSW every day in 1990, this number was 2.1 pounds in Japan and
Germany (101). Part of the reason for this difference is the greater emphasis on
recycling in those countries. On the other hand, the daily solid-waste generation in
Canada or in Mexico City is 6 pounds per person.

Solid-waste generation is not limited to MSW. Industry in the United States
alone generates some 250 million tons of toxic, corrosive, flammable, explosive, or
otherwise hazardous wastes. Over 90% of these hazardous industrial wastes are
treated at the plants that generate them. If one considers all the different types of
solid waste generated by industry, including agricultural, construction, and demolition debris, the total quantity for the United States is several billion tons.

Solid Waste Disposal Options and Costs

In the United States the average household generates about two tons of MSW
per year. The disposal costs vary from $10 to $20 per ton in the Midwest to $50 or
more in the Northeast. Such sums (up to $100 per household per year) are still

insufficient to motivate such basic changes in lifestyle as separating solid waste at the source to assist in recycling. As the disposal costs continue to rise, the "pocketbook motivations" for recycling will be felt more and more. Good examples are several towns on Long Island. In the town of Long Beach, MSW disposal cost in 1979 was the sixth largest item on the town's list of expenses; today it is the second. In Oyster Bay the yearly cost of MSW disposal in 1991 is estimated at $424 per household, which is more than half of the city taxes. East Hampton expects a 13% rise in taxes in 1991 and plans to recycle or compost up to 82% of its MSW to control the costs.

In Union, New Jersey, the annual dumping cost per household exceeds $400. As a consequence, the town is experimenting with a fairer distribution of the expenses, termed "billing by the bag." In High Ridge, New Jersey, the "pay by the bag" system was introduced in 1988. For a base fee of $200 a year, each household is given 52 town stickers, allowing them to set at curbside one 30-gallon bag or can per week. Stickers are also used on bulky items, requiring, for example, two stickers to dispose of a chair or six for a sofa. Additional stickers are sold at a unit cost of $1.65. As a result of this system, the amount of residential trash was reduced by 25%, from 4.3 to 3.3 pounds per person per day. In addition, glass, paper, and aluminum recycling has increased.

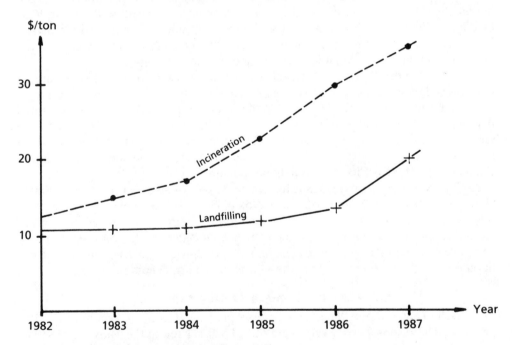

Fig. 1.6 Average tipping fees in the United States (110).

In the United States, a $20 billion industry collects and disposes of all MSW. The cost breakdown includes $12 billion for collection and transportation, $5 billion for landfilling, $2 billion for recycling, and $1 billion for incineration (110). The average "tipping fees" (the costs of unloading MSW at landfills or incinerators) have been rising continuously (Fig. 1.6). In the Northeast, the average tipping fees increased from $20/ton in 1986 to $40/ton in 1987 and have doubled again by 1991. In 1989 New York City raised the cost of dumping at its Fresh Kills landfill on Staten Island from under $50 to $100/ton (about $40/cubic yard). On the other hand, in order to stimulate recycling, it is planning to cut its dumping fee from $40 to $25 per cubic yard for garbage that is free of recyclables. In the rest of the country, the average tipping fee has risen more slowly, but this trend is also accelerating (110). Disposal costs in Europe and Japan tend to be higher; in Canada, they tend to be a little lower than in the U.S.

As the costs of disposal increase and the availability of landfill sites diminishes (Fig. 1.7), the emphasis on recycling is increasing. Actually, it is a natural process for landfills to fill up in 10 to 20 years; what is diminishing is not the space available for opening up new landfills, but the willingness of people to accept a landfill in the neighborhood. The consequence is that few new landfills are opening up and recycling is receiving more serious consideration. But relative to other industrialized countries, the U.S. has much catching up to do. Japan, for example, recycles about 50% of its

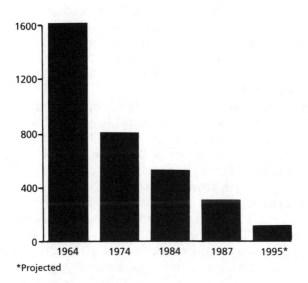

Fig. 1.7 Number of operating landfills in New York State. (New York State Department of Environmental Conservation [111].)

Table 1.14
CHANGING METHODS OF SOLID
WASTE DISPOSAL IN THE U.S.
(IN MILLIONS OF TONS/YEAR) (112)

	1960	1986	2000[a]
Landfill	81.7	131.2	136.8
Recycling	5.8	16.9	23.9
Incineration	0.0	9.6	32.0
Total	87.5	157.7	192.7

Source: Franklin Associates
[a] Projected

MSW, while the U.S. recycles only about 13% (Table 1.12). The highest recycling rate, 28%, has been achieved in Seattle, Washington, where disposal fees are based on the MSW volume generated by the household and where 60% of households participate in the recycling program. San Francisco is said to be recycling 25%, while New York City, Florida, and Washington, D.C., have set recycling goals of 25%, 30%, and 45% to be reached by the mid-1990s. It is anticipated that in the coming decades, both recycling and incineration will increase (Table 1.14), but the yearly demand for landfilling is not expected to decrease in this century. The likely scenario for the next decade is that the increase in recycling will compensate for the increase in population and the total quantity of MSW to be disposed of will remain constant at somewhere under 200 million tons per year.

To appreciate the trends of the last hundred years, one might look at the MSW of New York City in 1887 and in 1989 (190). During this century, the average New Yorker increased its daily MSW production from 3.25 to 5 pounds and the composition of this MSW also changed from being mostly (83%) coal ash to a more complex mixture, dominated by organics (37%) and paper (31%). The disposal methods have also changed. In 1887, 76% was dumped (73% in the sea, 3% in the rivers) and 24% was filled (18% by shoreline filling and 6% by landfilling). A century later, dumping has ended (with the exception of sludge dumping in the ocean) and most of the MSW (81%) is landfilled, while the rest is either incinerated (12.5%) or recycled (6.5%). The trend of the last century can be summed up by saying that ocean dumping has been replaced by landfilling. I hope that, a hundred years from now, another author can report that landfilling has been replaced by recycling and reuse.

Table 1.15
CLASSIFICATION OF WASTES (14)

Type 0—Trash, a mixture of highly combustible waste such as paper, cardboard, cartons, wood boxes and combustible floor sweepings from commercial and industrial activities. The mixtures contain up to 10% by weight of plastic bags, coated paper, laminated paper, treated corrugated cardboard, oily rags and plastic or rubber scraps.

This type of waste contains 10% moisture, 5% incombustible solids and has a heating value of 8500 BTU per pound as fired.

Type 1—Rubbish, a mixture of combustible waste such as paper, cardboard cartons, wood scrap, foliage and combustible floor sweepings from domestic, commercial and industrial activities. The mixture contains up to 20% by weight of restaurant or cafeteria waste, but contains little or no treated papers, plastic or rubber wastes.

This type of waste contains 25% moisture, 10% incombustible solids and has a heating value of 6500 BTU per pound as fired.

Type 2—Refuse, consisting of an approximately even mixture of rubbish and garbage by weight.

This type of waste is common to apartment and residential occupancy, consisting of up to 50% moisture, 7% incombustible solids and has a heating value of 4300 BTU per pound as fired.

Type 3—Garbage, consisting of animal and vegetable wastes from restaurants, cafeterias, hotels, hospitals, markets and like installations.

This type of waste contains up to 70% moisture, up to 5% incombustible solids and has a heating value of 2500 BTU per pound as fired.

Type 4—Human and animal remains, consisting of carcasses, organs and solid organic wastes from hospitals, laboratories, abattoirs, animal pounds and similar sources, consisting of up to 85% moisture, 5% incombustible solids and having a heating value of 1000 BTU per pound as fired.

Type 5—By-product waste, gaseous, liquid or semiliquid, such as tar, paints, solvents, sludge, fumes, etc., from industrial operations. BTU values must be determined by the individual materials to be destroyed.

Type 6—Solid by-product waste, such as rubber, plastics, wood waste, etc., from industrial operations. BTU values must be determined by the individual materials to be destroyed.

Table 1.16
CLASSIFICATION OF WASTES TO BE INCINERATED (15)

Classification of Wastes Type / Description	Principal Components	Approximate Composition, % by Weight	Moisture Content, %	Incombustible Solids, %	BTU Value/lb of Refuse as Fired	Requirement for Auxiliary Fuel BTU per lb of Waste	Recommended Min. BTU hr Burner Input per lb Waste	Density lb/ft³
0 Trash	Highly combustible waste, paper, wood, cardboard cartons and up to 10% treated papers, plastic or rubber scraps; commercial and industrial sources	Trash 100	10	5	8500	0	0	8–10
1 Rubbish	Combustible waste, paper, cartons, rags, wood scraps, combustible floor sweepings; domestic, commercial and industrial sources	Rubbish 80 Garbage 20	25	10	6500	0	0	8–10

2	Refuse	Rubbish and garbage; residential sources	Rubbish 50 Garbage 50	50	7	4300	0	1500	15–20
3	Garbage	Animal and vegetable wastes, restaurants, hotels, markets, institutional, commercial and club sources	Garbage 65 Rubbish 3	70	5	2500	1500	3000	30–35
4	Animal solids and organics	Carcasses, organs, solid organic wastes; hospital, laboratory, abattoirs, animal pounds and similar sources	Animal and human tissue 100	85	5	1000	3000	8000 (5000 primary) (3000 secondary)	45–55
5	Gaseous, liquid or semi-liquid	Industrial process wastes	Variable	Dependent on major components	Variable	Variable	Variable	Variable	Variable
6	Semisolid and solid	Combustibles requiring hearth, retort or grate equipment	Variable	Dependent on major components	Variable	Variable	Variable	Variable	Variable

Table 1.17
INDUSTRIAL WASTE TYPES (16)

Type of Wastes	Description
1	Mixed solid combustible materials, such as paper and wood
2	Pumpable, high heating value, moderately low ash, such as heavy ends, tank bottoms
3	Wet, semisolids, such as refuse and water treatment sludge
4	Uniform, solid burnables, such as off-spec or waste polymers
5	Pumpable, high ash, low heating value materials, such as acid or caustic sludges, or sulfonates
6	Difficult or hazardous materials, such as explosive, pyrophoric, toxic, radioactive or pesticide residues
7	Other materials to be described in detail

The Classification of Solid Waste

All waste can be defined as a mixture of combustibles, inerts, and moisture; the inerts include dirt, stones, glass, metals, and metal oxides. Metals tend to act as inerts if incineration temperatures and residence times are insufficient to burn (oxydize) them. The waste may also include appliances, furniture, tree limbs, construction debris, broken concrete, and other materials commonly classified as rubbish. The moisture content of the MSW is highly weather-dependent; after a storm the moisture content can reach a point where it will not burn.

The Incinerator Institute of America classifies solid waste into seven categories according to its moisture and inert content, heating value, and auxiliary fuel requirements (Table 1.15). Type 0 is also referred to as trash, type 1 as rubbish, type 2 as refuse, type 3 as garbage, etc. Having been generated by incinerator manufacturers, this listing does not cover some agricultural wastes, automobile scrap, or demolition or construction debris. Types 0, 1, 2, and 3 wastes, usually referred to as "domestic" wastes, can be burned in incinerators with pollution-abatement devices. Types 4, 5, and 6, industrial and agricultural wastes, require specially designed incinerators. Industrial wastes are also classified into seven categories as a function of their hazardous or explosive properties, combustible or pumpable nature, heating value, and other characteristics (Table 1.16).

Most classifications describe the composition, density, and heating value of the various types of solid wastes (Tables 1.18 and 1.19). The heating value of MSW tends to increase with the living standards of the society that generates it. Figure 1.8 shows the trends in MSW heating values generated by various cities around the world.

20

Table 1.18
AVERAGE WEIGHT (DENSITY) OF REFUSE[a] (14)

Type of Refuse	Refuse Density, lb/ft³
Type 0 waste	8 to 10
Type 1 waste	8 to 10
Type 2 waste	15 to 20
Type 3 waste	30 to 35
Type 4 waste	45 to 55
Garbage, 70% H_2O	40 to 45
Magazines and packaged paper	35 to 50
Loose paper	5 to 7
Scrap wood and sawdust	12 to 15
Wood shavings	6 to 8
Wood sawdust	10 to 12

[a] The weights given are general and are based on materials usually expected in refuse collection. Whenever the density of the material or its weight per cubic foot exceeds those listed, consideration must be given to the use of mechanical grates, or larger grate and hearth areas than the minimum. As an example, packaged paper, confidential records, bundled currency, etc., require special incinerator designs.

Table 1.19
APPROXIMATE CALORIFIC VALUES OF WASTES (13)

Material	BTU/lb	lbs/ft³
Animal fats	17,000	60
Brown paper	7,250	7
Citrus rinds	1,700	40
Coated milk cartons	11,330	5
Corn cobs	8,000	15
Corrugated boxes	7,040	7
Coffee grounds	10,000	30
Cotton seed hulls	8,600	30
Latex	10,000	45
Leather	7,250	20
Linoleum	11,000	90
Magazines	5,250	35
Newspapers	7,975	7
Polyethylene	20,000	50
Polyurethane (foamed)	13,000	2
Rags (silk and wool)	8,500	15
Rags (cotton & linen)	7,200	15
Rubber waste	10,000	100
Wood sawdust	8,500	12

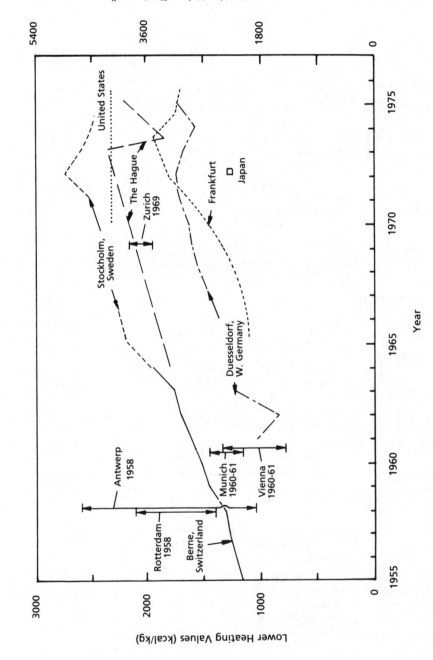

Fig. 1.8 Trends in lower heat values of MSW (31).

22

Chapter 2
Landfills

□
■ ■

CURRENT MSW DISPOSAL OPTIONS include recycling, burning, ocean dumping, and burial. The landfilling of solid waste, probably the oldest disposal method, grew from open or burning dumps to sanitary (earth-covered) landfills as concern over health effects increased. This is strip mining in reverse.

This chapter will discuss landfilling trends and practices and their effects on disposal costs. Regional and international trends will be noted for purposes of comparison. The second half of the chapter discusses the design, construction, and reclamation of landfills, with particular emphasis on the production of methane and leachate and on the prevention of groundwater pollution. The disposal of hazardous and toxic wastes is discussed in Chapter 3.

Existing Landfills

The excavation of landfills is a young and neglected field of research. Yet the recent findings of some researchers have already modified some old myths (179). One such myth involves the term "biodegradable." This term has been both misunderstood and exploited for commercial purposes in several ways. What we are gradually learning is that given the right conditions (temperature, moisture, oxygen) and the right, acclimated bioassay (bacteria that have been acclimated to break down the particular substance), everything, including plastics, crude oil, or even H_25 is biodegradable. On the other hand, we are also learning that in the compacted, dry environment of a landfill practically nothing is biodegradable. It has been found that after more than a decade in a landfill, such supposedly biodegradable materials as food and yard debris are only 25% to 50% degraded. Instead of being degraded, they

23

have been pickled and mummified by the anaerobic and acidic environment of the landfill. Other materials like newspapers and telephone books (while theoretically biodegradable) do not seem to degrade at all.

We have also learned that biodegradation itself is a mixed blessing. To most, biodegradation means letting nature do its work, and this sounds reassuring. In the case of landfills, this is not necessarily true. If paper products and plastics degraded in landfills, they would release all the toxic metals from the paints and inks into the groundwater. In the United States landfills contain some quarter million tons of PCBs (191). We are better off if the PCBs, as well as the toxic metals in ink and paint, stay in the landfills and do not leach into the groundwater.

The investigation of landfills has also placed a better perspective on the order of our priorities. Some have felt that fast food packaging, disposable diapers and other plastic products are a major load on our landfills. It turns out that fast food packaging represents a 0.1%, diapers a 1% and all the plastics combined a 12% volumetric load on our landfills (179). These same researchers tell us that while plastics are not as big a problem for landfills as we have thought, paper is an even greater problem than realized. Telephone books alone are a major problem in landfills, newspapers take up 10–18% of the landfill volume, and all paper products combined require 40–50% of all the landfill volume.

Research has also revealed that only 33% of the municipal landfills in the United States are provided with groundwater monitoring, only 15% of them have liners, and just 5% has leachate collection (191). On the other hand, 35% of the sanitary landfills are in counties containing geological faults, 14% are located on flood plains, and 6% are situated on wetlands. Because the most important ingredient of any solution is the clear understanding of what the problem is, such knowledge concerning the nature of existing landfills is very important.

Domestic and Overseas Trends

In the United States landfilling is still the dominant method of disposal for MSW, although many regions are rapidly running out of sites. In 1986 about 83% of the MSW generated in the United States was still being landfilled, while only 10% was being recycled. By 1990 landfilling had been slightly reduced, recycling increased to 13% and incineration amounted to 14% (200). In Japan landfilling amounts to only 15%, while recycling is about 50%. The reason Japan is a leader in recycling is its high population density and its shortage of space for landfills. By 1995 Tokyo's main landfill will be filled. This garbage heap is already 82 feet (25 meters) above sea level, and it will reach 98 feet (30 meters) by the time it is closed. Tokyo and Osaka are now considering extending the shoreline or creating islands from solid waste (159). The disposal practices of the densely populated European countries are similar to those of Japan.

In the United States the goal of the Environmental Protection Agency is to reduce landfilling to 55% and increase recycling to 25% by 1992 (a desirable, but excessively optimistic, goal). Unfortunately, the total amount of MSW generated is rising faster than the rate of recycling. With a daily MSW generation of 4.3 pounds per capita in the United States, a city of 10 million will cover a 1,000-acre area every year with an 8-foot layer of compacted waste. Even if the depth of the landfill is increased (Los Angeles has 100-foot-deep landfills), the United States is quickly running out of landfilling space (see Table 1.14). In the 1970s there were 18,000 operating landfills in the United States; by 1980 the number dropped to 10,000, and by 1988 there were 6,500. According to one study, all but four states are running out of suitable locations for landfills (101).

The need for new landfills is not a new development. Landfills fill up in 10 to 20 years and therefore new landfills have always been needed. What is new is the high cost and the public resistance to new landfills. Because of the strict state and Federal standards the cost of well-designed new landfills is nearing a half million dollars per acre.

The Northeast

The loss of landfill sites is most drastic in the more densely populated states of the Northeast. In the last twenty years the number of landfills in New York State has dropped from 1,600 to 300 (see Fig. 1.7). The state lost 53 landfills in 1988–89 alone, and only 12% of the remaining landfills have valid permits (129). Yet the amount of MSW generated in the state had risen 14% between 1986 and 1988, exceeding 20 million tons in 1988 (130). The situation has not been helped by the defeat of the $1.975 billion proposal on the 1990 ballot, the Environmental Quality Bond Act, which would have allocated $800 million for land acquisition and $525 million on closing landfills and for recycling.

In 1988 New York City generated 26,000 tons of MSW a day, of which 18,000 tons were landfilled at the Fresh Kills site on Staten Island (129). By 1990, total daily MSW generated by the city approached 30,000 tons (190). The only other operating landfill serving the city, at Edgemere in Queens, handles 1,200 tons a day. By 2010 the MSW production of the city is estimated to reach 38,000 tons per day (131). At this rate the Fresh Kills site, this great pyramid of unsafely stored garbage, will be full around the turn of the century. If the city builds five planned incinerators and accomplishes 15% recycling (today it stands at 6.5%), the daily MSW generation would be reduced to 14,000 tons, thus extending the life of Fresh Kills until 2007 (131). With the most optimistic scenario of 10% source reduction, 40% recycling, and two incinerators, the landfill space will still be depleted by 2011 (131).

In 1989 the officials of the state of New York filed a $76 million civil complaint against the city of New York for operating the Fresh Kills landfill in an unsafe manner, and discharging into the groundwaters a daily runoff of two million gallons of untreated leachate. Yet operation continues. In 1990 Mayor David Dinkins proposed an additional three years' delay in starting construction on five giant incinerators. Thus incineration will not reduce the load on Fresh Kills in the near future.

The mandatory recycling law passed in 1989 called for 25% recycling by 1994, but the program barely started when an austerity program was announced to delay it. For these reasons it might become necessary for New York City to follow the examples of Tokyo and Osaka and gain time by extending the island around lower Manhattan and into the East River. New York might also be forced to convert its incinerator bottom-ash into fused and stabilized building blocks to build the reefs needed for this shore-filling project before a permanent solution is found (201).

In New Jersey there were only 10 landfills still in operation in 1988 and some have since closed. In Connecticut the yearly MSW generation is nearly 3 million tons. There are 89 landfills in operation, but 42 of those are at or near capacity and 67 are causing environmental problems (132). "Vertical overfilling" tends to increase leaching and results in such unsafe operations as drivers driving up the sides of the landfills with their trucks in reverse.

In 1983 the New York State Department of Environmental Conservation ordered all landfills on Long Island to be closed by December 1990 because they were contaminating the underground aquifers that supply drinking water. Five towns—East Hampton, Southold, Riverhead, Smithtown, and Shelter Island—have not done so and have filed a suit to overturn this ruling as unconstitutional. In the meantime, the Department of Environmental Conservation has the right to take them to court and fine them $25,000 a day for noncompliance.

Landfilling Costs

Solid-waste disposal is a $20 billion industry in the United States, and $5 billion of that is spent on the operation of landfills (110). As the availability of landfills decreases, tipping fees increase (see Fig. 1.6).

In the Northeast the average tipping fees increased from $20/ton in 1986 to $40/ton in 1987, and doubled again by 1991. Under state law, the towns of Long Island were supposed to close all landfills by December 18, 1990; they could not, because they could not afford the even higher hauling fees ($110–$150 per ton, or $50 or more per cubic yard) to take the MSW off the island. At some landfills the tipping fee is based on volume instead of weight, which favors operators who shred and compact their MSW, increasing the weight of a cubic yard up to a half ton. At the Fresh Kills site on Staten Island, the tipping fee has increased from less than $50 to $100/ton in 1990 (129). In Philadelphia the cost of MSW disposal tripled in five years. In New Jersey half the household waste is trucked to out-of-state landfills up to 500 miles away at costs approaching or exceeding $100 per ton. Connecticut faces similar problems. The landfill used by the town of Waterford raised its tipping fee in 1988 from $20 to $60 per ton. This led the town to send some of its MSW to an incinerator in Hartford that charges $35 per ton plus $14 for transportation costs. As this incinerator reaches its capacity, the remaining option for Waterford is to send its MSW out of state at a cost of nearly $100 per ton (132).

In Europe and Japan the tipping fees are even higher than those in the North-east, while in Canada and in the other sections of the United States, they are lower, around $5 to $15 per ton; but the trend is unquestionably up.

Landfills and Incinerator Residue

Incineration does not eliminate the need for landfilling. In 1988 7 million tons of incinerator ash were produced in the United States, and there could be five times as much by the turn of the century if all planned incinerators are built. When un-separated MSW is "mass-burned," the residue is 25% by weight (10% by volume). If only the combustible portion of the MSW is burned (refuse-derived fuel, or RDF), the residue is half as much.

In most states incinerator residues are treated as nonhazardous solid wastes. Yet the dioxin, lead, mercury, cadmium, and other heavy-metal content of incin-erator ash frequently exceeds safe limits. The probability is that incinerator ash eventually will have to be disposed of in toxic-waste landfills provided with a double lining and continuous leachate collection and monitoring. This will be an expensive proposition, because as of now only 26 hazardous-waste dumps exist in the whole nation and none in the New York City area.

Transfer Stations

As landfills gradually become saturated in the more densely populated areas, "exporting" MSW becomes an option that can be considered at least as a temporary solution. Transfer stations are used for the long-distance transportation of MSW. At the transfer station (Figs. 2.1 and 2.2), garbage trucks unload the MSW, where

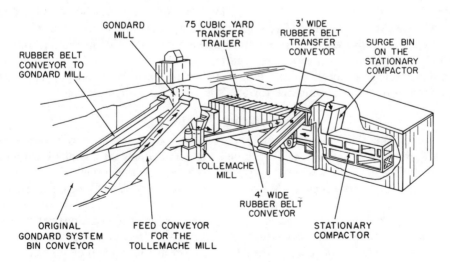

Fig. 2.1 Shredder transfer station with mill, compactor, and transfer trailer (81).

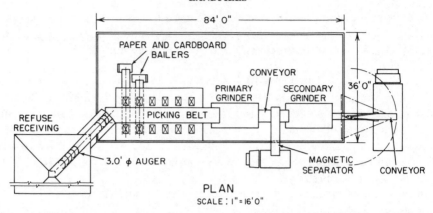

Fig. 2.2 A 300-ton/day refuse-recycling and transfer station (81).

it is shredded and compacted to about 1,000 pounds per cubic yard and is loaded on trailer trucks or railroad cars. Compacting reduces transportation costs as well as space requirements at the receiving landfill.

More and more states are using or building transfer stations. New Jersey, for example, ordered several counties to build transfer stations for the shipment of MSW out of state. These stations will operate until other, more permanent solutions can be found to the waste-disposal problem. Similar trends can also be observed in New York and Connecticut. In 1988 New York exported more than one million tons of its MSW, more than double the amount exported the previous year (130).

In the meantime, the costs of MSW exporting continue to rise. New York spent $800 million to export MSW in 1988 (130). In Newark the garbage-disposal budget rose from $6 million in 1987 to $30 million in 1988, with the MSW being trucked to Poland, Ohio. In Essex County (New Jersey), the tipping fees at the transfer station (which transports the waste to Pennsylvania) have reached $102 per ton. Pennsylvania has attempted to limit the "importation" of MSW, but this position appears to be illegal. In 1978, the U.S. Supreme Court ruled that it is unconstitutional and a violation of interstate commerce for one state to ban another's garbage.

International Practices

International law governing the exporting of solid waste to countries where landfill tipping fees are lower has yet to be formulated. Regulation is just beginning and mostly is aimed at the exporting of toxic wastes, although eventually it likely will be expanded to cover regular MSW. The Organization of Economic Cooperation (OEC), an association of twenty-four industrial nations, adopted a rule that holds the originating nation responsible for the waste if it is disposed of improperly. The European Community has also issued a directive to its twelve member nations requiring that no toxic waste be allowed to be exported unless the receiving country has agreed to receive it and is capable of disposing of it adequately.

The debate has yet to be resolved whether the disposal of waste should be regulated—and thus legitimized—or be prohibited outright. Some environmental organizations argue in favor of prohibition, claiming for example, that no developing country possesses a safe hazardous-waste incinerator or landfill. The United Nations environmental organization is also attempting to draft a treaty to regulate the international shipment of hazardous wastes. Countries that sign the treaty, which is still under discussion, would agree to take back the waste if it cannot safely be disposed of in the importing country. They would also recognize the right of nations to refuse all hazardous-waste imports and would not attempt to make shipments to such nations. Exporting countries would also agree to punish those who illegally ship toxic waste.

The problem of regulating waste disposal has been further complicated in the Northeast, where in some areas garbage collection and hauling has been controlled by the mob. On Long Island in 1989, federal prosecutors filed a civil racketeering suit against 100 hauling operators, alleging that they are controlled by the Luccese and Gambino crime families. Federal officials have also said that the Luccese crime family was involved in the operation of the garbage barge that went around the world in 1987 (192).

Sanitary Landfills

The city dumps of earlier centuries were open, frequently burning dumps. Few open dumps remain in operation in the industrialized world today. Open dumps are a source of vermin, odors, flying papers, cans, and other forms of irritation. Sanitary landfills should minimize such adverse effects. A sanitary landfill, as its name implies, tends to contain the solid waste in a safe and healthy manner. In a sanitary landfill the solid waste is spread in thin layers and compacted and covered daily by a six-inch layer of soil. Some operators feel that the daily covering is unnecessary if the MSW is shredded, since flies and rodents find it more difficult to survive on shredded, compacted waste. In open dumps there is no compaction, shredding, or covering.

The ideal site for a sanitary landfill must meet many requirements (Table 2.1). It should conform to the land-use plan of the area, have public approval, be reasonable in cost, adequate in area, and easily accessible, not pollute groundwater or the surrounding land, be protected from uncontrolled methane generation, and have a sufficient supply of earth cover.

Selecting a Landfill Site

The selection of a landfill site must receive public acceptance and therefore must be preceded by frank and open discussion. The community should be advised of the costs of various sites and the size in terms of years of available capacity, among other relevant factors. One option is to select an area that is zoned for industry. Another is to select a remote site that can serve several communities or a whole region. Yet another possibility is to select barren meadows or strip-mining sites near large cities and landscape them after they have reached capacity. The site

Table 2.1
SANITARY LANDFILL EVALUATION CHECKLIST (117)

	No	Yes
Requirements		
A. Open burning prohibited	___	___
B. Access limited	___	___
C. Spreading and compacting accomplished	___	___
D. Daily cover applied	___	___
E. Final cover applied	___	___
F. Environmental protection provided	___	___
G. Litter control provided	___	___
H. Salvage prohibited	___	___
I. Operational considerations	___	___
J. Special waste handling	___	___
Recommendations		
1. Operation instructions for users provided	___	___
2. Measurement provided	___	___
3. Communications available	___	___
4. Employee facilities provided	___	___
5. Equipment maintenance facilities provided	___	___
6. Unloading area and working face controlled	___	___
7. Fire protection provided	___	___
8. Bulky waste handling provided	___	___
9. Vector control provided	___	___
10. Dust control provided	___	___
11. Accident prevention and safety practiced	___	___
12. Drainage and grading provided	___	___
13. Planning, development and plan execution provided	___	___

selection process should always anticipate the "NIMBY" ("not in my backyard") factor.

Sufficient distance between landfills and inhabited areas should be provided to minimize noise. Fencing is needed to keep out children and prevent trash dispersal. The fences should be about ten feet tall and portable. The site should be accessible to vehicles under all weather conditions and access should be over high-speed roads with ready on-off exits at the entrance point in both directions. If possible, the site should be equidistant from the communities it serves.

The area required for the site will depend on the MSW density in the landfill, which ranges from 400 to 1,500 pounds per cubic yard, depending on the degree of shredding and compaction. If shredding is done at the site, the density should be no more than 1,000/lb. per cu.yd. The required area for a landfill can be calculated by assuming that 4–5 cubic yards of space will be needed per household per year, and by knowing the number of households, the depth of the fill, and the number of decades the landfill is expected to be in operation. If it grows at its present rate Fresh Kills landfill on Staten Island will reach 500 feet in depth, while the depth of a landfill in Los Angeles is already 100 feet.

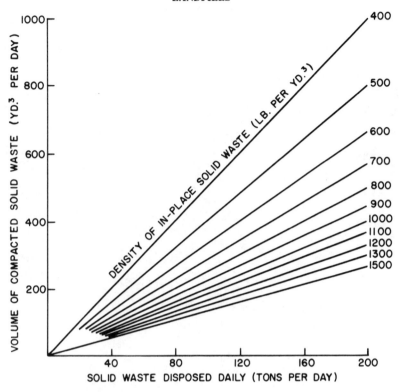

Fig. 2.3 Daily volume of compacted solid waste from small communities (113).

Knowing the rate at which MSW is generated and the density at which it is disposed of in the landfill, daily volume of landfill space required can be obtained from Figures 2.3 and 2.4. The yearly volume requirement in acre-feet can be obtained by multiplying the cubic-yard-per-day-value obtained from Figures 2.3 or 2.4 by 0.226 (113). A common rule of thumb is that 10,000 people will cover a one-acre area with an 8-foot-deep layer of MSW every year.

Leachate Production

To prevent groundwater pollution, surface or rainwater should be prevented from percolating through the refuse and into the groundwater aquifer. The optimum landfill design would eliminate all leachate formation. If the landfill is located near a river, the leachate can percolate directly into that stream in addition to the groundwater (Fig. 2.5). Depending on the permeability of the subsurface soil, water wells in the area can be contaminated. The leachate not only dissolves organic matter and salts but also chlorides and nitrates found in the decomposing wastes. Ammonia and organic nitrogen are gradually converted to nitrates, which in excess of 10 to 20 mg/1 are considered hazardous to newborn infants. Near landfills the iron content

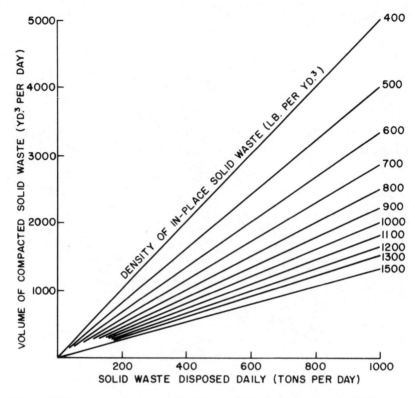

Fig. 2.4 Daily volume of compacted solid waste from large communities (113).

WASTES IN SOLUTION LEACH INTO THE GROUND
WATER AND EVENTUALLY THE RIVER

Fig. 2.5 Pollution resulting from sanitary landfills (114).

32

in the wellwater frequently rises, while the pH drops and the carbon dioxide concentration increases.

If a population of 10,000 yearly fills an area of one acre with an 8-foot-deep
landfill, and assuming that the annual rainfall of 36 inches per year (about one-third
of which becomes groundwater) is allowed to percolate through the landfill, the
leachate will be one-acre-foot per year at a BOD (biochemical oxygen demand)
concentration of about 10,000. This is equivalent to the yearly BOD loading from
the treated sewage of a population of 7,200 (115). Since the landfill remains active
for about twenty years, the BOD pollution load from an unsealed landfill can match
the total waste-water pollution load of the population. Dangerous concentrations of
leachate have been measured at distances exceeding a mile from the landfill (115).

Landfills continue to generate leachate for twenty or thirty years. Tables 2.2
and 2.3 show the characteristics of leachate samples taken at various landfill sites.
In some tests leachate BOD concentrations of 23,000 ppm and total solids concentrations of 33,000 ppm were measured, in addition to trace amounts of many chemicals (115). According to some surveys, 9% of the landfill sites in the United States
have serious leachate pollution problems (115). Actually, because in most communities the groundwater quality is not monitored, the percentage could be as high as
90%. The problems get worse as the landfill is lowered closer to the groundwater

Table 2.2

CHEMICAL CHARACTERISTICS OF SANITARY LANDFILL
LEACHATE CONTRASTED WITH DRINKING-WATER
QUALITY (113)

Constituent	Content of Leachate[a] (mg/l)		Level in Ground Drinking Water (mg/l)
	Low	High	
Alkalinity ($CaCO_3$)	730	9,500	5 to 100
Chloride	96	2,350	3 to 7
Calcium (Ca^{2+})	240	2,330	—[c]
Hardness ($CaCO_3$)	890	7,600	80
Iron, total (Fe)	6.5	220	0.3[b]
Magnesium (Mg^{2+})	64	410	—[c]
Nitrogen, organic (N)	2.4	465	—[c]
Nitrogen, ammonia (NH_4^+)	0.22	480	0.002 to 0.020
Potassium (K^+)	28	1,700	—[c]
Phosphate (PO_4^{3-})	0.3	29	1.0
Sodium (Na^-)	85	1,700	5
Sulfate (SO_4^{2-})	84	730	250[b]
Biochemical Oxygen Demand (BOD)	21,700	30,300	0
pH	6.0	6.5	7.6 to 8.5

[a] From a study by the California State Water Pollution Control Board, 1954. Average values in first 1.3 liters
of leachate per cu. foot of compacted, representative municipal solid waste.
[b] Not to exceed.
[c] Not usually specified.

Table 2.3
ANALYSIS OF LEACHATES FROM LANDFILL
REFUSE (115) (mg/l)

Characteristics	Test Bin[a]	Site 2 Test Fill[a]	Site 8 Test Well[b]	Site 11 Test Well[c]
		Sample Points		
pH	5.6	5.9	8.3	7.2
Total hardness as CaCO₃	8120	3260	537	1540
Calcium	2570	905	72	302
Magnesium	280	254	87	192
Iron. total	305	33.6	219	2.0
Sodium	1805	350	600	69
Potassium	1860	655	nr	13
Boron	nr	7.13	Nil	1.15
Sulfate	630	1220	99	615
Chloride	2240	nr	300	nr
Nitrate	nr	5	18	nr
Alkalinity as CaCO₃	8100	1710	1290	974
Ammonia nitrogen	845	141	nr	0.3
Organic nitrogen	550	152	nr	2.1
COD	nr	7130	nr	36
BOD	32400	7050	nr	nil
Total dissolved solids	nr	9190	2000	1960

Note: nr = no result
[a] Sample from sumps at bottom of test fills.
[b] Sample of water found in refuse during drilling to test well.
[c] Sample from test well located 50 feet from large refuse fill.

table. If the refuse is less than 5 feet above the groundwater, pollution is likely. An example is the otherwise well-designed North Hempstead landfill on Long Island, where in December 1990 the leachate pumping system broke down. The accumulated leachate level has risen and as a result has caused extensive odor emission in the area.

In Seattle, refuse was used to fill a swamp around Lake Washington. Despite precautions, the swamp became severely polluted (115). At the Fresh Kills landfill on Staten Island, some 4 million gallons of leachate are produced every day, a stream of polluted water that is working its way toward the beaches.

Groundwater Pollution

A most precious and irreplaceable resource, groundwater is stored beneath the earth's surface in geologic formations called aquifers. More than half the population of the United States receives its drinking water from acquifers. This water source can be endangered when pollutants seep into the aquifers. In addition to landfills, fertilizers, pesticides, leaking underground tanks, and chemical disposal wells all can contribute to groundwater pollution. The pesticide office of the EPA has identified some fifty toxic chemicals in the drinking-water wells in some thirty states. It has also found some nineteen pesticides in the groundwaters of twenty-four states.

In Iowa half the wells serving the drinking-water supplies of the state's cities were found to be contaminated (134).

Groundwater flows similarly to surface water. The Resource Conservation and Recovery Act requires that all land disposal facilities be monitored by placing one monitoring well upstream and several monitoring wells downstream but immediately adjacent to the disposal site. The EPA is responsible for limiting the increase in groundwater contamination caused by a disposal site.

The most important consideration in siting a landfill is the groundwater table (the level where the water stands in wells at atmospheric pressure). The capillary fringe or moisture zone that exceeds the water levels in wells and landfills should be kept above this moisture zone to make sure that leachate does not reach the groundwater table. The location of the moisture zone and the level of the groundwater table should be measured at its maximum (usually the spring). The bottom of the landfill should be located at a minimum of five feet, but ten feet *above* the groundwater table is preferable. The direction of water flow in the groundwater table should be established in order to determine the likelihood of leachate from the landfill entering the drinking-water supply.

Leachates are potent pollutants that also contain nitrates, which in high concentrations have toxic effects on humans (see Table 2.2). The landfill must be so designed that rainwater and surface runoff will not percolate through the landfill and produce leachates with high concentrations of toxic organics and harmful salts. Before a site is selected or approved, the soil and underlying rock formations must be examined. The type of soil affects drainage and also the availability of suitable cover material. The earth-cover material should be easily workable, compactible, and free of large objects or organic matter that are conducive to the harborage and breeding of vectors.

The Soil Conservation Service of the U.S. Department of Agriculture will examine soil samples and determine the soil composition at the site. The soil classifications used by the USDA and the draining, permeability, and compaction characteristics of different types of soils are listed in Table 2.4. Sandy landfill covers do not prevent rain or surface waters from entering the compacted refuse. Clay is more impervious when moist but in dry summer periods tends to crack, allowing the rainwater runoff to seep through the cracks. The best cover is a layer of clay covered by a second layer of moisture-holding sand or compost.

Methane and Other Gases

When MSW is compacted and buried, the environment become anaerobic. The anaerobic decomposition process consists of an acidic phase and a methane-gas-producing phase. In landfills the concentration of volatile acids rises, reducing the organic matter to strong, putrid acids, which tend to "pickle" the waste and keep it in a putrid state almost indefinitely. One can dig up landfills that are more than twenty years old and still find a malodorous mess, with visibly unchanged, pickled orange peels, showing that decomposition has stopped.

Table 2.4
UNIFIED SOIL-CLASSIFICATION SYSTEM AND CHARACTERISTICS PERTINENT TO SANITARY LANDFILLS (13)

Major Divisions	Name	Potential Frost Action	Drainage Characteristics[a]	Value for Embankments	Permeability cm. per sec.	Compaction Characteristics[b]	Std. AASHO Max Unit Dry Weight lb. per cu. ft.[c]	Requirements for Seepage Control
Coarse-Grained Soils	Well-graded gravels or gravel-sand mixtures; little or no fines	None to very slight	1	Very stable; pervious shells of dikes and dams	$k > 10^{-2}$	Good, tractor, rubber-tired steel-wheeled roller	125–135	Positive cutoff
	Poorly graded gravels or gravel-sand mixtures; little or no fines	None to very slight	1	Reasonably stable; pervious shells of dikes and dams	$k > 10^{-2}$	Good, tractor, rubber-tired steel-wheeled roller	115–125	Positive cutoff
	Silty gravels, gravel-sand-silt mixtures	Slight to medium	2; 4	Reasonably stable; not particularly suited to shells, but may be used for impervious cores or blankets	$k = 10^{-3}$ to 10^{-6}	Good, with close control, rubber-tired, sheepsfoot roller	120–135	Toe trench to none
	Clayey gravels, gravel-sand-clay mixtures	Slight to medium	4	Fairly stable; may be used for impervious core	$k = 10^{-6}$ to 10^{-8}	Fair, rubber-tired, sheepsfoot roller	115–130	None
	Well-graded sands or gravelly sands; little or no fines	None to very slight	1	Very stable; pervious sections slope protection required	$k > 10^{-3}$	Good, tractor	110–130	Upstream blanket and toe drainage or wells
	Poorly graded sands or gravelly sands; little or no fines	None to very slight	1	Reasonably stable; may be used in dike section with flat slopes	$k > 10^{-3}$	Good, tractor	100–120	Upstream blanket and toe drainage or wells
	Silty sands; sand-silt mixtures	Slight to high	2; 4	Fairly stable; not particularly suited to shells, but may be used for impervious cores or dikes	$k = 10^{-3}$ to 10^{-6}	Good, with close control, rubber-tired, sheepsfoot roller	110–125	Upstream blanket and toe drainage or wells
	Clayey sands; sand-clay mixtures	Slight to high	4	Fairly stable; use for impervious core for flood control structures	$k = 10^{-6}$ to 10^{-8}	Fair, sheepsfoot roller, rubber-tired	105–125	None

	Inorganic silts and very fine sands, rock flour, silty or clayey fine sands or clayey silts with slight plasticity	Medium to very high	2	Poor stability; may be used for embankments with proper control	$k = 10^{-3}$ to 10^{-6}	Good to poor, close control essential, rubber-tired roller, sheepsfoot roller	95–120	Toe trench to none
	Inorganic clays of low to medium plasticity, gravelly clays, sandy clays, silty clays, lean clays	Medium to high	5	Stable; impervious cores and blankets	$k = 10^{-6}$ to 10^{-8}	Fair to good, sheepsfoot roller, rubber-tired	95–120	None
Fine-Grained Soils	Organic silts and organic silt-clays of low plasticity	Medium to high	3	Not suitable for embankments	$k = 10^{-4}$ to 10^{-6}	Fair to poor, sheepsfoot roller	80–100	None
	Inorganic silts, micaceous or diatomaceous fine sandy or silty soils, elastic silts	Medium to very high	2	Poor stability, core of hydraulic dam, not desirable in rolled fill construction	$k = 10^{-4}$ to 10^{-6}	Poor to very poor, sheepsfoot roller	70–95	None
	Inorganic clays of high plasticity; fat clays	Medium	5	Fair stability with flat slopes; thin cores, blankets and dike sections	$k = 10^{-6}$ to 10^{-8}	Fair to poor, sheepsfoot roller	75–105	None
	Organic clays of medium to high plasticity; organic silts	Medium	5	Not suitable for embankments	$k = 10^{-6}$ to 10^{-8}	Poor to very poor, sheepsfoot roller	65–100	None
Highly Organic Soils	Peat and other highly organic soils			Not recommended for sanitary landfill construction				

[a] Values are for guidance only; design should be based on test results. *Key:* 1 = excellent; 2 = fair to poor; 3 = poor; 4 = poor to practically impervious; 5 = practically impervious

[b] The equipment listed will usually produce the desired densities after a reasonable number of passes when moisture conditions and thickness of the lift are properly controlled.

[c] Compacted soil at optimum moisture content for Standard AASHO (Standard Proctor) compactive effort.

37

As the organic matter in the MSW decomposes under anaerobic conditions, hazardous gases are generated: CO_2, CH_4, H_2, and H_2S. Of these gases CH_4 and H_2 are explosive, CO_2 is acidic and can cause groundwater corrosion, and H_2S is malodorous, corrosive, and poisonous in higher concentrations. If methane escapes from the landfill, it can cause explosions and severe damage. Horizontal movement of methane gas has been noted at distances as great as 800 feet from the fill.

Landfill gas production is tested by portable instrumentation or by pushing hundreds of probes into the fill at different locations. Based on the detected gas concentrations, contour maps are prepared and vent or collection pipes laid. Most landfills generate enough methane to justify pumping it into pipelines and marketing it.

The heating value of the "dirty" gas from a sanitary landfill is about half that of natural gas. The quantity of gas produced drops off after the landfill is closed for a decade or two. In Pittsburgh, for example, the Chambers landfill produces 3,400 cubic feet of methane every minute. As a typical single-family house requires about a quarter cubic feet of natural gas per minute, the "dirty" methane gas from Chambers could supply nearly 10,000 households (194).

The regulations for landfill operations require that methane be pumped and vented. Compliance with these regulations requires a per-site capital investment of about $1.5 million. For a total per-site capital cost of $3 million, the methane can be vented, collected, and pumped into pipelines. The payback period for this investment is estimated as 3.5 years (133).

Landfill Construction

A detailed map should be prepared for the site showing all homes, buildings, wells, watercourses, dry runs, rock outcroppings, and roads. In addition, a plot plan with a scale not greater than 200 feet/inch should be prepared to show all locations of soil and rock borings, proposed trenches, winter-cover stockpiles, and fencing. Cross sections should show the original ground and the proposed fill elevations. The maps should show the proposed fill area, any borrow sections, access roads, grades for proper drainage of each lift, and a typical cross section of a lift (Fig. 2.6), together with drainage and gas-control devices. The access to the site must be controlled to keep out unauthorized persons and to prevent salvaging or scavenging at the working face of the fill.

Drainage from surrounding areas should be isolated from the landfill by ditches. The landfill covers should be sloped sharply into these ditches to guarantee that water will drain *away* from the fill. The top surface should be graded (slope of 2% to 4%) to drain the surface runoff water. The side slopes should not be so steep as to cause erosion problems. Runoff water from higher areas around the site should be diverted to prevent infiltration and cover erosion.

The sanitary landfill site should include an all-weather building for the purposes of weighing trucks and for record keeping. A weighing-scale platform of 10 by 34

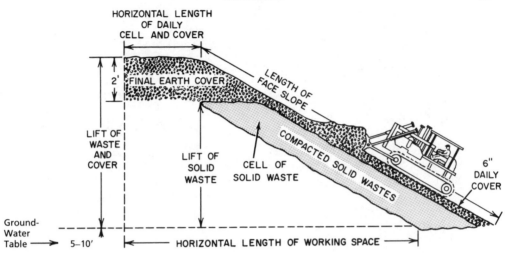

Fig. 2.6 Sectional view of a sanitary landfill (113).

feet is sufficient for most collection vehicles, but a 50-foot platform is needed for truck and trailer units. The equipment requirements vary with the size of the population being served. A village of 10,000 might require only a tractor with a front-end loader, while a town of 200,000 will require two such trucks, plus such optional equipment as a compactor, road grader, water truck, scraper, and drag line (Table 2.5). The equipment should be sufficient for spreading, compacting, and covering operations. In case of equipment breakdown, backup equipment should be available within 24 hours.

Landfilling Procedures

Several procedures are recommended during landfilling (116): The MSW should be spread in thin layers, not more than 2 feet prior to compaction. The number of layers in a cell depends on cell depth, usually about 8 feet. Landfills can be constructed based on the *area method* (Fig. 2.7), where MSW covers the total area, or the *trench method* (Fig. 2.8), where 6 feet of undistured earth is left between trenches. With either method, the working face should be as small as possible for safe and efficient equipment operation.

The solid waste should be covered daily by a 6-inch layer of compacted soil. In addition to aesthetic reasons, this will prevent the blowing of litter, will provide fire breaks between cells, and ensure insect and rodent control. Once the final grade is reached, a minimum layer of two feet of compacted soil should be added as the final cover. This thickness can be increased to protect nearby trees or shrubs. The final cover should be added within a week of completing a section and grass should

Table 2.5

AVERAGE EQUIPMENT REQUIREMENTS FOR THE OPERATION OF
SANITARY LANDFILL SITES (113)

Population	Daily Tonnage	Equipment No.	Equipment Type	Size in lbs.	Accessory[a]
0 to 15,000	0 to 40	1	Tractor crawler or rubber-tired	10,000 to 30,000	Dozer blade Front-end loader (1 to 2 yd) Trash blade
15,000 to 50,000	40 to 130	1	Tractor crawler or rubber-tired	30,000 to 60,000	Dozer blade Front-end loader (2 to 4 yd) Bullclam Trash blade
		[a]	Scraper Dragline Water truck		
50,000 to 100,000	130 to 260	1 to 2	Tractor crawler or rubber-tired	30,000 or more	Dozer blade Front-end loader (2 to 5 yd) Bullclam Trash blade
		[a]	Scraper Dragline Water truck		
100,000 or more	260 or more	2 or more	Tractor crawler or rubber-tired	45,000 or more	Dozer blade Front-end loader Bullclam Trash blade
		[a]	Scraper Dragline Steel wheel compactor Road grader Water truck		

Source: U.S. Department of Health, Education and Welfare, Public Health Service, National Center for Urban and Industrial Health, Solid Wastes Program Publication Number 1792 Sanitary Landfill Facts, Government Printing Office, Washington, D.C., p. 17, 1968.
[a] Optional, dependent on individual need.

be planted to protect it from erosion or surface deterioration. The 2-foot layer of compacted soil should consist of a 12- to 18-inch bottom layer of firmly compacted clay covered by a 6- to 12-inch layer of moisture-holding topsoil, sand, or compost. During droughts, water should be sprayed onto the moisture-holding layer to prevent cracking.

Burning waste should not be allowed and the blowing of paper should be prevented. Daily records should be kept of the quantities and types of MSW received,

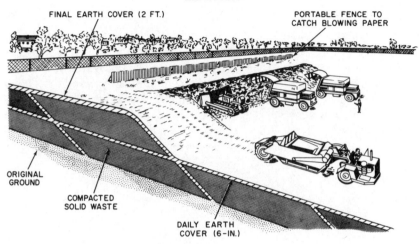

Fig. 2.7 Area method of sanitary landfill (113).

of cover materials used, of the portion of landfill used, and of equipment maintenance and costs. No domestic or wild animals should be allowed on the site. The site should be inspected regularly for rodents and insects and certified reports submitted by independent pest-control firms. A detailed plot plan of the completed fill site should be filed with the local land-record agency to provide information to future users of the site.

The appropriate regulatory agency should periodically inspect the landfill operation. All cracks and eroded or uneven areas in the final cover should be repaired after the completion of the fill and inspected again.

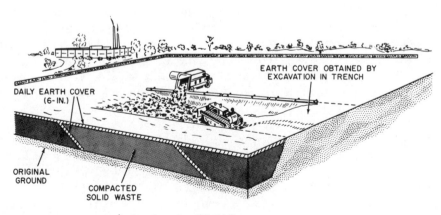

Fig. 2.8 Trench method of sanitary landfill (113).

Landfill Record Keeping

Record keeping should be consistent with the accounting system recommended for sanitary landfill operations by the Solid Waste Management Office, U.S. Public Health Service: Publication No. 2007. The costs associated with building a landfill include clearing, excavation, drainage, and tax and interest, referred to as concealed costs (Table 2.6), and fencing, grading, equipment, and utility and shed building, referred to as site-improvement costs (Table 2.7). Both concealed and site-improvement costs are given in Table 2.8. To obtain the overall yearly cost of operating a landfill, the following calculation has to be made (114):

$$i_2\left(C_1 + \frac{C_1}{(1 + i_1)^{n_1} - 1} + C_s + \frac{C_s}{(1 + i_2)^{n_2} - 1}\right.$$

$$\left. + \frac{C_h}{i_2} + \frac{C_{om}}{i_2} + \frac{C_e}{(1 + i_3)^{n_3} - 1} + \frac{C_{an}}{i_2}\right) = C_a$$

where:

C_a = Annual total cost of landfill operation
C_1 = Cost of land
C_s = Cost of site improvement
C_h = Yearly cost of hauling
C_{om} = Operating and maintenance costs
C_e = Cost of equipment for hauling and operation
C_{an} = Administrative and nuisance costs
i_1 = Interest rate of bonds floated to purchase land (in fractions). If no bonds were issued, use i_2
n_1 = Life of the landfill in years
i_2 = Going discount rate obtained from bank in fractions
n_2 = Average life of sheds, etc., in years
n_3 = Average life of equipment in years

Stabilizing Landfills

Sanitary landfills pickle waste and allow the slow leaching of pollutants into the groundwater. This anaerobic process can be prevented by shredding the MSW and blowing air through the fill for six to twelve months. (The equipment used in shredding is described in Chapter 10.) Shredding doubles the density of the waste and thereby allows more MSW to be disposed of in the same space. Shredding with oxidation stabilizes the MSW and makes it nonodorous and unattractive to flies and rodents. This can eliminate the need for the daily earth covering.

Composting landfills may also be built. (Composting processes are discussed in Chapter 9.) In these fills compressed air is forced through low-cost, 4-inch corrugated polyethylene pipes prelaid under the shredded MSW. As air is blown through the fill, oxidation raises the temperature of the inner layers well above the pasteurization point of 142 F°, killing all pathogenic organisms (99). Some hazardous wastes can also be biologically oxidized and thereby rendered less hazardous or almost

Table 2.6
CONCEALED COSTS OF SANITARY LANDFILL DEPENDENT ON SITE SELECTION (114)

Cost Item	Comments
Clearing	Trees, brush and debris on a site must be cleared. Excessive quantities adversely affect costs.
Excavation	Since the sanitary fill will raise the level of the land, a sizable expense may be the amount of initial excavation needed and haulage costs for excess fill material.
Type of soil	A sandy loam will be easier and, thus, cheaper to work than a dense clay or marsh land.
Drainage	If extensive drainage revisions are needed to prevent pollution or flooding, the installation can be costly.
Tax loss	Purchase of a landfill site will remove it from the property tax rolls which might adversely affect a poorer district.
Interest rates	As with most public improvements, purchase of real estate depends on timing to obtain a favorable rate if the funds are raised by bond issue. Interest rates must also be included when amortizing the land investment.

inert. Aerobic stabilization tests at landfills have resulted in good stabilization and increased compaction.

Composting landfills are more expensive to operate than conventional landfills, but they have the advantage of making earth covering unnecessary, thus minimizing the likelihood of water pollution. They require about one-third the fill area and, in addition, generate a compost product, which can be sold if a market exists for it in the area.

Using Landfills in Land Reclamation

Ideally, landfilling could be used to convert eroded, barren areas into farmlands, parking lots, runways for light airplanes, playgrounds, fair grounds, golf courses, parks, or similar areas where settlement is not critical. The value of the land is

Table 2.7
ITEMS INCLUDED IN THE SITE IMPROVEMENT COST (114)

Fencing	Scraper (if used)
Equipment shed	End loader (if used)
Personnel shed	Haul road
Scales	Grading
Bulldozer(s)	Utilities

Table 2.8
FORM FOR ITEMIZING SANITARY LANDFILL
COSTS (114)

Cost Item	Total Units	Total Cost
Population		
solid waste quantity per person per year	_____	_____
quantity of solid waste per year in tons	_____	_____
Real estate		
number of acres	_____	_____
price per acre	_____	_____
total initial cost	_____	_____
interest rate for bonds	_____	_____
life of landfill until completion	_____	_____
amortized cost of land	_____	_____
Site improvements		
cost of sheds	_____	_____
cost of scales	_____	_____
cost of utilities	_____	_____
other costs	_____	_____
current interest rate	_____	_____
average life of improvements	_____	_____
amortized cost of site improvements	_____	_____
Hauling		
distance of landfill from collection site 1	_____	_____
distance of landfill from collection site 2	_____	_____
average cost per mile	_____	_____
cost per trip	_____	_____
capacity of trucks in tons	_____	_____
number of trips required per year	_____	_____
total cost of hauling	_____	_____
interest rate	_____	_____
life of hauling equipment	_____	_____
amortized cost of hauling equipment	_____	_____
Operation and maintenance		
equipment items	_____	_____
cost to operate (fuel, maintenance, etc.)	_____	_____
hours of operation	_____	_____
number of operators	_____	_____
wage rates for each operator	_____	_____
Total cost of operation		
interest rate	_____	_____
life of operating equipment	_____	_____
amortized cost of operating equipment	_____	_____
Other costs (same procedure to determine the cost of hauling the cover material, etc.)	_____	_____
Administrative costs		
engineering fees	_____	_____
nuisance costs, percentage of total	_____	_____
Total cost	_____	_____
Annual cost	_____	_____

usually increased by the landfilling operation. MSW was used to extend Manhattan's shoreline by a quarter mile into the East River, and valuable land was created through similar means on the lake shore of Chicago.

Lightweight one-story structures such as warehouses can be built on reclaimed landfills. Ninety percent of the landfill settlement is expected to occur in the first five years, but this varies with the rate of decomposition, which in turn depends on many factors, including the weather. Therefore, on reclaimed landfills, no construction should take place in the first five years after completion. Even after five years, methane venting might be necessary (114). In Los Angeles, the amount of settling in landfills averaging 100 feet in depth was 4 feet in three years.

Well-designed landfills can be converted into parks and thereby improve the quality of life in congested urban areas, or barren or eroded rural areas can be converted into farmlands. When reclaimed landfills are used for agricultural purposes, plants with shallow roots are most successful because they are not harmed by the acids formed by the decomposing wastes. As areas suitable for landfilling become more difficult to find, the cost of this method of waste disposal will continue to rise to the point where it will be a more expensive solution than recycling.

Chapter 3
Hazardous Waste Disposal

□
■ ■

TOXIC WASTES include thousands of chemicals and metals—gasoline and other hydrocarbon products, solvents, paints, pesticides, insecticides, heavy metals—that can cause health problems ranging from birth defects to cancer. In the United States, for example, some 7,000 tons of mercury ends up in landfills, mostly household and vehicle batteries (191). The various disposal techniques include landfilling, inciner-ation, deepwell and ocean dumping, solidification, separation/recycling, and biolog-ical degradation. This chapter will cover the domestic and international trends and practices concerning hazardous wastes, the evolving regulations, and the various cleanup efforts in progress. Also discussed will be some of the newer methods of toxic-waste disposal and the design requirements of hazardous-waste landfills.

Types and Sources of Toxic Wastes

Wastes can be grouped into six categories according to their degree of hazard as listed in Table 3.1 (115). The EPA maintains a list of nearly 500 regulated hazardous wastes, including solvents, paints, inks, disinfectants, oils, sludges, and toxicants. The EPA considers a waste hazardous if it is ignitable, corrosive to steel, has a heavy metal content, or is reactive in the sense that it can generate toxic gases or fumes, or can cause illness or death.

Nearly 50% of all hazardous wastes in the United States are generated by the chemical industry, 25% by the metals industry, and 10% by the petroleum industry; the remaining percentages are distributed among all other industries, including ve-hicle maintenance and construction. The total amount of hazardous waste generated

46

Table 3.1
TOXIC WASTES GROUPED INTO TYPES ACCORDING TO THEIR DEGREE OF HAZARD (115)

I. *Inert or relatively inert substances* include municipal trash, tree, shrub and grass trimmings, street sweepings, demolition and construction material, many manufacturing wastes, abandoned vehicles and excavated fill.

II. *Nonpoisonous chemicals and strong, unstable organic wastes* include all of the normal wastes destined for the sanitary landfill that do not fall into category I, including ordinary municipal refuse, *digested* sewage sludge and industrial wastes (many acids, caustics and salts), which are nontoxic.

III. *More hazardous organic wastes* include raw sewage sludge and more hazardous industrial organic wastes, cesspool waste, dead animals, hospital wastes and manures, more hazardous acids, caustics and salts.

IV. *Oils, solvents and volatile sludges,* which may be safely burned under properly controlled conditions.

V. *Pesticides, herbicides and other poisonous liquid and solid chemical wastes,* which also include the very low-grade radioactive wastes.

VI. *Moderate to strong radioactive wastes.*

in the United States exceeds 250 million tons per year (137). This represents only about 3% of all wastes generated, which amounts to 8 billion tons a year (Fig. 3.1).

International law governing the exporting of toxic wastes is just beginning to be formulated. On 22 March 1989 in Basel, Switzerland, more than 100 nations adopted a treaty restricting the shipments of hazardous wastes across international boundaries. The purpose of the treaty is to prevent the exportation of hazardous waste to unsafe or inadequate sites and to require the waste-exporting countries to obtain the consent of the receiving countries prior to shipment. It is the responsibility of each country to ensure safe disposal. The Basel Convention on the Control of Transboundary Movements of Hazardous Waste defines as hazardous the wastes from hospitals and pharmaceutical factories, polychlorinated biphenyls (PCBs), compounds containing mercury or lead, and wastes from the production or use of dyes, paints, and wood-preserving chemicals.

According to the Organization for Economic Cooperation and Development (OECD), 100,000 to 120,000 international shipments of toxic waste, amounting to 2.5 million tons originate in Western Europe alone (10% of total Western European production). Of these shipments 80% are sent to other Western European countries, 15% to Eastern Europe, and 5% to the third world (137). West Germany alone sends more than 300,000 tons of hazardous waste to what was East Germany every year. From the United States less than 1% of the toxic waste is exported over its borders

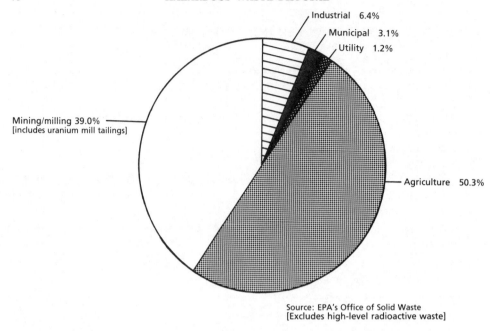

Fig. 3.1 The sources of solid waste generated in the United States at a rate of 8 billion tons per year (136).

to Canada and Mexico. The OECD, an organization of twenty-four industrial nations, adopted a rule that holds the waste-originating nation responsible if the waste is disposed of improperly.

Some European nations have additional rules concerning toxic wastes. In Austria, Switzerland, and the Netherlands, producers of toxic waste must declare types and quantities generated. In Denmark, Finland, and Sweden, toxic wastes must be deposited in central repositories for controlled disposal. Sweden and West Germany have banned wood preservatives containing dioxin (pentachlorophenols), while others have outlawed the manufacture of PCBs.

The economic motivations for the illegal exporting of toxic wastes can be substantial. Landfilling in Africa can cost as little as $20/ton, but disposal through a properly controlled high-temperature incinerator can reach or even exceed $1,000/ton. The case of the Italian garbage barge, the "Karin B," illustrates how the OECD rule is beginning to take effect. In 1988 the "Karin B" carried some 2,100 tons of toxic waste originating from three or four European countries, to Koko, Nigeria. After Nigeria protested and pointed to the OECD rule, the "Karin B" returned to Europe. It first attempted to dispose of the waste in various Western European countries and, after being rejected in each port, finally docked in Livorno, Italy. Because of this unfortunate experience, Italy has now banned the export of toxic waste to the third world. In 1988, in a similar incident, a barge of MSW originating in Islip, Long Island, spent months attempting disposal in the ports of several nations but was forced to return to New York.

Other recent incidents were not resolved within the law. The "Pelicano," for example, a ship carrying toxic incinerator ash from Philadelphia, managed to lose its load in 1988 at the end of a two-year journey. The "Pelicano" arrived empty in Singapore, its load of aluminum, arsenic, chromium, copper, lead, mercury, nickel, zinc, and toxic dioxins probably having been dumped in the sea. Unfortunately, the case of the "Pelicano" may be representative of many unpublished cases of illegal dumping of toxic waste, both in the ocean and in the landfills of the third world.

Toxic Waste Cleanup in the United States

The United States has 30,000 toxic waste dumps. In 1980 the Comprehensive Environmental Response, Compensation, and Liability Act (CERCLA) was passed with the goal of cleaning up these dumps. Cleanup could eventually involve 22,000 sites at a potential cost of more than $100 billion (138). The Hazardous Waste Trust Fund, nicknamed "Superfund," was established at an original budget of $1.6 billion to finance remedial cleanup and emergency situations. In the first six years Superfund completed 6,484 site investigations, initiated cleanups at 156 sites, took civil action at 91 sites, and actually completed the cleanup of 14 sites (136). Since 1980, responsible parties entered into 372 settlements with the EPA worth $619 million in cleanup expenditures. In addition, the EPA recovered $37 million in compensation for cleanups it performed (136). The number of major toxic waste sites selected for urgent cleanup in 1981 was 1,077. As of May, 1991, 1,047 of these have *not* yet been cleaned up (191).

The first decade of Superfund efforts did not bring the anticipated results; instead of cleaning up the dumps, the toxic wastes were buried in landfills or transported to other waste sites, which themselves were leaking. Because it takes an average of five years or more to complete the cleanup of a site, only fourteen sites nationwide had been cleaned up by 1986 and only 30 by 1990 (191).

There is a notable exception to the evenhanded treatment of all industries in the United States: the oil industry is *not* responsible for cleaning up the results of its operation. One such side effect is the radium from the earth's crust that has been brought to the surface by the drilling of millions of wells. Oil drilling equipment, piping and buildings have been reported to emit 8 millirems per hour of radiation, while OSHA requires special protection for workers when the radiation intensity in the workplace exceeds 2 millirems (195). For naturally occurring radioactive materials, the United States has only one repository in the desert near Salt Lake City, but the cost of disposal there is $300–$500 per barrel (195).

For these reasons Congress, in October 1986, renewed the toxic-waste law for another five-year period, increased the funding to $8.6 billion, and provided an additional $500 million to regulate leaking underground storage tanks. The 1986 Superfund Amendments and Reauthorization Act (SARA) will have spent some $9 billion on cleanup between 1986 and 1991. By 1989 only twenty-six properly designed hazardous-waste sites were in operation in the United States.

50

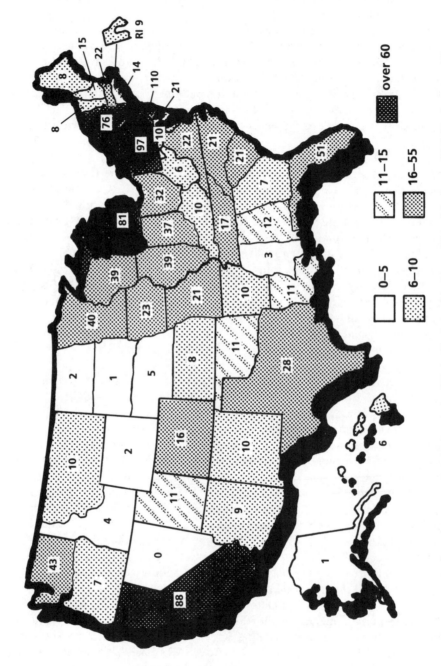

Fig. 3.2 Hazardous waste sites, June 1988. Total of 1,177 sites includes 9 in Puerto Rico, 1 in Guam. (U.S. Environmental Protection Agency, National Priorities List Fact Book [178].)

0–5

6–10

11–15

16–55

over 60

In 1989 the federal government forced states to submit their plans for disposing of hazardous wastes over the next twenty years. To participate in Superfund, the states have to prove either that they will be able to dispose of their toxic wastes within their own borders or that they will have entered into regional agreements with other states to do so. The effectiveness of this approach was limited because in 1990 only twelve states were receiving Superfund money.

The Financial and Legal Aspects

The cost of hazardous-waste disposal is much higher than that of MSW. Tipping fees in hazardous-waste landfills can exceed $100/ton, while disposal costs in special high-temperature incinerators can reach $1,000/ton. The cleanup costs of existing toxic dump sites are also high. In 1989 the Army and the Shell Oil Company agreed jointly to pay about $1 billion to clean up the Army's 27-square-mile Rocky Mountain Arsenal next to Denver's Stapelton Airport. Shell produced the pesticides aldrin and dieldrin at the site, and the cleanup became unavoidable when trichlorethylene was detected in the drinking-water supplies of the area. The goal under the agreement is to complete the cleanup by the year 2000.

Another new development in the toxic-waste controversy is the flurry of lawsuits. The nearly 30,000 toxic-waste sites represent a cleanup liability of more than $100 billion. Faced with such liabilities and with possible toxic-exposure litigations, producers of toxic waste are turning to their insurance companies to find ways to reduce their liabilities. Today, the laws regulating toxic cleanup remain unsettled. For example, the Fourth Circuit of the United States Court of Appeals ruled that property and casualty insurance does not cover toxic-waste cleanup and related liabilities. This ruling, supporting the insurance companies' arguments, was allowed to stand by the U.S. Supreme Court. Courts in New York and New Jersey have ruled that insurance should cover such liabilities. Because of all the uncertainty, manufacturers and insurance companies are filing lawsuits against each other on a "sue before you are sued" basis over the issue of who must pay for toxic-waste cleanup (138).

An important new development is the prosecution of companies participating in illegal dumping. In May 1989 the General Marine Transport Corporation of Bayonne, New Jersey, its president, and five employees were indicted on criminal charges. The indictment names twenty-five occasions when sewage sludge containing pathogenic bacteria and viruses, as well as hazardous, toxic, and acidic wastes, was discharged into the coastal waterways and ocean off New Jersey. A civil suit has also been filed to seize two of the company's tugboats and three of its barges. New York State later shut down the company's barging operation and in 1990 the Superior Court in Newark sentenced the president of the company to five years probation and the company itself to $1 million in fines.

Changing Regulations and Practices

The Resource Conservation and Recovery Act (RCRA) and CERCLA and its 1986 amendments (SARA) have imposed strict guidelines for new hazardous-waste landfills and have placed the responsibility for identifying and controlling hazardous wastes on the EPA. The EPA controls the movement of hazardous waste by assigning identification numbers to all firms that are authorized to generate, transport, treat, store, or dispose of toxic waste. Each shipment is assigned a manifest that identifies the waste, its quantity, and its source and destination. All parties involved in the shipment are required to sign the manifest.

The relationship between industries using hazardous materials and the communities in which they operate is changing. In 1986 Congress passed legislation requiring all industrial plants to disclose if they are using or storing any of more than 300 hazardous substances. About 1.5 million farms and factories are subject to the new provisions (139) and all must disclose to the states and the EPA the quantities of hazardous substances they release annually into the atmosphere, water, or ground. This information will be given to local emergency committees, fire departments, and the public. Armed with such information, citizens would be able to pressure local governments and industries to correct the practices that threaten the health of the community or the environment.

Interstate trafficking of hazardous wastes is also fraught with problems. The largest hazardous-waste landfill in the United States is in Emelle, Alabama, while the second largest is in Pinewood, South Carolina. The amendments to SARA require states to certify that they have a twenty-year capacity for disposal. This disposal capacity can be either within or outside a state's borders. The EPA also considers illegal any interference with the interstate transportation of hazardous waste, but this concept is being tested. South Carolina has banned toxic wastes (poisonous chemicals like arsenic and mercury, waste oils, pesticides, and flammable or corrosive chemicals) from those states that do not allow disposal or treatment within their own borders. This ban applies to twenty-three states and Puerto Rico and, if upheld in the courts, will require new rules governing the interstate transport of toxic wastes (140).

Reducing and Recycling Hazardous Waste

As the federal government moves to curtail the land disposal of the 298 million tons of hazardous wastes generated every year, a whole new industry will emerge. By 1994, $13 billion a year will be spent on recycling, recovering, reducing, and treating toxic wastes (141). Some 250,000 sites—from large chemical plants to small photo processors—now generate toxic wastes. Added to this are the garden chemicals, used oil, paints, pool chemicals, antifreeze, solvents, and cleaning agents generated by the average household.

Some toxic-waste generation can be reduced at the source, while other wastes can be reused or recycled. Thus the challenge involves more than just disposal. It is likely that some chemicals will be replaced by less hazardous or nontoxic substances as the costs of disposal and liability insurance rise. Substances for which nontoxic alternatives are urgently needed include solvents and pesticides, cyanide wastes in metal extraction, and heavy metal and hydrocarbon contaminants. There are many signs that some replacement and redesign are already taking place. The toxic waste produced in Connecticut, for example, dropped by 15% between 1983 and 1987 simply because some corporations redesigned their manufacturing processes to cut wastes, while others introduced evaporative techniques to recapture and reuse metallic wastes (142).

A number of alternative technologies are aimed at reducing toxic-waste generation. Ion-exchange methods are widely used to remove metals from solutions, allowing them to be reused (141). Other waste-separation technologies include stripping by air or steam to remove volatile chemicals, coal adsorption, precipitation and flocking, and soil washing and flushing (136).

Hazardous-Waste Disposal Techniques

In the past most toxic wastes were disposed of in conventional landfills, injected underground (about 25% of hazardous wastes in the United States), or dumped in the oceans. These disposal options have already been banned in many states (141). The thermal destruction of hazardous wastes has been used in 13% of all Superfund removal actions through the use of mobile incinerators, fluidized bed or infrared treatment, plasma arc, and pyrolysis (136). High-temperature incineration—multiple-heart and fluidized bed incinerators (discussed in more detail in chapter 5)—is the leading technology for destroying toxic wastes. In the multiple-heart design, the waste is burned and reburned to ensure complete destruction, while in the fluidized-bed units superheated wastes are passed over a bed of sand to hasten combustion.

Offshore incineration in international waters has also been considered as a means of disposing hazardous waste. It has been argued that safety is enhanced when the incineration of dioxins and polycarbonated biphenyls (PCBs) is carried out at remote locations far from population centers. On the other hand, it has also been argued that spills or accidents that might occur during offshore incineration of toxic wastes will destroy marine life and threaten coastal communities.

The EPA has selected some sites as being potentially suitable for offshore incineration. In the 1970s it conducted tests on burning PCBs and dioxin-containing herbicides such as Agent Orange in the Gulf of Mexico. These tests, conducted in part by Waste Management Inc. of Illinois, which has two incinerator ships, have been terminated because of public opposition. Ocean incineration in the North Sea is also expected to be terminated by the mid-1990s.

Another method of toxic-waste disposal is immobilization or solidification, a technique that disposes of indestructible metals by fixation or immobilization in flyash or cement blocks. (The storage of immobilized wastes is more convenient than the storage of liquids.) One of the most promising techniques is biodegradation. It has been found that bacteria species can be developed to degrade just about any toxic substance (143). With this method microorganisms are introduced into the contaminated area to consume the hazardous waste. In most cases the microbes already exist in the contaminated soil and groundwater (this is nature's way of cleanup anyway), and engineers need only to make the conditions more conducive to the growth of these organisms.

In the future engineers may be able to "genetically design" microbes for the task of degrading specific hazardous wastes. Remedial Technologies of Boston specializes in biodegradation processes, while Detox Industries of Sugarland, Texas, promotes the use of naturally occurring microbes to metabolize hazardous waste (143). The unit costs of these methods can be under $50/ton, which compares favorably with high-temperature incineration (143).

Incinerator Ash

Incineration does not eliminate the need for landfilling. The 3,000-ton/day incinerator proposed for the Brooklyn Navy Yard will produce 900 tons of daily residue, containing lead, cadmium, and possibly dioxins. In 1988, 7 million tons of incinerator ash was produced in the United States and there could be five times as much by the turn of the century. When unseparated MSW is "mass-burned," the residue is 25% of the feed by weight (10% by volume). If only the combustible portion of the MSW is burned (RDF), the residue is half as much.

Most states are treating municipal incinerator ash as nonhazardous. More recent investigations show that, while bottom ash is non-hazardous, fly-ash is not. Since the heavy metal (lead, cadmium, etc.) and dioxin content of this residue frequently exceeds federal limits, incinerator fly-ash should be placed in toxic-waste dumps. Unfortunately, there are only twenty-six such sites nationwide and none exists in the New York City area. Still, the probability is that eventually incinerator fly-ash will be required to be disposed of in landfills provided with double lining and with continuous leachate collection and monitoring. Of course, the high tipping fees at these hazardous-waste dumps will increase the cost of MSW incineration.

Another possible way to dispose of incinerator bottom ash is to mix it with 15% concrete and use it as building blocks. Scientists working at the Marine Sciences Research Center at Stony Brook used such blocks to build a reef on Long Island Sound in 1987 and a boathouse in 1991 (201). The heavy metal or other toxic leaching from these blocks is reported to be nondetectable. Incinerator bottom ash has also been used in this manner in Japan and it is possible that regular or bottom ash can eventually be disposed of in this manner.

Landfilling Hazardous Waste

In the past it was acceptable to place category I, II, and III hazardous wastes (as defined in Table 3.1) in sanitary landfills if the landfills were provided with all-weather impervious covers and were located in sandy subsoil at an elevation of at least 10 feet above the highest groundwater table. If category V material was disposed of in sandy soil, a two-foot layer of clay was placed over the sand and sloped toward a perforated tile covered with sand or gravel. The tiles brought the leachate to a roofed-over monitoring basin for testing and evaluation. Most large landfills used to have a "hazardous waste section" that was constructed according to these criteria and subject to stricter rules. Today, the distinction between hazardous and sanitary landfills is being drawn more precisely.

In 1985 the EPA reduced by one-third the types of hazardous wastes that are allowed to be placed in conventional sanitary landfills. For example, it banned liquid

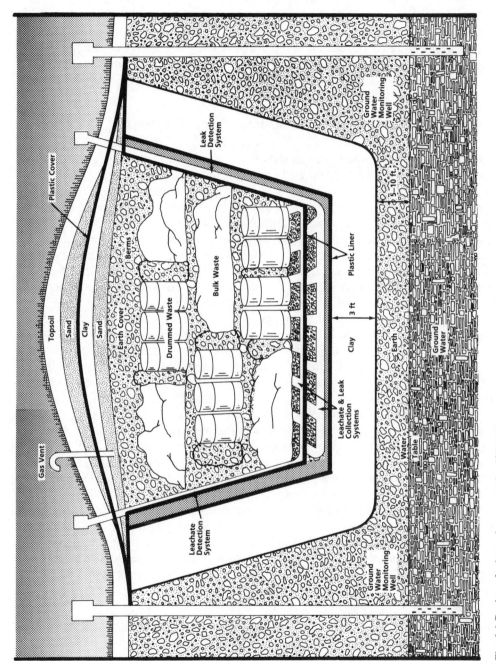

Fig. 3.3 Design of a hazardous-waste landfill (117).

55

wastes from land disposal. A similar ban is expected shortly on deepwelling or underground injection of hazardous wastes. The EPA is also considering banning the land disposal of all solvent and dioxin-containing wastes. Unfortunately, while progress has been made in regulating the design and operation of new landfills, few existing municipal landfills have been upgraded. Regarding the upgrading of the operation of existing landfills, some states, including Ohio, now require the monitoring of methane formation.

The Design of Hazardous-Waste Landfills

The design of a compact hazardous-waste landfill is illustrated in Figure 3.3. The landfill itself must be located at an elevation of 5 to 10 feet above the groundwater level and groundwater monitoring wells must be provided around the landfill. The outer shell of the landfill consists of a three-foot layer of compacted clay to prevent the seeping out of leachate. Two high-density polyethylene plastic liners are placed above the layer of clay. Between these two liners is aggregate, honeycombed with a leakage collection system consisting of perforated collection pipes connected to a leachate detection and pump-out system. The collected leachate is treated in a couple of tanks, one to degrade the organic content, the other to remove ammonia, before discharging the residue. In addition, the operation is continuously monitored. The cost of building such a landfill is around $500,000 an acre.

When it is full, the landfill is capped with a sloped multilayered cover. The bottom layer of the cap is made of a porous material such as sand. Porous pipes are placed in this layer to collect the methane and thereby eliminate the fire hazard. The collected gas is either vented, burned on the site, or piped to users. The next layer is clay covered by a plastic cover, which protects it from drying out and also seals it from the infiltration of rain or runoff water. Another layer of sand is placed on top of the plastic cover, followed by a layer of topsoil for planting grass.

The operator of the landfill is responsible for checking each shipment of waste for properties like pH and density and to make sure the material is correctly labeled. If there is a possibility of mislabeling, further checks and analysis are performed. After checking, the wastes are placed in the landfill according to a preapproved layering plan (Fig. 3.3).

A separate book could be devoted to the topic of hazardous-waste reduction, reuse, and disposal. This chapter has discussed only the general trends, practices, and emerging technologies. Society is just beginning to become aware of the scale of the toxic-waste problem, and it will take decades before the halfway measures of burning and burying can be replaced by safer and more permanent solutions, such as replacement, recycling, and biodegradation.

Chapter 4
Sewage Sludge Disposal

□
■■

SEWAGE SLUDGE, the stabilized and digested solid-waste product of the waste-water treatment process, can be disposed of by landfilling, incineration, composting, or ocean dumping. Nature returns organic materials to the soil as fertilizers (Table 4.1). Organic materials become wastes when they are not returned to the soil but instead are burned, buried, or dumped in the ocean. These unhealthy practices began when chemical fertilizers took away the market from sludge-based compost and when industrial wastes began to contaminate sewage sludge with toxic metals (lead, cadmium), making it unusable for agricultural purposes (Table 4.2). Until recently, the bulk of the sewage sludge generated by metropolitan areas has either been landfilled or dumped in the ocean. These options are gradually disappearing and as a result municipalities will be forced to make some hard decisions.

In late 1988 Congress passed a bill banning the dumping of sludge at sea by 1992. The bill calls for dumpers to pay disposal fees of $100 per dry ton in 1989 and $200 by 1991; beginning in 1992, civil penalties of $600 per dry ton will be imposed. This penalty is to increase 11% each year, with part of the penalties collected deposited in a fund for the development of alternate disposal methods.

The Sludge Problem in the Northeast

Ocean dumping of sewage sludge has been going on since 1924. In 1972 Congress authorized the EPA to regulate this form of disposal. Until 1987 New York and New Jersey were dumping the bulk of their sewage sludge in 80-foot-deep water twelve miles off Sandy Hook, New Jersey. When that area turned into a virtual dead sea, the EPA designated the 6,000-foot-deep waters over the edge of the Continental

Table 4.1
DRY BASE COMPOSITION OF POTENTIAL ORGANIC
FERTILIZERS[a] (150)

Type of Potential Fertilizer	Total Nitrogen (%)	Phosphorus P_2O_5 (%)	Potash (%)	Volatile Solids (%)
Cow manure[b]	3	1	3	66
Poultry manure[b]	5	2	1	64
Human excreta[b]	12	5	4	72
Raw sludge	5	1	—	70
Anaerobically digested sludge	2	0.5–4	0.5	45–60
Activated sludge	6	3–4	1	60–75

[a] Approximate values are given.
[b] Nitrogen is expressed as N and phosphorus is expressed as phosphoric acid.

Table 4.2
CHEMICAL CONSTITUENTS IN SLUDGES
(PERCENT, DRY BASIS) (147)

Chemical Constituents	Chemical Symbol	Digested (Toledo)	Activated (Milwaukee)
Silicon	SiO_2	15.6	8.452
Iron	Fe_2O_3	5.43	7.148
Aluminum	Al_2O_3	6.8	3.211
Calcium	CaO	6.64	1.675
Magnesium	MgO	1.83	1.810
Potassium	K_2O	0.42	0.862
Sodium	Na_2O	—	0.988
Titanium	TiO_2	—	0.083
Manganese	MnO	0.78	0.0327
Copper	CuO	1.0	0.0465
Barium	BaO	0.06	0.0611
Zinc	ZnO	1.24	0.01627
Lead	PbO	0.81	0.225
Nickel	NiO	—	0.00561
Cobalt	CoO	—	0.0002
Sulfur	SO_3	8.12	2.900
Chlorine	Cl_2	—	0.501
Chromium	Cr_2O_3	0.37	0.219
Arsenic	As_2O_3	0.13	0.01347
Boron	B_2O_3	—	0.00426
Iodine	I	—	0.00113
Phosphorus	P_2O_5	2.02	3.089
Ignition loss	—	44.4	68.55
Total		95.65	99.9036

Shelf (Fig. 4.1) at a location called the "106 Mile Site" for the dumping of the 9 million yearly tons of sludge. (This name was given to the 130-square-mile site because it is 106 miles from the closest landfall at Cape May.) As of this writing, the only agencies still dumping sludge off the American coast are the nine sewer agencies of New York and New Jersey. Partially treated sewage is also being discharged into the sea by Boston and Los Angeles (187).

The issue of ocean dumping is still not fully resolved in New York. Federal and state edicts require that both New Jersey and the New York counties of Westchester and Nassau halt their ocean dumping in 1991. New York City claims that it cannot meet such an early target and wants to continue dumping until 1998. (In the 1970s the mayor of New York had promised to stop ocean dumping by the early 1980s, but economic considerations prevented him from meeting that target.)

The landfilling of sewage sludge varies from state to state. New York and Connecticut landfill about 25% of their sludge regardless of its metal content. The land-disposal regulations in New Jersey do not allow that, so some 18% of New Jersey's sludge is shipped to less regulated, out-of-state landfills (144). The plans for the future also vary from state to state. New York and Connecticut are trying to make useful products such as fertilizers and composts out of their sludge. New composting plants are being built in Schenectady, Binghamton, and Plattsburg, New York, and in Hartford and Bristol, Connecticut (144). Middlesex County, New Jersey, also hopes to produce a soil-substituting landfill-cover material from its sludge

Fig. 4.1 Diseased lobsters and crabs have been found in undersea canyons along the Continental Shelf. (The *New York Times,* May 22, 1988 [124].)

by chemically binding the cadmium, zinc, and copper in the sludge. Other counties in New Jersey hope to solve their sludge-disposal problem by building a regional incinerator in Bayonne (144). New York City has not yet decided if it will turn to incineration or composting in the future and for the next few years plans to continue dumping its sludge in the ocean.

The Cost of Sludge Disposal

The cost of sewage-sludge disposal has risen substantially during the last quarter century. In the 1960s the cost of composting, landfilling, or ocean dumping seldom exceeded \$10/dry ton (120 and 153), and even incineration could be accomplished for under \$20/ton (Tables 4.3, 4.4, and 4.5). Considering that the daily per capita sludge production on a dry basis is only 0.2 pounds, the cost of sludge disposal did not represent a major element in a municipality's budget. Even when all the sludge preparation, pretreatment, and ash-disposal costs were included in the total sludge-disposal costs, the costs remained reasonable (Table 4.6).

In 1980 the city of Baltimore still estimated its sludge-disposal options as ranging from only \$22/ton for barging over a round-trip distance of 300 miles to \$38/ton for incineration (120). Today, the economic picture is quite different. The utility cost of sludge incineration alone is reported to range between \$20 and \$42/ton (145). The tipping fee at the Fresh Kills landfill on Staten Island has reached \$100/ton and the ocean disposal fee set by Congress for 1991 is \$200/dry ton.

As the options of ocean dumping and landfilling disappear, cities will be faced with major new capital investments for building sludge-handling facilities. The new composting plant in Hartford, Connecticut, is estimated at \$20 million, while the new sludge incinerator for Newark, New Jersey, will cost \$300 million (144). The cost of sludge disposal is rapidly becoming a considerable cost in the total budget of municipalities.

Ocean Dumping

In 1970 Jacques Cousteau, the underwater explorer, said: "In 30 years of diving I have seen this slow death everywhere under the water. In the last 20 years, life in our oceans has diminished by 40 percent. If it continues, I predict that man has only

Table 4.3
INCINERATION AND DRYING COSTS (152)

Equipment	Type of Operation	Deodorizing Equipment Included	Total Annual Cost ($/ton of dry combustible solids)
Multiple-hearth or other	Incineration	no	15.51
type of incinerator		yes	18.97
Flash	Incineration	no	18.19
dryer		yes	21.94
incinerator	Drying	no	29.09
		yes	37.02

Table 4.4

ESTIMATED SEWAGE TREATMENT COSTS BASED ON A 7.5 MGD RAW SEWAGE FLOW (146)

	Conventional Treatment Steps			Treatment Steps	Advanced Treatment Steps		
Cost Items	Primary and Secondary Treatment $/MG ($/1000 m³.)	Odor Control and Disinfection, $/MG ($/1000 m³.)	Organic Sludge Handling, Incineration and Ash Disposal $/MG ($/1000 m³.)	Phosphorus Removal, Lime Costs and Recalcination Costs $/MG ($/1000 m³.)	Nitrogen Removal and Recarbonation $/MG ($/1000 m³.)	Filtration $/MG ($/1000 m³.)	Carbon Adsorption Regeneration $/MG ($/1000 m³.)
Capital amortization	56.00 (14.75)	2.50 (0.658)	15.50 (4.08)	25.50 (6.72)	18.00 (4.74)	19.50 (5.13)	23.50 (6.18)
Operating labor	22.00 (5.79)	—	6.39 (1.682)	13.94 (3.67)	2.62 (0.69)	2.96 (0.779)	13.39 (3.52)
Power	18.20 (4.79)	—	3.16 (0.832)	2.70 (0.711)	5.13 (1.35)	13.55 (3.57)	0.30 (0.079)
Fuel	—	—	18.70 (4.92)	24.52 (6.45)	—	—	0.80 (0.21)
Chemicals	—	1.83 (0.482)	17.54 (4.62)	29.50 (7.76)[a]	—	12.89 (3.39)	8.06 (2.12)[b]
Maintenance labor	1.35 (0.355)	0.18 (0.047)	4.07 (1.07)	2.76 (0.727)	1.46 (0.384)	0.40 (0.105)	1.39 (0.366)
Equipment replacement	1.65 (0.434)	0.02 (0.005)	2.94 (0.774)	1.99 (0.523)	1.04 (0.274)	0.15 (0.040)	0.17 (0.004)
Total Cost/MG	$99.20 (26.12)	$4.53 (1.19)	$68.30 (17.99)	$100.91 (26.56)	$28.25 (7.44)	$49.45 (13.01)	$47.61 (12.49)

Note: Total conventional treatment costs: $172.03 ($45.4); total tertiary treatment costs: $226.22 ($59.6); total all treatment costs: $398.25 ($105.0).
a Cost of lime makeup.
b Cost of carbon makeup.

Table 4.5

ESTIMATED OPERATING DATA AND COSTS OF MULTIPLE-HEARTH SLUDGE INCINERATORS (146)

Operating and Cost Factors		Equivalent City Population Used in Estimate				
		10,000	20,000	50,000	100,000	1,000,000
Sludge incinerated lbs. per day (wet basis)[a]		8,000	16,000	40,000[b]	80,000[b]	800,000[b]
Sludge incinerated lbs. per day (dry basis)		2,000	4,000	9,928	19,857	198,570
Operating schedule hrs. per week		35	35	70	70	168
Furnace feed lbs. per hour (wet basis)		1,600	3,200	3,971	7,943	33,095
Furnace feed lbs. per hour (dry basis)		400	800	993	1,986	8,274
Furnace required: number of hearths outside diameter		6 hearth 10.75' OD	6 hearth 12.75' OD	6 hearth 14.25' OD	6 hearth 18.75' OD	12 hearth 22.25' OD
Installed cost	$	357,500	429,000	494,000	624,000	1,007,500
Annual fuel cost	$	536	823	1,987	3,385	27,346
Annual power cost	$	302	352	813	1,077	7,069
Annual labor cost	$	5,460	5,460	10,920	10,920	26,208
Total annual operating cost	$	6,298	6,634	13,721	15,382	60,623
Annual amortization cost 25 yrs., 6% interest	$	27,966	33,559	38,644	48,813	78,813
Total annual system cost	$	34,264	40,194	52,365	64,196	139,436
Cost per ton dry solids:						
fuel	$	1.47	1.13	1.10	0.94	0.76
power	$	0.83	0.48	0.45	0.30	0.20
labor	$	15.00	7.50	6.04	3.02	0.73
total operating cost	$	17.30	9.11	7.59	4.26	1.68
amortization cost	$	77.00	46.20	21.40	13.50	2.17
Total system cost	$	94.30	55.31	28.99	17.76	3.85
Total cost per ton wet sludge	$	23.58	13.83	7.25	4.44	0.96

[a] Cake moisture: 75%
Volatile content: 65%, 10,000 BTU/lbs of volatile solids
[b] Rounded off number, scaled up from line below.

Table 4.6
COST OF SLUDGE DISPOSAL METHODS (151)

Method of Disposal[a] or of Volume Reduction	Total of Capital and Operating Costs ($ per Dry Ton)	
	Average	Range
Composting with refuse	8[b]	5–15[b]
Heat drying	50[c]	40–55[c]
Incineration		
wet combustion	42	26–50
multiple-hearth and fluidized bed	30	10–50
Landfilling dewatered sludge	25	10–50
Disposal as a soil conditioner w/o heat drying (dewatered)	25[c]	10–50[c]
Disposal on land as a liquid soil conditioner	15[c]	8–50[c]
Lagooning	12	6–25
Barging to sea	12	5–25
Underground disposal	Unknown, potentially inexpensive	
Pipeline to sea	11	—

[a] Ultimate disposal includes cost of preparation, such as dewatering, digestion and ash disposal.
[b] Costs added by author omitted in reference.
[c] Gross cost; does not account for money received from sale of sludge.

50 more years on this planet'' (121). Cousteau's concerns, while possibly overstated, are certainly justified. For many centuries the oceans were considered to be infinite (a mile and a half deep on average, covering most of the globe), inexhaustible, and everlasting. Today, we know that the oceans are a limited, sensitive, and valuable resource that must be protected not only from dumping but also from polluted rivers and land runoffs. On the other hand, we are beginning to learn that the damage caused by sludge dumping in shallow waters will not necessarily occur in deeper regions.

Some calculations suggest that the cycle time of the oceans (the time it takes for an ocean's waters to evaporate, form clouds, return to the land in the form of rain, percolate down into the groundwaters, and eventually return to the rivers and back to the ocean) is about 2,000 years. Cousteau was reporting only the first visible consequences of ocean pollution. It will take 2,000 years to learn the total impact of the pollution to date.

The Effects of Ocean Dumping

Some authorities argue that the addition of organic nutrients benefit the sea, citing statistics on the increased yield of fish (120). The increase in fish yield in controlled environments such as nutrient-rich fish ponds tends to support this view.

Others argue that the ocean dumping of sludge is safe because sludge contains only treated and stabilized biodegradable substances without any floatables.

Some will also argue that as long as there is enough dissolved oxygen in the water to support animal life and decompose the organic wastes, sludge dumping will not upset the ecological balance of the receiving waters. With this logic one could view the ocean as a great sink, capable of absorbing almost anything that is thrown into it. If this view were correct, San Francisco Bay could handle the wastes of 200 million people, since tidal action in the bay replaces the water twice a day (120).

It is frequently argued that sludge, the end product of the sewage treatment process, is a relatively benign substance. This is not completely true. Unfortunately the sludge from a city like New York also contains toxic industrial wastes, because industrial plants frequently dump their wastes in the municipal sewage system. The synergistic effects of the many synthetic chemicals, toxic substances, pesticides, PCBs, heavy metals, and medical wastes containing viruses and bacteria are not fully understood and are likely to be very harmful to the ecology of the receiving waters. Many experts feel that neither the dissolved oxygen level nor the organic-waste-assimilating capacity of the oceans is a safe criterion for waste disposal. The effects and interrelationships are much more complex and the consequences are much less well understood to accept such simplistic arguments.

Since dumping started at the 106 Mile Site, which receives New York's sludge, Rhode Island fishermen and lobstermen have reported diseased lobsters and crabs and a general drop in their catch of bottom-dwelling fish. There is some controversy concerning the fishermen's reports that sludge is driving the fish away or that it causes shell disease (burn-spot disease) in lobsters and crabs. The National Oceanic and Atmospheric Administration (NOAA) has indicated that the dilution of the ocean is sufficient to eliminate the harmful effects, but it made that statement without studying the fish in the area (124).

Scientists have reported a proliferation of certain forms of sea life at the 106-Mile Site (187). This discovery could raise questions about the wisdom of banning ocean dumping by the end of 1991. Part of the reason why the damage to the receiving waters is reduced, if not eliminated, is the great depth and large area of this site. The heavy fraction of the sludge takes 3–4 days to sink to the bottom, while the lighter particles take up to a year. This allows more time for microorganisms to decompose the sludge. While the scientists reported that the ocean bottom at the dump site teems with sea life, they have not yet determined if this sea life is contaminated with bacteria, viruses, or heavy metals that might enter the food chain (187).

Regulating Ocean Dumping

Ocean dumping of sludge and other solid wastes is widely practiced in Europe and Japan. For many decades in the United States, sewage sludge was barged to so-called approved areas on the Gulf and Atlantic coasts (122). Toxic wastes dumped in the ocean were usually put in containers and shipped to more remote locations.

While many cities, such as San Diego and San Francisco, have banned ocean dumping, others like New York City continue to barge their sludge into the sea.

The marine-disposal of radioactive wastes was terminated in the United States in 1967. Yet in 1968 the yearly quantity of other types of wastes dumped in the sea (Table 4.7) was still close to 50 million tons (120). Even though Congress has banned the ocean dumping of sewage sludge by 1992, it will probably take until the turn of the century before this form of waste disposal is stopped completely.

Unregulated disposal beyond the boundaries of the Territorial Sea imperils the waters, resources, and beaches of the maritime nations. Specific legislation is needed to give national and international authorities the responsibility for preventing ocean pollution and protecting ocean resources. The creation of such authorities and enforcement methods is proving to be a slow, difficult process, and no effective policy for ocean management has yet evolved on either the national or international level.

Illegal Ocean Dumping

The incident of the "Khian Sea" illustrates the state of international controls on ocean dumping. In September 1986 the "Khian Sea," owned by a Bahamian company, the Amalgamated Shipping Corporation, loaded up in Philadelphia with 28 million pounds of toxic incinerator ash. The ship attempted, unsuccessfully, to discharge its load in the Bahamas, the Dominican Republic, Honduras, Costa Rica, Guinea-Bissau, and the Cape Verde Islands. In February 1987 it did discharge 4 million of its 28 million pounds of ash in Haiti but then was ordered out of that country.

In September 1988 the "Khian Sea" was sighted in the Suez Canal with a new name, "Felicia," and a new owner, Romo Shipping, Inc. In October 1988 Captain Abdel Hakim, vice-president of Romo Shipping, sent a message to Amalgamated Shipping (which has since gone out of business) indicating that the ash had been discharged; Hakim did not say *where* it had been discharged. It is not clear how the case will be resolved and who will answer for what in front of which legal authority. But this incident illustrates the chaotic state of international controls over dumping in international waters. Illegal dumping is not limited to international waters. In December 1988 the State Attorney General of New York accused the General Marine Transport Corporation of illegal dumping in the Raritan River, in the Hudson Bay, and in the coastal waters of New York City. The lawsuit also names four officers of the corporation and recommends placing environmental police on the barges.

Some sludge haulers did not take their loads to designated areas, but dumped them closer inshore. To control this situation, the EPA now requires that each load of sludge be accompanied by a "black box," which is dumped with the load (187). This allows the EPA to protect against cheating. Another EPA requirement is that the barge must dump the load slowly, so that dispersal is maximized.

Illegal dumping also includes medical wastes, which has caused the closure of the New York area beaches. Until this occurred, no government agency was charged with the task of tracking the safe disposal of hospital wastes from the point of

Table 4.7

ESTIMATED MARINE DISPOSAL FOR 1968 (153)

Type of Waste	Pacific Coast		Atlantic Coast		Gulf Coast		Total		Percent of Total	
	Annual Tonnage	Estimated Cost, $	Annual Tonnage	Estimated Cost, $	Annual Tonnage	Estimated Cost, $	Annual Tonnage	Estimated Cost, $	Tonnage	Cost
Dredging spoils	7,320,000	3,175,000	15,808,000[a]	8,608,000	15,300,000	3,800,000	38,428,000	15,583,000	80%	53%
Industrial wastes	981,000	991,000	3,011,000	5,406,000	690,000	1,592,000	4,682,000	7,989,000	10%	27%
containerized	300	16,000	2,200	17,000	6,000	171,000	8,500	204,000	<1%	1%
Refuse[b]	26,000	392,000					26,000	392,000	<1%	<1%
Sludge[c]			4,477,000	4,433,000			4,477,000	4,433,000	9%	15%
Miscellaneous	200	3,000					200	3,000	<1%	<1%
Construction and demolition debris			574,000	430,000			574,000	430,000	1%	2%
Explosives			15,200	235,000			15,200	235,000	<1%	<1%
Total wastes[d]	8,327,500	4,577,000	23,887,400	19,129,000	15,986,000	5,563,000	48,210,090	29,269,000	100%	100%

[a] Includes 200,000 tons of flyash
[b] At San Diego, 4,700 tons of vessel garbage at $280,000 dumped in 1968 (discontinued in Nov. 1968).
[c] Tonnage on wet basis. Assuming average 4.5% dry solids, this amounts to about 200,000 tons/yr. of dry solids being barged to sea.
[d] Radioactive wastes omitted because sea-disposal operations were terminated in 1967.

generation to the point of disposal. After the beach closings, bills were introduced requiring the EPA to create a "paper trail" in order to control the disposal of medical wastes. New regulations in New York State have been issued requiring that hazardous and infectious medical debris, including needles, be placed in strong, moisture-resistant "red bags" conspicuously labeled "infectious." Medical practitioners must either carry their infectious waste to an approved hospital incinerator or deliver it to a certified trucker. An elaborate system of record keeping has also been set up to track the waste from source to disposal.

Sludge Pipelines and Marine Fills

Some coastal cities have found that piping sewage sludge into the sea is less expensive than barging it. In the late 1960s Los Angeles, for example, reduced its sludge-handling costs from $14 to $2/dry ton by constructing a 7-mile-long, 22-inch-diameter pipe and discharging the sludge through the pipe at a depth of 320 feet on the edge of a submarine canyon (120). Today, Los Angeles mixes some of its sludge with sawdust and sells it as compost, but it pipes the rest into the sea. Boston and Los Angeles are the only two cities in the United States that still pipe their sewage sludge into the ocean.

In designing outfalls in the ocean, the goal is to achieve good mixing between the heavier saltwater and the lighter sewage effluent. Good mixing is essential to ensure that the waste does not surface like an oil slick and pollute the shoreline and beaches. The following equation has been suggested in designing an outfall (120):

$$N = (K)(Q^2)/(Y)(X^2)$$

Where:

N = the maximum tolerable shore pollution expressed as the arithmetic mean or 80 percentile

Q = the average sludge flow

Y = the depth at which the sludge is discharged into the ocean

X = the distance of the discharge point from the shore

K = a constant that varies from 5 to 10 million, if the units are given in feet and gallons

Another method of sludge and MSW disposal is to build ocean landfills. Hong Kong has built "marine fills" as a means of disposing of its refuse in the estuaries of the bay. The marine fill is surrounded by a solid rock dyke and refuse is loaded and compacted into it. Tidal action causes some leaching of the marine fill. After about three years the refuse converts into compost and is used on nearby farms.

Sludge Composting

Part of the difficulty in enforcing bans on ocean dumping is the prohibitive cost of other options for sludge disposal. The alternatives to ocean dumping include landfilling, incineration, and composting. If sludge is deposited in conventional sanitary landfills, the environmental damage is transferred from the ocean to the land.

   The most desirable alternative to ocean dumping of digested sludge is to use the sludge as an organic fertilizer on croplands (123). In fact, the most natural method of disposal for organic wastes in general is through composting (discussed in more detail in chapter 9). The economics of composting are dependent on the markets for the soil conditioner produced and on the toxic-metal content of the sludge. This is a solution to the sludge-disposal problem only if toxic metals are prevented from entering the sewer system and if the market for sludge-based compost is supported by the government.

   Composting sewage sludge is frequently accomplished in conjunction with shredded municipal refuse. The shredded refuse is usually mixed in a proportion of about 50% to the sludge. The refuse provides the needed body or structure to improve aeration, which is essential for high-quality compost. When sludge is composted without shredded refuse, some of the dryer, already-composted sludge solids must be recycled with the raw, dewatered sludge to obtain a drier, more porous sludge, which ensures good conditions for composting.

   Nature maintains the organic content of the soil by returning the residue of dead vegetation to it. When crops are removed from the land, the soil must be replenished by other means. The traditional solution has been to add manure-based organic fertilizers to improve friability and porosity, increase moisture retention, reduce the leaching of chemical fertilizers, stimulate root growth, and make the otherwise insoluble minerals available to the vegetation. Organic fertilizers must be

Table 4.8
FERTILIZER VALUES OF SEWAGE SLUDGE AND VARIOUS
MANURES (147)

| Material | Percent Moisture | Percent on Dry Basis | | | |
		Nitrogen as N	Phosphoric Acid	Potash	Volatile Solids
Digested primary sludge	52.7	2.30	1.4	—	46.7
Digested primary with trickling filter sludge	41.0	2.54	1.8	—	49.0
Digested activated sludge	35.0	3.03	2.5	—	50.5
Heat-dried activated sludge	—	5.13	3.27	0.40	63.5
Human urine	96.3	16.2	4.6	5.4	65.0
Human feces	77.2	4.38	4.82	1.1	86.8
Human excreta (std.)	94.5	12.2	4.66	3.9	71.8
Dried poultry manure	—	5.0	1.95	1.16	64
Dried goat manure	—[a]	1.35	1.00	3.00	—
Dried sheep manure	—	1.5–3.0	1.0–2.5	0.3–2.2	48
Horse manure	59	1.7	0.6	2.0	—
Cow manure	79	2.9	1.1	2.9	66
Steer manure	78	3.3	2.4	2.4	66
Hog manure	87	3.8	2.7	3.0	—

[a] Not available

free of weed seeds or pathogens and must contain a minimum amount of nitrogen, phosphorous, and potassium (NPK). As shown in Table 4.8, the fertilizer values (NPK) of sewage sludges and manures tend to be low, while their organic values tend to be high.

Sludge Processing and Preparation

The first step in stabilizing sludge is aerobic or anaerobic digestion. Aerobic digestion (activated sludge process) produces a better fertilizer value because it concentrates and traps the nitrogen in the microorganism cells. The longer the period of aeration, the higher the percentage of nitrogen incorporated. In anaerobically digested sludge a large percentage of the nitrogen is lost as it becomes soluble and is lost with the sludge supernatant (147).

Dewatering sludge (Fig. 4.2) is accomplished through vacuum filtration, centrifugation, sand-bed drying, prefreezing, and heat drying. The relative costs of these processes are given in Table 4.9. Heat drying sterilizes the organic fertilizer without changing its chemical content. In open drying beds, lagoons, or stockpiles, the loss in nitrogens, phosphorous, and other organic compounds can amount to 25% to 50% due to rainfall leaching and anaerobic digestion by bacteria. Smaller communities favor open sand beds for drying sludge, while larger ones reduce the labor and land requirement by using vacuum filters. In vacuum filtering, minute quantities of polyelectrolytes are added to increase the efficiency of filtration.

Sand-Bed Drying

In smaller communities air drying digested sludge is an accepted method of dewatering. More than half of communities with fewer than 5,000 inhabitants use it; this proportion drops to 25% for municipalities between 5,000 and 25,000 and to 5% for cities with more than 25,000 inhabitants (149). The installation is simple and requires little investment, and the operation requires no special skills. In smaller communities these considerations overshadow the disadvantages in that air drying requires large areas (1 to 3 square foot per capita) and the labor costs of removing the sludge after drying are high. Other drawbacks include effects of adverse weather, the likelihood of odor, the breeding of insects, and the inability of the process to destroy pathogens and weed seeds.

The drying beds are constructed of a smooth layer of 9 to 18 inches of 0.3 to 0.75 mm sand with 4″-diameter underdrain pipes spaced 8 to 10 feet apart. The beds are partially paved with a sloping asphalt surface with a slope of 1 to 2 inches per foot toward the bed to allow front-end loaders to remove the dried sludge. The advantage of drainable sand beds is the superior drying rate. A reasonable compromise is to use an open asphalt bed with a thin layer of sand over it for drainage (149).

A well-digested primary sludge of 6% to 8% solids content is spread on an open sand bed in a thickness of 7 or 8 inches. The dried sludge is removed when it begins to peel away from the sand, but before it starts to crumble. The dried sludge still has a moisture content of 60% to 70%. Figure 4.3 shows the relationship between

Fig. 4.2 Dried sludge in a treatment plant (149).

Table 4.9
COST OF SLUDGE-HANDLING
PROCESSES (151)

Method of Sludge Handling[a] and Pretreatment	Total of Capital and Operating Costs ($ per Dry Ton)	
	Average	Range
Thickening		
(1) gravity	—	1–5
(2) air floatation[b]	—	6–15
(3) centrifugation[b]	—	3–20
Dewatering		
(1) vacuum filtration	15	8–50
(2) centrifugation	12	5–35
(3) sand bed drying	—	3–20
Anaerobic digestion	—	4–18
Elutriation	—	2–5
Lagooning	2	1–5
Landfilling	—	1–5[c]
Pipeline transportation	5	—[d]
Liquid sludge disposal on land as a soil conditioner	10	4–30
Heat drying	35	25–40
Incineration	20	8–40
Barging to sea	10	4–25

[a] Sludge handling (specific costs of individual treatment or disposal processes).
[b] Varies greatly, depending on the need for chemicals.
[c] Long hauls would be higher.
[d] For moderate distances, cost varies with length.

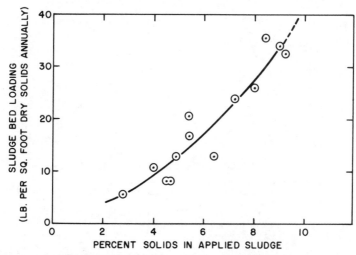

Fig. 4.3 Sludge volume reduces at a variable rate during drying, with the greatest volume reduction occurring at 60 percent to 70 percent moisture content (149).

the solid content of the applied sludge and the annual production of dried solids in pounds per square foot of bed area. The beds often can be left open, except in a cold climate or if odor becomes a problem.

Sludge Incineration

Incinerating sewage sludge has been practiced for the last sixty years. Early designs were either "flash drying" or "multiple-hearth" types, while in recent years fluidized-bed incinerators have also been used. The flash-drying process has a low

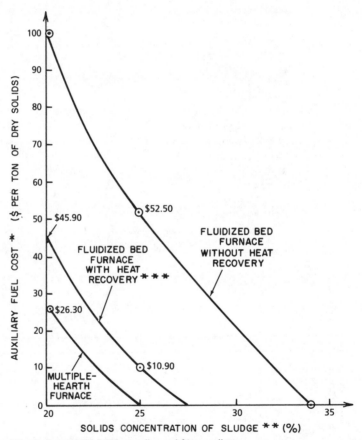

° No. 2 oil at 135,000 BTU per gallon and $1 per gallon
°° Sludge solids are assumed to be 75 percent volatile and to have a heating value of 9500 BTU per pound.
°°° Air is assumed to be preheated to 1000°F, whereas exit air temperature is assumed to be 1500°F; the calculations are based on the use of 20 percent air. For the multiple-hearth furnace an 800°F exit air temperature has been assumed.

Fig. 4.4 Auxiliary fuel costs of sludge incineration processes as a function of sludge moisture content (146).

capital cost and is flexible in the sense that it can produce as much dried sludge as there is a market for; the remainder can be incinerated. Its limitations are the added cost of pay-fuel and the associated odor and pollution problems. The multiple-hearth design, the most widely used for sludge incineration today, reduces odor and pollution but provides less operating flexibility because it cannot dry the sludge without also incinerating it. The most recent and, in the judgment of some, the most advanced design is the fluid-bed sludge incinerator, which can operate in either the combustion or the pyrolysis mode. The exhaust temperature from the fluid-bed incinerator is higher than from a multiple-hearth furnace, so afterburners are less likely to be needed to control odor.

As shown in Figure 4.4, the auxiliary fuel cost of sludge incineration is higher with fluidized-bed incinerators than with multiple hearths. The cost varies according to the moisture content of the sludge and with the degree of heat recovery (146). To eliminate the need for auxiliary fuel, the dry-solid content must exceed 25% for multiple-hearth and 32% for fluidized-bed incinerators (146). In some recent fluidized-bed installations in Japan, operating costs have been cut in half through heat recovery (145).

In a multiple-hearth incinerator with a feed containing 10% solids, the ash is about 10% of the feed. The compositions of incinerator ashes are given in Table 4.10. The ash is either landfilled or marketed as soil conditioner. Table 4.11 gives the composition of Vitalin, the ash from Tokyo's Odai plant. (The Japanese word "lin" means phosphorus). The term "soil conditioner" is used instead of "fertilizer" because the phosphate content is less than 12%, the nitrogen content is under 6%, and the total NPK content is less than 20%.

Table 4.10

TYPICAL ANALYSIS OF ASH FROM TERTIARY QUALITY ADVANCED WASTE TREATMENT SYSTEM (100)

Content	Percent of Total Sample			
	Lake Tahoe 11/19/69	Lake Tahoe 11/25/69	Minn.–St. Paul 9/30/69	Cleveland 3/2/70
Silica (SiO$_2$)	23.85	23.72	24.87	28.85
Alumina (Al$_2$O$_3$)	16.34	22.10	13.48	10.20
Iron Oxide (Fe$_2$O$_3$)	3.44	2.65	10.81	14.37
Magnesium oxide (MgO)	2.12	2.17	2.61	2.13
Total calcium oxide (CaO)	29.76	24.47	33.35	27.37
Available (free) calcium oxide (CaO)	1.16	1.37	1.06	0.29
Sodium (Na)	0.73	0.35	0.26	0.18
Potassium (K)	0.14	0.11	0.12	0.25
Boron (B)	0.02	0.02	0.006	0.01
Phosphorus pentoxide (P$_2$O$_5$)	6.87	15.35	9.88	9.22
Sulfate ion (SO$_4$)	2.79	2.84	2.71	5.04
Loss on ignition	2.59	2.24	1.62	1.94

Table 4.11
COMPOSITION OF MHI ASH FROM ODAI PLANT
IN TOKYO

Silica oxide	30.00	Potassium	1.00
Magnesium oxide	3.30	Nitrogen	0.20
Calcium oxide	30.00	Manganese	0.06
Phosphoric oxide	6.20	Copper	0.61
Ferric oxide	18.20	Boron	200.00 ppm

Note: Ash is marketed under the trade name Vitalin.

Flash-Dryer Incineration

The flash-dryer incinerator process was first introduced in the 1930s as a relatively low-capital-cost, space-saving alternative to air drying sludge on sand beds. At the time, this method of drying was advantageous because the resulting heat-dried sludge was virtually free of pathogens and weed seeds and also because the process was flexible enough to produce only that amount of dried sludge that could be marketed. The disadvantages of this process are dust and odor. These problems, while manageable through the use of dust collectors and afterburners, make the flash-dryer less popular than the multiple-hearth and fluidized-bed incinerators.

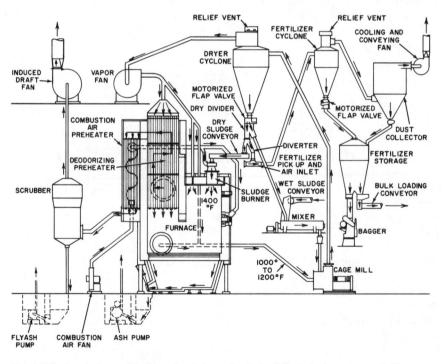

Fig. 4.5 Sludge drying and incineration using a deodorized flash-drying process (148).

In the flash-drying process (Fig. 4.5), the wet, dewatered sludge is mixed with dry sludge from the dryer cyclone. This preconditioned mixture comes in contact with a stream of 1000° to 1200° F gas, which moves it at a velocity of several thousand feet per minute. In this turbulent, high-temperature zone, the moisture content of the sludge is reduced to 10% or less in only a few seconds. As the mixture enters the dryer cyclone, the hot gases are separated from the fine, fluffy heat-dried sludge. Depending on the mode of operation, the flash-dried sludge is either sent to the sludge burner and incinerated at 1400°F, or it is sent to the fertilizer cyclone and recovered as a saleable fertilizer product. When the incoming wet sludge contains about 18% solids, it takes about 6500 BTUs (half a pound of coal plus one cubic foot of natural gas) to produce a pound of dry sludge (15,000 kJ/kg) (148).

Multiple-Hearth Incineration

Multiple-hearth incineration was developed in 1889 and was first applied to sludge incineration in the 1930s. It is the most widely used method of sludge incineration (26). The multiple-hearth furnace consists of a steel shell lined with a refractory (Figs. 4.6 and 4.7). The interior is separated into compartments by horizontal brick arches. The sludge is fed through the roof by a screw feeder or a belt feeder

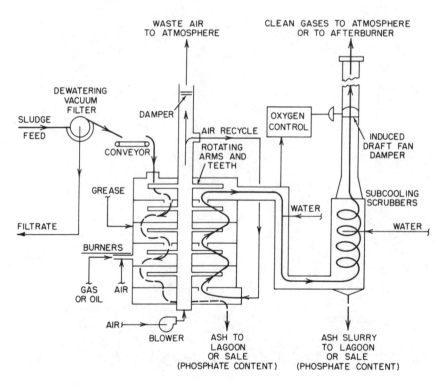

Fig. 4.6 Multiple-hearth incineration of sludge (26).

Fig. 4.7 Multiple-hearth furnace system. Odai treatment plant, Tokyo, Japan. Feed conveyors are shown from vacuum filter to storage bin (top right) and from sludge storage bins to furnace feed (top left). Exhaust gas pipes at right lead to air scrubbing devices and to stack (100).

and flapgate at a rate of about 7 to 12 pounds per square foot. Rotating rabble arms push the sludge across the hearth to drop holes, where it falls to the next hearth. As the sludge travels downward through the furnace, it turns into a phosphate-laden ash (see Tables 4.10 and 4.11).

The sludge is dried in the upper, or first, operating zone of the incinerator. In the second zone, it is incinerated at a temperature of 1400° to 1800° F (760° to 982° C) and deodorized. In the third zone, the ash is cooled by the incoming combustion air. The air, which travels in counterflow with the sludge, is first preheated by the ash, then participates in the combustion, and finally sweeps over the cold incoming sludge, drying it until the moisture content is about 48%. At this percentage of moisture content, a phenomenon called "thermal jump" occurs as the sludge enters the combustion zone. The thermal jump allows the sludge to bypass the temperature zone where the odor is distilled. The exhaust gases are 500° to 1100° F (260° to 593° C) and are usually odor-free. The sludge temperature profile across the furnace is shown in Table 4.12.

The pollution-control equipment usually includes three-stage impingement-type scrubbers for particulate and sulfur dioxide removal (Chapter 6) and standby afterburners, which destroy malodorous substances such as butyric and caproic acids (26). The need for afterburners is a function of exhaust-gas temperature. Usually at temperatures above 700° to 800° F (371° to 427° C) in a well-controlled incinerator, where the combustion process is complete, the afterburners are not necessary to guarantee odor-free operation. If combustion is not complete, however, the exhaust-gas temperature might need to rise to 1350° F (732° C) before the odor is distilled. In such cases it is less expensive to install an afterburner than to use auxiliary fuel to achieve such high exhaust temperatures.

If the incoming sludge contains 75% moisture and if 70% of the sludge solids are volatile, the incineration process will produce about 10% ash. The ash can be used as a soil conditioner (Fig. 4.8) and as the raw material for bricks, concrete blocks (Fig. 4.9), and road fills, or it can be landfilled. In the United States the

Table 4.12
SLUDGE FURNACE
TEMPERATURE
PROFILE (26)

Hearth No.	Approximately at Half Capacity (°F)	Nominal Design Capacity (°F)
1	670	800
2	1380	1200
3	1560	1650
4	1450	1450
5	1200	1200
6	325	300

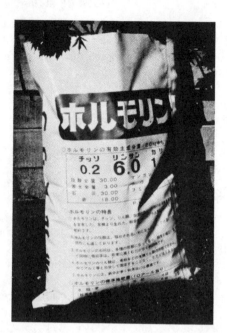

Fig. 4.8 Incinerator ash marketed in Japan as soil conditioner (100).

Fig 4.9 Experimental block from sludge ash (Nakahama Treatment Plant, Osaka, Japan). The numbers 60, 20, and 20 refer to ash, slag and silicas content, respectively (100).

78

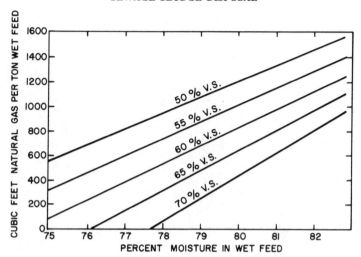

Fig. 4.10 Multiple-hearth incinerator fuel consumption as a function of moisture content in the feed and percentage of volatile solids. *Note:* 1. Curves not applicable for feed rates below 4 tons per hour; 2. Curves do not include allowance for lime as filter aid; 3. To correct for lime, downgrade volatile solids according to lime dosage. Assume lime forms calcium hydroxide, therefore each pound of CaO forms 1.32 pound calcium hydroxide; 4. Natural gas calorific value is assumed to be 100 BTU per ft.³; 5. Heat content of sludge is based on 10,000 BTU per pound of volatile solids; 6. %VS represents percentage of volatile solids in the feed (146).

supply of phosphates is sufficient for less than a century (100), so the phosphate content of sludge ash is important. If the ash also contains zinc or chromium, it can damage certain crops, although it will not damage cereals or grass (100). The harmful effects are more likely to occur in acidic soils and can be offset by adding lime to the sludge.

The main advantages of multiple-hearth incinerators include their long life and low operating and maintenance costs (146), their ability to handle sludges with a moisture content of up to 75% without requiring auxiliary fuel, their ability to incinerate or pyrolize hard-to-handle substances such as scum or grease, their ability to reclaim chemical additives such as lime in combination with or separately from incinerating the sludge, and their flexibility, in that they can be operated intermittently or continuously at varying feed rates and exit-gas temperatures (26). The auxiliary fuel requirement varies with the dry-solids content of the sludge and with the percentage of volatiles in the solids (Fig. 4.10).

Fluidized-Bed Incineration

Fluidized-bed incineration can handle sewage sludge containing as much as 35% solids. The sludge is injected into a fluidized bed of heated sand. The incinerator (Fig. 4.11) is a vertical cylinder with an air distribution plate near the bottom, which

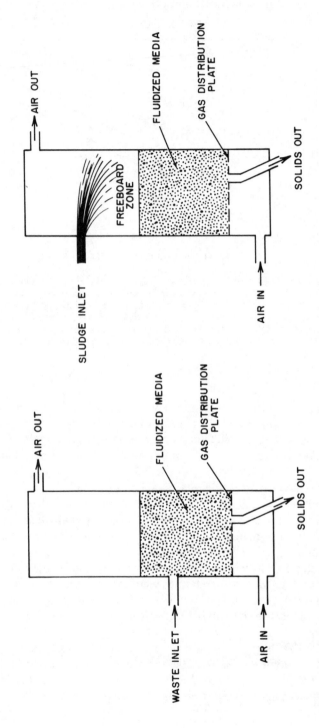

Fig. 4.11 Fluidized-bed incinerator (28).

allows the air to enter the sand bed while also supporting it. As the air flow increases, the bed expands and becomes fluidized. The solid waste in the sludge can be destructed by either combustion or pyrolysis. During combustion the organic materials are turned into carbon dioxide:

$$\text{Sewage Sludge} + \text{Oxygen} \rightarrow CO_2 + H_2O + \text{Ash} + \text{Heat}$$

The heat of combustion contributes to maintaining the fluidized bed at a temperature of about 1400° F (760° C). In the pyrolysis process the sludge is decomposed in the presence of inert gases at 1400° F (760° C), which yields hydrogen, methane, carbon monoxide, and carbon dioxide. For pyrolysis, auxiliary heat is required to maintain the fluidized bed at the high reaction temperature:

$$\text{Sewage Sludge} + \text{Inert Gas} \rightarrow H_2 + CO + CO_2 + CH_4 + \text{Auxiliary Heat}$$

The operation of the fluidized-bed incinerator is optimized by the control of the air-flow rate and of the bed temperature. Since reaction rates are related to bed mixing and the source of agitation is the fluidizing air, reaction rates can be adjusted by changing the airflow supply. Bed temperature is usually maintained between 1300° and 1500° F (704° to 815° C). For complete combustion (odor-free operation), about 25% excess air is needed (28).

The organic materials can be deposited on the sand particles (agglomerative mode) and removed by continuous or intermittent withdrawal of the excess bed material. An alternative mode of operation (nonagglomerative) is to combine the organic ashes with the exhaust gases and collect them downstream with dust collectors.

The main advantages of fluidized-bed incinerators include the uniformity of the bed, the elimination of stratification and hot or cold spots, the high rate of heat transfer for rapid combustion, the elimination of odor and the need for afterburners, and the low maintenance requirements of the process. The disadvantages include the high operating-power requirement, the need for auxiliary fuel if the dry solid concentration is less then 32% (see Fig. 4.4), and the need for dust-collection devices.

Fluidized-Bed Incinerators with Heat-Recovery

The addition of heat-recovery equipment can increase the capacity of fluidized-bed incinerators by about 20% (145). The plant shown in Figure 4.12 has been in operation in Tokyo since 1984 (145). Heat recovery is achieved by inserting a heat exchanger into the stream of the exiting hot gases, which serves to generate a supply of hot thermal oil. The oil is then used as the energy source to heat the sludge cake dryers.

The hot oil passes through the hollow inside of the motor-driven screws (Fig. 4.13), while the sludge cake is both moved and heated by these screws. As shown in Table 4.13, the cake is dried to a substantial degree (some 20% of the inlet flow

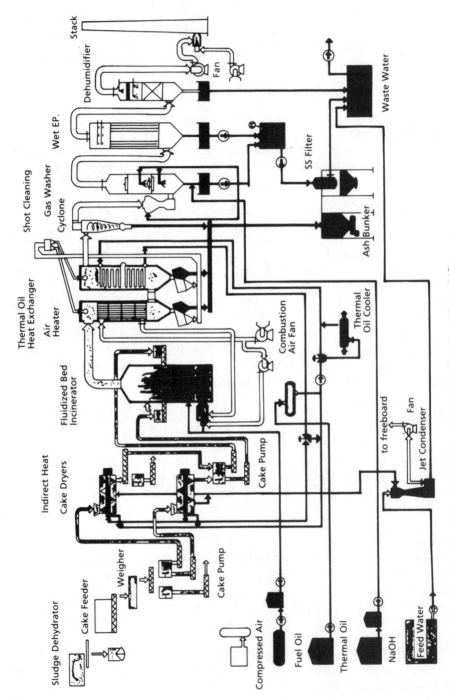

Fig. 4.12 Flow diagram of sewage sludge incineration plant with indirect heat dryer (145).

82

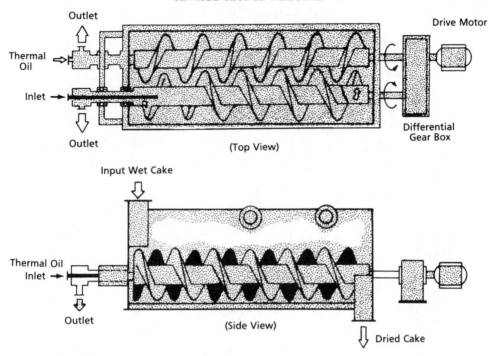

Fig. 4.13 Top and side view of indirect heat cake dryer (145).

is evaporated) as the sludge cake passes through the cake dryer. The heating oil circulates in a closed cycle and is maintained at about 480° F (250° C) inside the screw-conveyor dryers by the throttling of two three-way valves. One valve can increase the outlet oil temperature from the cake dryers by blending in the warmer inlet oil; the other can lower the outlet temperature by sending some of the oil through an oil cooler.

The operators of the Tokyo incinerator feel that the total capital cost of the plant is unaffected by the addition of the heat-recovery feature because the cost of the heat-transfer equipment is balanced by the reduced capacity requirement. The operating costs, on the other hand, are cut in half with the heat-recovery system (Table 4.14).

Another interesting feature of this system is the method of cleaning the accumulation of ash and soot from the heat-transfer surfaces. This automated system utilizes 3- to 5-mm-diameter steel-shot balls that are dropped every three to six hours from the top of the hot-air heaters. The random movement of the balls removes the

Table 4.13
OPERATIONAL DATA OF FLUIDIZED BED SLUDGE INCINERATOR WITH AND WITHOUT INDIRECT HEATING (145)

	Direct Incineration (without drying)			Incineration with Dried Cake					
	Design	Run 1	Run 2	Design	Run 3	Run 4	Run 5	Run 6
Input cake (tpd)	80	60	80	96	60	80	96	70
Cake property								
moisture (%)	78	83–86 (84.7)	←	75–80 (78)	83–86 (84.8)	←	←	78–80 (79)
combustibles (%ds)	56–64 (60)	80.2–80.7 (80.5)	←	56–64 (60)	80.1–80.2 (80.2)	79.1–79.4 (79.3)	79.3–80.4 (84.7)	(80)
lower heating value (MJ/kg ds)	12.6	18.5–18.9 (18.7)	←	12.6	18.7	18.8	18.5	18.8
Dried cake								
moisture (%)	same	same	same	66–73 (70)	80.1–80.9 (80.6)	79.8–80.6 (80.2)	79.2–80.2 (79.7)	70–72 (71)
weight (tpd)	80	60	80	70	46.7	61.8	72.4	43
Supporting fuel								
(liter/h)	312	238	265	47	92	94	96	0
(liter/ton cake)	94	95	80	12	37	28	24	0
Furnace temperature								
sand bed (°C)	800	790	780–820 (800)	800	770–810 (780)	770–810 (780)	760–810 (780)	(780)
outlet (°C)	850	860	850–950 (900)	800	760–800 (780)	760–800 (780)	760–810 (780)	(780)
Fluidizing air (m³ Normal/h)	7500	6400	7500	5500	5500	5500	5500	5500
Excess air ratio	1.40	1.62	1.52	1.56	2.02	1.61	1.41	(1.8)

Note: Values in parentheses are average values. All units are in metric (SI).

Table 4.14
OPERATING COSTS OF FLUIDIZED-BED SLUDGE INCINERATION WITH AND WITHOUT HEAT RECOVERY (145)

Items	Utility Unit Cost	Normal Undried Cake Incineration		Newly Developed Indirectly Dried Cake Incineration	
		Amount	Cost	Amount	Cost
Plant capacity		80 tons/d		96 tons/d	
Operation cost			US$/d		US$/d
1. Supporting fuel	0.35 US$/L	6360 L/d	2226	2304 L/d	806
2. Electricity	0.1 US$/kWh	10800 kWh/d	1080	10300 kWh/d	1030
3. Chemical (NaOH)	0.35 US$/kg	220 kg/d	77	151 kg/d	53
4. Lubrications	2.6 US$/L	0.8 L/d	2	0.8 L/d	2
Total per day	—	—	3385 US$/d	—	1891 US$/d
Unit treatment cost per input cake vol.	—	—	42.3 US$/ton	—	19.7 US$/ton

Note: All units are in metric (SI).

Table 4.15
CONCENTRATION OF POLLUTANTS IN ATMOSPHERIC EMISSIONS AND IN ASH PRODUCED BY A FLUIDIZED-BED TYPE SLUDGE INCINERATOR (145)

Items	Regulations	Run 1	Run 4
Exhaust gas			
sulfur oxides (SO_x)	292 ppm	4 ppm	25 ppm
nitrogen oxides (NO_x)	250 ppm	—	41 ppm
hydrogen chloride (HCl)	93 ppm	3 ppm	—
dust density	0.05 g/m³N	0.006 g/m³N	0.002 g/m³N
Residual ash			
amount	7.0 tons/d	2.4 tons/d	2.9 tons/d
ignition loss	<15%	0.6%	0.6%
dissolution to water			
alkyl mercury (Hg)	N.D.	<0.0005	—
total mercury (Hg)	0.05	<0.0005	—
cadmium (Cd)	0.1	<0.05	—
lead (Pb)	1	<0.2	—
phosphorus (P)	0.2	<0.01	—
chromium 6+ (Cr[VI])	0.5	<0.05	—
arsenic (As)	0.5	<0.19	—
cyanide (—CN)	1	<0.05	—
PCB	0.003	<0.0005	—

Notes: Values marked (—) were not measured. Suffix "N" means the value converted at normal condition of 273 K, 1 atm.

Table 4.16

CONCENTRATION OF WASTEWATER
GENERATED BY FLUIDIZED-BED
INCINERATOR (145)

Content (mg/l)	Feed Water	Dryer Condensate	Wet E.P. Effluent	Dehumidifier Effluent
Tkj—N	24.71	29.67	24.97	24.52
NH4$^+$—N	21.14	24.40	21.24	20.75
Org—N	3.57	5.27	3.73	3.77
NO2$^-$—N	0.18	0.16	N.D.	0.11
NO3$^-$—N	0.63	0.60	0.98	0.62
T—N	25.52	30.43	25.95	25.25
BOD	6.86	41.5	2.24	3.44
CODmn	12.2	21.4	38.8	12.5
SS	3.3	15.7	84.0	2.3
Cl$^-$	74.3	74.5	58.3	76.1

dust from the heater tubes. The dust is removed at the bottom, while the balls are collected and returned to the top.

Table 4.15 gives the composition of the ash residue and the stack gases (after they have been cleaned by wet electrostatic precipitation); both meet Japanese regulations. Table 4.16 gives the composition of the waste water produced by this process. According to the operators, the process produces almost no odor.

Chapter 5
The Incineration Process

IN THE UNITED STATES the use of incinerators fluctuated over the years. By 1920 incinerators were in operation in a dozen American cities, and on the eve of World War II some 700 incinerators were in operation (179). In the 1950s, mostly for aesthetic reasons, many municipalities converted from incineration to landfilling. When the cost of oil skyrocketed in the 1970s, incinerators were rediscovered and renamed "resource-recovery plants." By 1977 twenty such plants were in operation and another forty-five were in various stages of completion (179). Later, some of these plants were shut down because of operation and maintenance problems and incinerator manufacturers started to promote the "mass-burning" incinerator. Today the United States has 130 heat recovery and 50 regular incinerators in operation and another 105 planned or under construction (185 and 200).

Not too long ago it appeared that nuclear energy would replace oil as the main energy source of industry. Today this is no longer a certainty. Similarly, it seems that as space for landfills is exhausted, more incinerators will need to be built. The landfill-incinerator transition is similar to the oil-to-nuclear transition in that both require public acceptance. Opponents of incineration point to the health hazard, the noise and odor, the increase in truck traffic, and the potential for air, land, and water pollution. The NIMBY (not in my backyard) syndrome, fueled by health and pollution concerns (which stopped nuclear plant construction in many parts of the world), could lead to the same experience in the incineration industry. The purpose of this book is to help avoid such repetition.

The incineration of hazardous wastes involves additional concerns. First is the potential health risks that would result from accidental spillage or release of toxic substances, or the dangers of incomplete burning. Second, the presence of a toxic-waste incinerator could lead to the development of "dirty industry rights" in the area. Some municipalities are fighting incinerator construction, claiming that incinerators constitute a substantial detriment to public health, safety, and welfare. Officials have tried to overcome such public opposition by proposing to locate incinerators on prison grounds (Florida), on already contaminated industrial "brown field" lands (Linden, NJ), and on isolated rural "green fields" (Millstone, NJ). But the opposition is far from over. More stringent regulations and better enforcement, while increasing costs and lengthening schedules, would serve to reassure the public.

Incineration Statistics

Because of the indestructibility of matter, incineration does not eliminate waste; it just changes its form and volume. The residue continues to exist in the flue gases and ash. Therefore incineration is only a waste-reduction process, not a waste-disposal process.

As was discussed in Chapter 2, the availability of landfills (101) in the United States has diminished drastically (Fig. 1.7). This explains why so many incinerators are being planned, particularly in the Northeast: 22 in New York State, 5 in New York City, one for each borough. The capacities of the planned incinerators are up to 3,000 tons/day, more than a million tons per year, and their unit costs approach half a billion dollars (179). In 1970 only 1% of the MSW in the United States was being incinerated. At the end of 1990, it was 14% (200). By 1992, according to a prediction of the National Solid Wastes Management Association, MSW incineration will reach 25% (185).

Some 80% of MSW is still being landfilled, while recycling and incineration amount to 13% and 14%, respectively (200). The 1992 target of the EPA is to reduce landfilling to 55% and increase recycling and incineration to 25% and 20%, respectively. These goals might seem unrealistic, although Japan, for example, already recycles 50% of its MSW, while incineration and landfilling accounts for 35% and 15%, respectively.

If we only consider the tonnage of the MSW that remains in Japan, after recycling has reduced its quantity by 50%, Japan incinerates 68% of its "remaining" MSW. Europe incinerates 30% to 50% of its MSW, while the United States incinerates about 14%. Yet, as the availability of landfill sites diminishes, the pressure to build new incinerators will increase. Since the 1960s Japan has built 1,915 incinerators, 361 of them being the heat-recovery type (108).

In 1988, 111 solid waste incinerators operated in the United States, and an additional 210 were being planned at an estimated cost of $15 to $30 billion. By the end of 1989, 122 heat recovery and 50 conventional incinerators were in operation, 31 were under construction, and 74 were in the planning stage (185). By the end of 1990, the number of heat recovery incinerators in operation in the U.S. reached 130 (200). In 1987–88, 31 incinerator projects were put on hold, representing a total

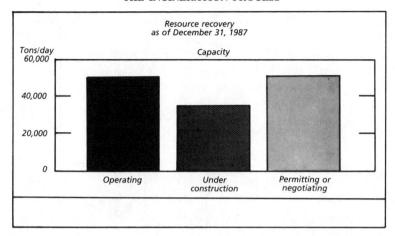

Fig. 5.1 Capacity of resource-recovery plants in operation and under development. (Kidder, Peabody & Co. Inc. [108].)

estimated budget of about $3 billion. In the U.S., 100 of the operating incinerators are of the "mass burning" type and 130 recover some of the energy content of the MSW in the form of steam or electricity. A few dozen of the planned incinerators are also provided with "front-end" processing for recovering some of the more valuable MSW constituents prior to burning. In the coming years this trend is likely to gain momentum as municipalities mandate more recycling prior to burning.

The main suppliers of heat-recovery incinerators (HRIs) in the United States are Ogden Martin (29 plants, 20% market share), Wheelabrator Environmental (25 plants, 18% market share), American Ref-Fuel (8% market share), Combustion Engineering (7%), and Westinghouse Electric (5%). In 1987 Ogden and Wheelabrator received 40% and 15%, respectively, of all orders (108). In terms of new orders for HRI units in the United States, in 1987 a total of 20,585 tons per day (TPD) new capacity was ordered, while 35,655 TPD capacity from previous orders was canceled, and 11,607 TPD of new capacity was put into commercial operation (108). Figure 5.1 shows the overall status of the HRI industry in the United States and Figure 5.2 shows a typical HRI plant. (For a detailed discussion of HRI plants, see Chapter 8.)

Incineration Costs

Incineration used to be inexpensive. In the 1960s the cost of incinerating a ton of MSW was around $5 (Table 5.1). Even the cost of sewage-sludge incineration, including drying and deodorizing, seldom reached $30/ton of combustible solids (Table 5.2). In 1977 the average operating cost of mass-burning a ton of MSW was

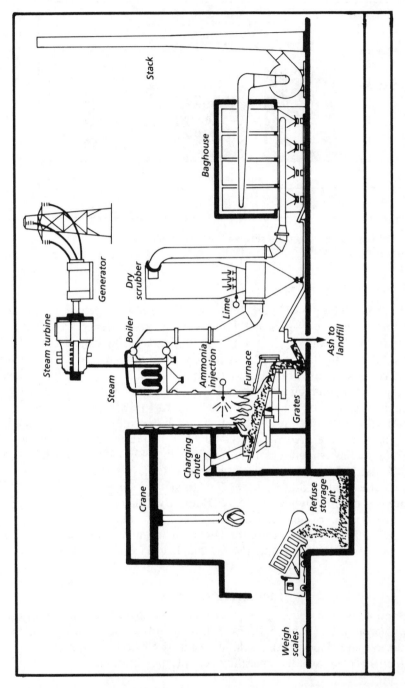

Fig. 5.2 Schematic of Commerce waste-to-energy plant in southern California. (Commerce Refuse-to-Energy Authority [108].)

90

Table 5.1
OPERATING COSTS
FOR MUNICIPAL
INCINERATORS IN
SIX U.S. CITIES (6)

City	Cost per Ton of Refuse Processed
Philadelphia	$4.24
Washington, D.C.[a]	2.28
Detroit	4.30
Milwaukee	6.49
New York City[b]	5.55
Los Angeles	3.13

Note: Costs are for one plant in each city in 1959, except for New York, where the cost is the 1958 average for three plants.
[a] Does not include amortization costs.
[b] Cost computed on basis of tons burned (amount charged minus residue).

$26, and after the sale of the produced energy this cost was reduced to $16. Capital costs of building incinerators were also low, about $5,000 per TPD of capacity (Table 5.3). These figures do not reflect the social costs of incineration. As designs were improved to achieve better material recovery and more complete combustion at higher temperatures, or when scubbers and bag filters were added to improve pollution abatement, the costs increased substantially (Table 5.4). Capital costs of advanced incinerators in 1988 were around $100,000 per TPD of capacity (31) and by 1990 they approached $133,000. A large, 3,000 TPD incinerator installation costs about $500,000. Tipping fees have also increased substantially.

The addition of pollution-control equipment alone can greatly increase capital costs. In Saugus, Massachusetts, adding scrubbers, for example, increased the tipping fees from $25 to $50 per ton. In Detroit, the retrofit addition of scrubbers and

Table 5.2
INCINERATION AND DRYING COSTS (5)

Equipment	Type of Operation	Deodorizing Equipment Included	Total Annual Cost ($/ton of dry combustible solids)
Multiple-hearth or other type of incinerator	Incineration	no	15.51
		yes	18.97
Flash dryer incinerator	Incineration	no	18.19
		yes	21.94
	Drying	no	29.09
		yes	37.02

THE INCINERATION PROCESS

Table 5.3
COMPARATIVE COSTS OF TWO TYPES OF
INCINERATORS IN NEW YORK CITY (7)

	Cost	
Cost Item	Mechanized Continuous Type, Average for 3	Manually Stoked Batch Type Average for 4
Total construction costs per ton per day of capacity (including engineering but exclusive of land)	$5,500.00[a]	$3,750.00
Total operating costs per ton of refuse destroyed	5.55	7.50
Operating less residue disposal	2.40	4.20
Maintenance and repair	1.05	1.05
Administration and supervision	0.50	0.65
Pension	0.60	0.90
Fuel and utilities	0.05	0.05
Amortization	0.95	0.95

[a] Two plants have since been constructed elsewhere for $3,600 per ton per day.

other pollution-control devices will cost some $30 million (185). Higher levels of technology and more expensive construction materials are also required when the MSW is burned at temperatures as high as 2500° F. When the residence time of MSW reaches 45 minutes in the high-temperature combustion zone, special grates are needed. New and specialized designs are also needed because the increased concentration of plastics in the MSW increases the heating value of the waste. Special equipment is also needed to handle the molten metal and glass slug. All these requirements increase the complexity, sophistication, and therefore the capital cost of modern incinerators.

If we define an "average" incinerator as one that serves a population of 500,000, generating about 1,000 tons of MSW per day, then an "average" American incinerator will cost about $125 million in 1991. Costs are naturally lower in rural areas and higher in metropolitan areas, but the trends point in the same direction. The anticipated added regulations on dioxin abatement and ash disposal are likely to increase these costs further.

Incinerator Types

Two types of incinerators are being built today. "Mass burn" incinerators are operated at high enough temperatures (up to 2500° F) to burn *everything*, including the slagging of metals; therefore the MSW needs no pretreatment or separation prior to incineration (Fig. 5.3). Incinerator manufacturers favor this technique because it

Table 5.4

RECENT AND PROPOSED LARGE U.S. WASTE-TO-ENERGY PLANTS (31)

	Year of Start-Up	Daily Capacity (tons)	Type	Costs (millions $)	Generating Capacity MW	Expected Energy Recovery (kW/ton)	Cost Factor ($/ton of capacity)
Pinellas County, Florida	83	2100	Mass Burn	160	51	583	76,190
Westchester County, New York	84	2250	Mass Burn	237	47	501	105,333
Tampa	85	1000	Mass Burn	118	23	552	118,000
Baltimore	85	2010	Mass Burn	254	50	597	126,368
North Andover, Mass.	85	1500	Mass Burn	197	40	640	131,333
Central Mass. (Worcester)	87	1500	Mass Burn	200	38	610	133,333
Detroit	88	3300	RDF	—	65	473	—
Columbus Orig. Est. (1979)	83	3000	RDF-Coal	118	90	720	39,333 (Est.)
Nominal (1985)		2000	RDF	208	60	720	104,000

94

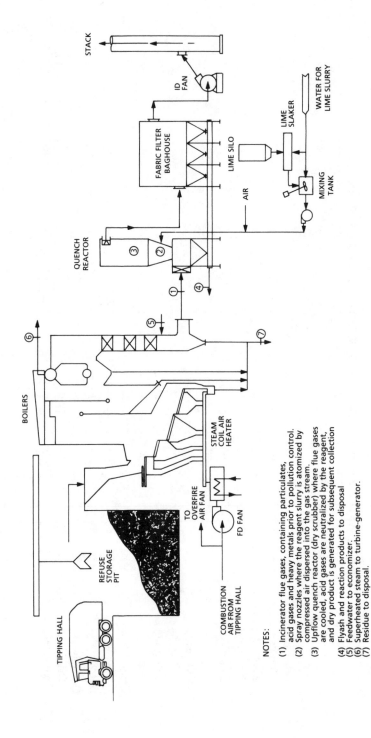

NOTES:

(1) Incinerator flue gases, containing particulates, acid gases and heavy metals prior to pollution control.
(2) Spray nozzles where the reagent slurry is atomized by compressed air dispersed into the gas stream.
(3) Upflow quench reactor (dry scrubber) where flue gases are cooled, acid gases are neutralized by the reagent, and dry product is generated for subsequent collection
(4) Flyash and reaction products to disposal
(5) Feedwater to economizer.
(6) Superheated steam to turbine-generator.
(7) Residue to disposal.

Fig. 5.3 Schematic of a typical mass-burning facility with dry scrubber and baghouse (106).

eliminates all the design and operational problems associated with recovering non-combustibles prior to burning. This technology has been imported from Europe, where mass-burn incinerators usually generate steam, which in turn drive turbines to generate electricity. The atmospheric emissions are cleaned by electrostatic precipitators or by acid scrubbers followed by fabric filter baghouses. There are some 500 "mass burning" incinerators in operation around the world (31), 100 of them in the United States (179). They also can handle the corrosive, plugging, damp, explosive, and adhesive nature of MSW. "Mass burn" technology has been more successful in Europe than in the United States partly because our MSW plastic and toxic content is higher (resulting in more pollution) and because European operators tend to be better trained. The largest such U.S. incinerator, in Detroit, was shut down in 1990 due to excessive mercury emissions.

A competing incinerator design converts MSW into refuse-derived fuel (RDF) through "front-end processing," thereby burning only the combustible portion of the MSW stream. (The design of RDF plants is discussed in Chapter 10.) Front-end processing removes noncombustibles (glass, metals) from the MSW and converts the combustible portion into a more uniform, pelletized fuel through particle-size reduction (usually 4- to 6-inch pellets).

RDF technology is preferred by recycling-oriented users, partly for economic reasons (income from the sale of recovered aluminum, for example) and partly because it cuts the incinerator residue to less than half and thereby reduces the amount of leftover material that needs to be landfilled. RDF-fired boilers can also respond faster to load variations, require less excess air, and are able to operate at higher efficiencies. On the other hand, RDF technology is still in the developmental stage and its first installations included some major disappointments.

The choice between "mass-burn" and RDF is an important one. It makes little sense to burn everything, including noncombustibles. It also seems that the future of incineration in general and of heat-recovery incineration (HRI) in particular is dim without recycling. As recycling increases, it will reduce the combustible fraction of the MSW because of the removal of newsprint and other paper products. This in turn will make the remaining MSW less suited for mass burning. So it would seem that the future belongs to RDF. Yet the majority of incinerators under construction are mass-burn. Part of the reason is the complex nature of the RDF process, which remains an expensive and maintenance-intensive alternative to "mass burning."

Incinerator Ash Residue

The amount of ash residue depends on the content of the refuse and might fluctuate from 5% to 15% by volume or from 20% to 30% by weight. The density of ash residue from conventional incinerators is about 1,200 pounds per cubic yard, while from high-temperature slagging incinerators it can be twice as much or more. The residue from mass-burning incinerators contains fine and light fly ash and heavier bottom ash, mixed with partly melted and burned cans, glass, and pieces of metal.

On average, the ash from mass-burning plants is about 25% of the weight of the MSW feed. When refuse-derived fuel (RDF) is burned and the metal, glass, and other noncombustibles removed, the quantity of ash drops to about 10% by weight. In mass-burning plants, fly ash makes up about 5% of the total (35), while in RDF plants it is somewhat higher (31). In modern installations, 99% of the fly ash is recaptured in the particulate-collection devices. Incinerator bottom ash is beginning to be used as concrete filler, fertilizer base, and road-base material. Additional uses for bottom ash are being developed (201).

Because of the heavy metal and dioxin content of the incinerator fly-ash (see Chapter 6), the disposal of incinerator ash in regular landfills likely will be banned in the near future. If a decision is made to dispose of incinerator fly-ash only in Class I landfills, which are designed for toxic-waste disposal, the impact on incinerator economics, and thus on their general acceptance, will be significant. As discussed in Chapter 3, landfills built for the disposal of hazardous solid wastes must be provided with double lining and with drainage to capture the leachate. This is expensive (about $500,000 per acre). In 1988, 7 million tons of incinerator ash was produced in the United States, and that could increase five times by the turn of the century. It does not seem feasible to dispose of such quantities of ash in hazardous-waste landfills, and it has not been proved that it is safe to dispose of it in regular landfills.

In addition to heavy metals and dioxins, incinerators release nitrous oxides and acid gases, which contribute to the acid rain problem affecting forests. The release of carbon dioxide adds to the global greenhouse effect, which in turn affects the global climate. As will be discussed in Chapter 6, the quantities of pollutants released by incinerators depend on their design. The more advanced designs minimize the quantities of harmful emissions released into the environment, but few "advanced" incinerators are in operation (101) and their costs are high.

Practical Experience: Delays and Overruns

In 1982 a modest HRI in Rutland, Vermont, was budgeted for $22 million. It was to burn MSW at a unit cost $16.50/ton. As of this writing, plant start-up has been delayed because of emission concerns. The investment now stands at $35 million and the estimated "tipping fee" at $40/ton. This is not an isolated example. At the Brooklyn Navy Yard, a 3,000 ton/day incinerator was supposed to start operation in 1984; but the unresolved problems with recycling, ash disposal, and pollutant emissions prevented even the start of construction. The capital cost is now estimated at $535 million, or $178,000 per TPD of capacity.

The advantages anticipated from waste-heat recovery have not fully materialized either. Connecticut passed a law requiring the Connecticut Light and Power Company to purchase electricity from the HRI units at the "municipal rate." Based on this estimate, the tipping fees would have been $45/ton in 1990 and $74/ton in 2000; by 2014, when the original investment is fully paid, it would drop to $22/ton. In 1988 the company filed a suit to have the law overturned. If the utility has its way, the tipping fee will rise to $147/ton by 2014.

The cost of incinerating toxic and hazardous wastes can reach $1,000/ton. High disposal costs are not necessarily undesirable because they act as effective incentives to modify the process that produces such wastes and thereby reduce or eliminate toxic emissions. Just as energy conservation became a beneficial side effect of the increase in oil prices, so the high costs of toxic-waste disposal may result in a cleaner, healthier environment. It will be more profitable to revise the polluting or hazardous processes than to pay the disposal costs required to meet stringent regulations.

The Components of Incinerators

Incinerators range from 50 to more than 3,000 TPD of capacity, with the majority of them in the 200- to 400-TPD range. Incinerator terminology is given in Figure 5.4. Figures 5.5 and 5.6 illustrate a small and a large incinerator, respectively. Incinerators perform best when operated continuously at steady loads. Small plants (under 100 TPD) operate only one shift a day, while those above 400 TPD usually

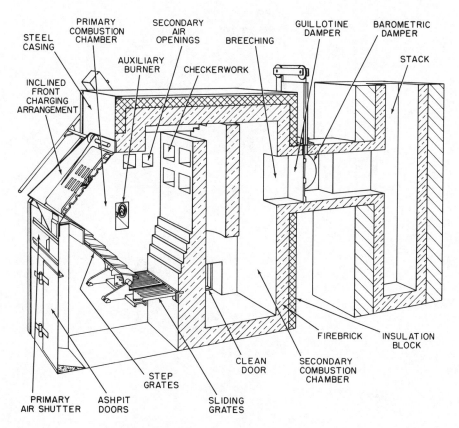

Fig. 5.4 Direct-charge incinerator, class III or IV (14).

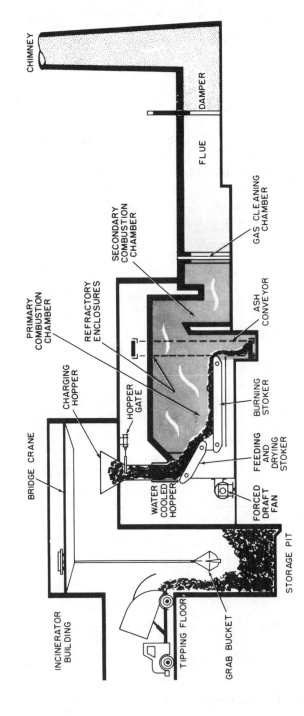

Fig. 5.5 Layout of a municipal incinerator plant. (Courtesy of M. H. Detrick Co. [17].)

98

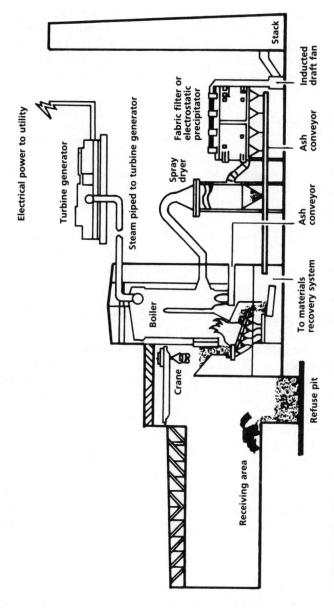

Fig. 5.6 Waterwall-type heat-recovery incinerator. (Wheelabrator-Fryer-Environmental Systems, Inc.)

operate in three shifts. On average, municipal incinerators operate two or three shifts a day five or six days a week.

The by-products of incineration include exhaust gases containing particulates, chlorine, and oxides of sulfur and nitrogen, ash residue potentially contaminated with heavy metals and dioxin, and waste water from wet scrubbers and other sources. Designers of incineration plants must be familiar with federal and state regulations for water and air quality, sanitation, zoning, and building and electrical codes. In addition, they must follow EPA and OSHA regulations regarding public health and safety.

Receiving and Storage Facilities

Site selection depends on a number of factors, including public acceptance, traffic, building restrictions, site foundations, drainage, and prevailing winds. These considerations frequently result in combining sewage-treatment and incineration facilities on the same site. Hillsides are desirable sites because trucks can deliver the waste to the highest point while the residue trucks remove the ash from a lower elevation.

The MSW is usually delivered by trucks to the incinerator plant during the day shift. The loaded trucks are weighed before the waste is dumped on the tipping floor. The tipping floor in smaller installations should provide about 150 sq.ft. (15.4 sq. m) per TPD of plant capacity. Ideally, the tipping floor should be large enough to accommodate more than one truck in the dumping position simultaneously. Storage space is provided through bins or pits. The MSW is discharged from the tipping floor into the storage pit or directly into the furnace.

The floors of the storage pits are pitched to facilitate drainage. They are constructed of reinforced concrete with steel plates or rails along the sides, which protects them against damage by the crane bucket (20). The storage pit usually is sized to hold about 36 hours of collection, but this varies with collection practices. The storage pit would need to be larger if the incinerator operates 24 hours a day, 7 days a week, with collections limited to one 8-hour shift per day. Pit size is usually based on an MSW density of 350 pounds per cubic yard of pit volume.

Storage pits are usually long, deep, and narrow. One pit might be located in front of the furnace or a pit might be situated on each side of the furnace. The pit area is usually enclosed by the refuse-storage building; the escape of odors or dust is prevented by drawing in combustion air for the furnace from the refuse-storage building. This creates a slight vacuum inside the building, which draws in atmospheric air and prevents the escape of odors or dust.

Charging Cranes

In larger, continuously-operating plants, refuse is transferred from the storage pit into the charging hopper of the furnace by a crane equipped with a grapple. The pits should be designed so that a monorail crane can operate as the trucks release

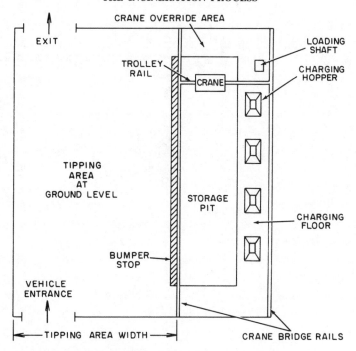

Fig. 5.7 Municipal incinerator layout (20).

their loads (Fig. 5.7). Incinerator facilities with more than a 300-TPD capacity use bridge-type cranes. Both crane types are provided with either a clam-shell grapple or with an abrasion-resistant steel bucket with a capacity between 30 and 150 cubic feet (1–4 cubic meters); spare buckets are recommended. The monorail crane is designed to move in a straight line over the storage pit at one end to lift the MSW into several charging hoppers at the other end. The crane is controlled from a stationary position and can move in only one direction. The storage pit serving a monorail crane must be no more than one meter wider than an open bucket if it is to move the refuse along the sides of the pit.

On a bridge-type crane (Fig. 5.8), the bridge itself travels across the length of the storage pit while a trolley moves the bucket over the length of the bridge. The pit can be as wide as 30 feet (10 meters), but since the time it takes for the trolley to traverse the pit affects efficiency (and therefore capacity), wide pits with long bridges are not economical. Bridges, trolleys, and hoists travel at a speed of about 6 feet per second (100 meters per minute). Modern designs provide air-conditioned crane control cabs to improve visibility and safety.

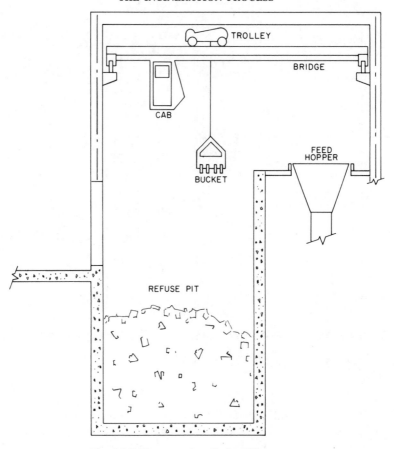

Fig. 5.8 Bridge-crane installation (20).

Charging Hoppers and Gates

In plants under 100 TPD capacity, the MSW is loaded directly from the trucks into the charging hoppers. In larger plants, if the storage area and the charge hoppers are at the same elevation, the MSW is transferred into the charge hoppers by ram feeders (Fig. 5.9) or by front-end loaders. Where the hoppers are located above the storage pits, the MSW is lifted by cranes or, in specialized applications such as sawdust burning, transported by screw-type or pneumatic conveyors (Fig. 5.10). The purpose of the charge hopper is to provide some holdup volume while guaranteeing a reasonably uniform waste flow into the incinerator. In multicell furnaces, each furnace cell usually has one charge hopper (Fig. 5.11). Most hoppers have an "angle of slide" surface of 30 to 60 degrees from the vertical. Depending on the furnace design, the bottom opening can be rectangular or circular and range from 20 to 50 square feet (2 to 4 square meters).

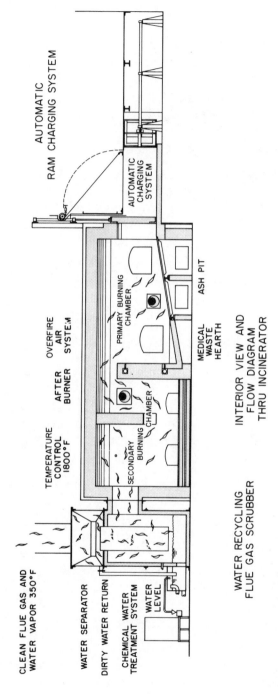

Fig. 5.9 Incinerator with ram-feed system (14).

103

104

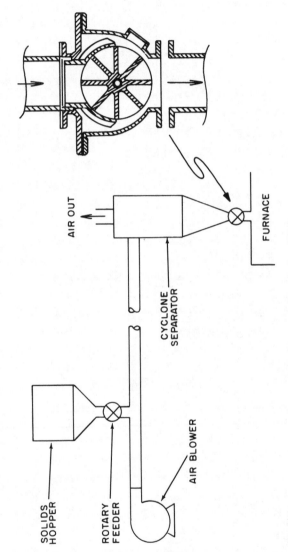

Fig. 5.10 Pneumatic conveying system (20), with rotary valve (50).

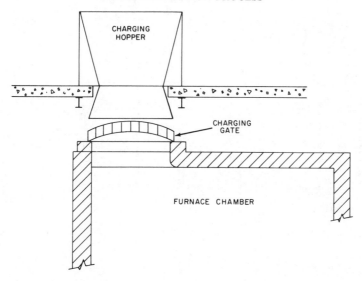

Fig. 5.11 Charging hopper for rectangular batch-feed incinerator (20).

The charging hopper can be continuous or batch type (Fig. 5.12). The capacity of batch-type units is usually limited to 100 TPD. In the batch-type hopper the waste falls by gravity, with its flow rate regulated only by the opening of the hopper outlet gate. The charging gates are made of refractory-lined steel and serve to seal the MSW-feed opening at the top of the furnace. They can be designed to slide (Fig. 5.12), rotate (Fig. 5.10), flap or tip. The tipping-gate valves open and drop their charge into the furnace when the weight of MSW on the flaps exceeds the counter-weight (Fig. 5.13). In applications where a tight pressure seal is needed between the furnace and the hopper, installing two spherical-segment tipping valves in series will ensure a positive seal.

In a continuous-charging hopper, the outlet gate is kept open, the air seal is maintained by the MSW, and the movement of the mechanical grate below (Fig. 5.14) charges the furnace. The vertical charging chute extends from 12 to 14 feet (3 to 4 meters) from the hopper to the front of the furnace. Because of its close proximity to the combustion zone, the continuous hopper is usually water-cooled.

The use of batch hoppers results in uneven operation because large quantities of MSW are dumped intermittently. They also expose the hot refractory to a blast of damaging cold air during each charging cycle. The continuous-charge hopper, however, allows for better furnace temperature control and thereby reduces the need for refractory maintenance. It also spreads the MSW more evenly over the grate, in a relatively uniform and thin layer, while sealing the furnace from cold air.

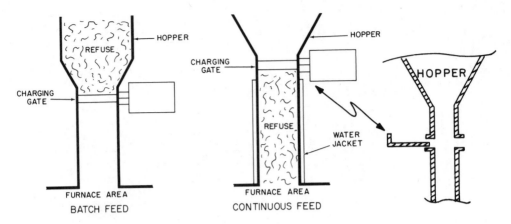

Fig. 5.12 Charging hoppers (17) with slide gate valve (50).

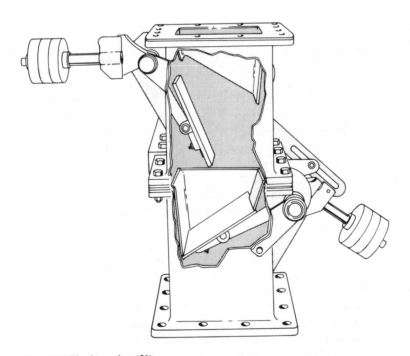

Fig. 5.13 Tipping valve (50).

106

Furnace Designs

Incinerator furnaces can be circular or rectangular in shape and single or multicell in design and can also be distinguished according to the type of grates used (Figs. 5.15 to 5.19). Furnaces are usually sized on the basis of a heat-release rate of 18,000 to 20,000 BTU per cubic foot. Therefore a continuous 100-TPD furnace would require a volume of approximately 3,000 cubic feet. (Some of the key parameters of incinerator design are listed in Table 5.5.)

In the primary combustion chamber of the furnace (Fig. 5.14), the MSW is dried, ignited, and burned at temperatures between 1300° and 2500° F. In the more automated incinerators, the combustion temperature is usually controlled at around 1800° F by varying the amount of air that is allowed to enter the combustion zone. The so-called underfire (or primary) air is introduced beneath the MSW and forced upward through the burning bed by a forced-air fan. This air stream controls the rate of combustion. The overfire (or secondary) air is blown over the top of the

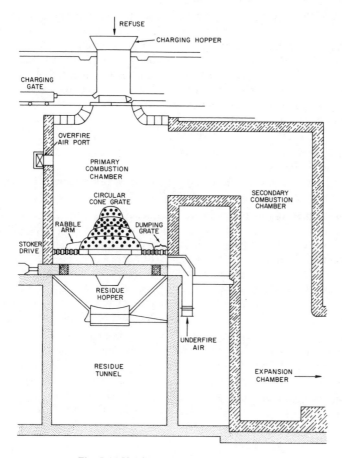

Fig. 5.14 Vertical circular furnace (20).

Table 5.5
INCINERATOR DESIGN
PARAMETERS (19)

Capacity and chamber size: The capacity or duty of an incinerator is usually predetermined; existing practice tends to base size on past experience: combustion intensity is often used.

Combustion intensity: This is the mean volumetric burning rate: specifications mostly call for intensities less than 25,000 Btu/hr cu ft, which in practice ranges from 5,000 to 50,000, depending on waste type; if combustion intensity and incinerator capacity are specified, chamber size is also determined.

Grate area: For a specified capacity, grate area is determined by permitted firing densities, which may range from 100,000 to 400,00 Btu/sq ft hr, depending on the nature of the waste and unit capacity; permitted firing densities still tend to be based on experience.

Bed depth: For a selected firing density, bed depth is determined by the refuse density: at higher firing densities bed depths can be as much as 3.5 ft.

Air proportioning: The ratio of overfire air determines degree of combustion or gasification of the solid bed, bed temperature, slagging potential in the bed and stability of the overbed combustion; the trend is for this ratio to rise, thereby cutting down bed velocities and ash pickup.

Overfire air momentum: High momentum is required for jet penetration without which mixing is inadequate; to obtain penetration, air pressures in excess of 6 inches H_2O are increasingly common.

Overfire air jet configuration: The jet locations, spacing, number and directions determine degree of backmix and dimensions of backmix (stirred reactor) region; this region may have to be 50% or more of the total combustion chamber volume; optimum configuration is probably the biggest unknown at present; current mixing practice is poor.

Excess air: High excess air, ranging from 50 to 400% is common for temperature control to prevent slagging; resulting low temperatures can cause terminal flame quenching, resulting in pollutant emissions.

Gas velocity: High velocity favors pickup and retention of flyash; specifications often call for 35 fps overall and less than 9 fps in any one settling chamber; such specifications will set chamber and gas pass dimensions but should not void overfire air momentum requirements.

Chamber configuration: Current practice favors two or more chambers (exclusive of settling chambers); the practice could be an alternative to purely aerodynamic means of generating a backmix region in the flow pattern.

Grate type: Small units generally have fixed grates and are fired as batch or semi-batch; movable grates can determine supply rate of waste, evenness of combustion and completeness of burnout; gentle stirring is advantageous to break up accumulations tending to clog bed; too active stirring can increase flyash pickup.

Wall type and refractory selection: Refractories must be resistant to slag attack and abrasion; resistance is improved by air cooling; if waste is to be used as fuel for steam raising, water walls can be used, allowing reduction of excess air.

burning solid waste pile by overfire fans and serves to promote the complete combustion of all unburned gases and particulates. The combustion gases travel from the primary to the secondary combustion chamber, while the solid residue is deposited into the ash pit.

Grate Designs

The key working part of the furnace is its grate. The purpose of the grate is to move the MSW from the feed point through the combustion zone to the point of ash discharge. The grate also controls the depth and residence time of the MSW in the combustion zone. The grate should also mix and tumble the MSW to allow it to burn uniformly. In large modern incinerators, the rectangular, movable mechanical grates can be operated at variable speeds to match the variable load conditions. Rectangular furnaces have rectangular grates; the furnace height is determined by the hourly rate of heat release (18,000 to 20,000 BTU/cu.ft.)

The grates support the burning MSW while allowing air to pass through and ash residue to fall through. Grates can be stationary or movable, manual or automatic, horizontal or inclined. Because they operate at such high temperatures, they are made of castable refractory materials or of cast iron. For burning MSW that contains large quantities of wood scraps or plastics, heavier grates made of castable refractory materials are needed to accommodate the higher temperatures. For sawdust applications, solid grates with 0.5- to 1.0-inch-diameter holes (1 to 2 cm) are necessary. Smaller incinerators use horizontal or circular, stationary cast-iron grates (Fig. 5.15). In large municipal incinerators the grates are of the tee or channel types and are inclined to facilitate charging (Figs. 5.16 and 5.17).

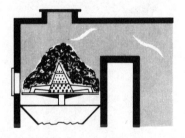

Fig. 5.15 Batch-feed cylindrical furnace. Refuse is dumped over a cone in the center of a circular grate; the cone provides the underfire air for combustion (17).

Fig. 5.16 Batch-feed rectangular furnace. Inclined grates provide burning surface (17).

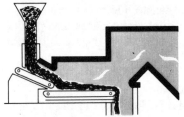

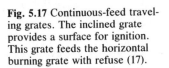

Fig. 5.17 Continuous-feed traveling grates. The inclined grate provides a surface for ignition. This grate feeds the horizontal burning grate with refuse (17).

Older, smaller plants use stationary and circular grates. The furnace is shaped like a vertical cylinder. In a batch-type circular furnace (Fig. 5.14), the outer areas (called dumping grates) are hinged to dump the residue. The solid waste is batch-charged through an opening in the roof and drops onto a central cone grate surrounded by a circular grate. The MSW is slowly stirred during burning by a pair of revolving arms that extend from the center cone. As the cone and arms rotate, the residue is moved to the perimeter, where the discharge gate is located. The furnace has doors for manual stoking or for dumping the residue. In a manual design, the grates can be raised or lowered hydraulically or manually. When the grate is lowered, the accumulated residue can be raked into the ashpit. These cylindrical furnaces are refractory lined and provided with overfire (secondary) air in their upper regions to support secondary combustion.

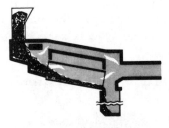

Fig. 5.18 Continuous-feed rotary kiln. The incline grate ignites refuse and feeds it to a rotary kiln where rotation provides turbulence and air for combustion (17).

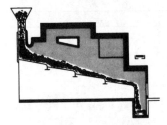

Fig. 5.19 Continuous-feed rocking grates. Alternate mechanical movement of the grates provides tumbling action for the forward movement of refuse (17).

Fig. 5.20 Ram feed. The ram and reciprocating action force and move the refuse through the furnace (17).

In modern rectangular furnaces, MSW movement is provided by gravity on the inclined grates (Fig. 5.16), multiple traveling grates (Fig. 5.17), rotary kilns (Fig. 5.18), rocking grates (Fig. 5.19), or ram feeders (Fig. 5.20). Continuous municipal incinerators have traveling grates (Fig. 5.17), designed either as a moving chain or as a rocking, reciprocating grate. These moving grates carry the burning refuse from the charging end to the residue-dumping outlet, with two or more grates configured in tiers to allow the waste to be mixed as it falls from one level to the next (Fig. 5.21). In the rotating-drum grate design (Fig. 5.22), the drying and burning sections are not separated but are intermixed on a single, continuously sloping surface. Rocking grates (see Fig. 5.19) stir and tumble the burning MSW as the grate elements rock and reciprocate. The rotary kiln or Volund (Danish) incinerator tumbles the burning waste as it passes through a revolving, refractory-lined cylinder (Fig. 5.23). In some rotary-kiln designs, the wastes are partially burned in a rectangular furnace before being fed into the kiln, which exposes the unburnt wastes to further combustion. Beyond the kiln outlet a secondary combustion chamber completes the combustion process. In other designs, the drying and ignition of the MSW are done on reciprocating stokers ahead of the kiln (Fig. 5.24). The "Düsseldorf incinerator" contains a series of rotating cylindrical grates installed in an inclined arrangement (18).

Figure 5.25 illustrates a spreader stoker of the horizontal traveling-grate design, which is used in the firing of refuse-derived fuels (RDF). As the RDF falls from the feed chute, it is distributed in the furnace by an adjustable, high-pressure air jet. The grate travels from the rear to the front of the furnace and dumps the ash at the front. The air distribution is matched to the fuel pattern on the grate by multiple

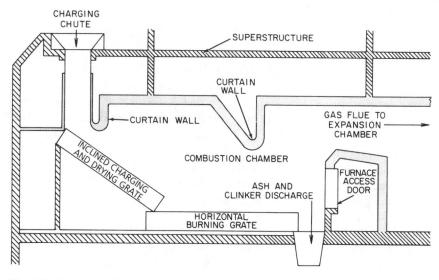

Fig. 5.21 Rectangular furnace (20).

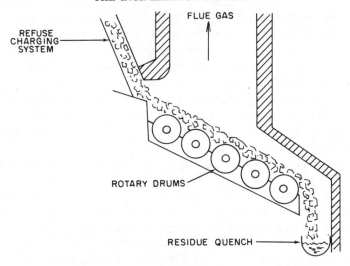

Fig. 5.22 Rotating-drum grate (20).

underground compartments, which control the airflow into each plenum. Secondary air is admitted by tangential overfire air-ports at an elevation well above the point where the fuel is injected.

In multicell furnaces the cells are arranged side by side (Fig. 5.26). Each cell has its own charge hopper (Fig. 5.7) and grate to move the MSW across it. Several furnace cells may share the same residue hopper and combustion chamber. The cells themselves are refractory-lined or, if the incinerator is designed for heat recovery, can be water-cooled.

Grate Sizing

The hourly burning rate (Fa) varies from 60 to 90 pounds of MSW per square foot of grate area (18). To reduce refractory maintenance and to provide a safety margin, an hourly rate of 60 lb/sq. ft. is recommended. In coal-burning furnaces, the grates are usually covered to a depth of 6 inches, corresponding to an hourly coal loading of 30 to 40 lb/sq.ft. The heating values and densities of uncompacted MSW are both less than half that of coal. Thus, to obtain the same firing densities on a BTU basis as produced by coal, the MSW should be supplied at an hourly rate of 60 lb/sq.ft. and should cover the grate to a depth of 3 to 4 feet. The required grate area in square feet is directly proportional to the maximum charging rate F (lb/hr) and is inversely proportional to Fa (lb/hr/sq.ft.):

$$A = F/Fa$$

Essenhigh (19) suggests another method of grate sizing. He would keep constant the ratio of 2/3 power of the combustion chamber volume and the grate area:

$$V^{2/3}/A = \text{Constant}$$

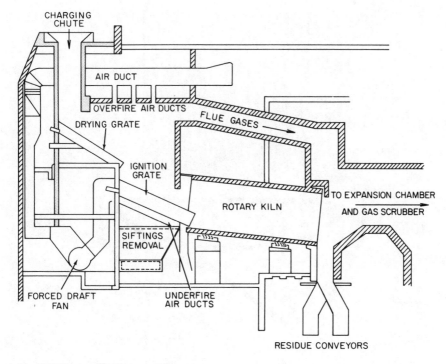

CHARGING
CHUTE

AIR DUCT

OVERFIRE AIR DUCTS

DRYING GRATE

FLUE GASES

IGNITION
GRATE

ROTARY KILN

SIFTINGS
REMOVAL

TO EXPANSION CHAMBER
AND GAS SCRUBBER

FORCED DRAFT
FAN

UNDERFIRE
AIR DUCTS

RESIDUE CONVEYORS

Fig. 5.23 Rotary-kiln furnace (20).

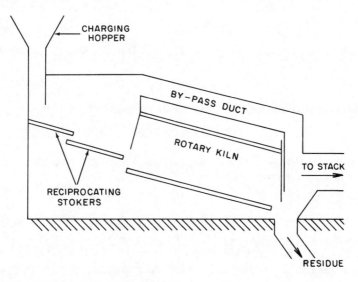

CHARGING
HOPPER

BY–PASS DUCT

ROTARY KILN

TO STACK

RECIPROCATING
STOKERS

RESIDUE

Fig. 5.24 Rotary-kiln incinerator (20).

113

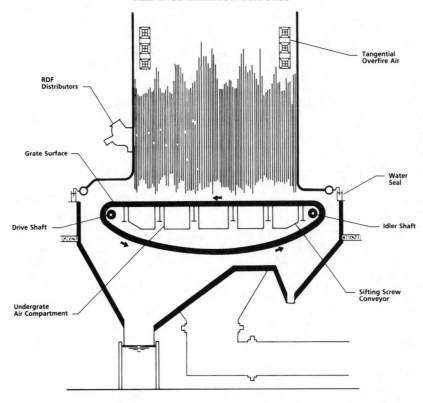

Fig. 5.25 Combustion Engineering's continuous ash-discharge type RC stoker for RDF (105).

He would also express the burning rate (*Fa*) as a function of the refuse-charging rate (*F*) as:

$$Fa = (K_1)(F^{1/3})$$

where K_1 is the so-called waste factor, having a value of:

$$K_1 = (V^{2/3}/A)(I/B)^{2/3}$$

Where:

I = combustion intensity
B = heating value of waste

Furnace Sizing

The furnace capacity is a function of its grate area and volume. The furnace volume is usually determined on the basis of an hourly heat release of 20,000 BTU/cu.ft. As the moisture and noncombustible content rises in the MSW, its heat release tends to drop. Auxiliary fuels are usually needed to complete the combustion when the hourly heat release drops to about 10,000 BTU/cu.ft. or when the combustibles content in the MSW is less than one-third. For dry and inert-free wastes, the hourly heat release can reach 40,000 BTU/cu.ft. If the hourly heat release is 20,000 BTU/cu.ft.

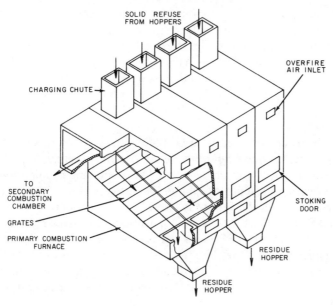

SOLID REFUSE
FROM HOPPERS

OVERFIRE
AIR INLET

CHARGING CHUTE

TO
SECONDARY
COMBUSTION
CHAMBER

STOKING
DOOR

GRATES

PRIMARY COMBUSTION
FURNACE

RESIDUE
HOPPER

RESIDUE
HOPPER

Fig. 5.26 Multicell furnace (20).

and the heating value of the MSW is 5,000 BTU/lb, the hourly burning rate is 4 lb/cu.ft. of furnace volume. A typical design basis is to provide 30 or 35 cubic feet of furnace volume for each TPD of incinerator capacity (18).

The combustion intensity I (BTU/hr/cu. ft.) can be calculated as the product of the maximum MSW charging rate F (lb/hr) and the heating value of the waste B (BTU/lb) divided by the flame-filled combustion chamber volume V (cu.ft.):

$$I = (F \times B)/V$$

The mean volumetric burning rate Rv (lb/hr/cu.ft.) is defined as:

$$Rv = I/B = F/V$$

On similarly sized incinerators, higher values of I and Rv indicate a better-quality design. Knowing the value of I of an existing incinerator makes it possible to determine the required combustion-chamber volume (V) for a new incinerator, if the values of F and B are known.

The combustion space within the furnace can be separated into the solid-bed reaction zone and the gas-phase reaction zone. One important furnace-design parameter is the ratio of the volumes of these two zones. As most of the reaction takes place in the gas phase, that zone is always the larger one. Another important design criterion is the amount of underfire and overfire air flows sent to these zones and their ratio to each other. The general flow pattern through the gas-phase zone is designed to obtain good turbulence as the gas passes through several interconnected chambers.

In most incinerators the flue gases first pass through the "stirred reaction region" above the burning waste; there they are "backmixed" through the introduction of the overfire air. The number, size, and location of the overfire-air inlet ports determine the amount of turbulence and backmixing in this region. A frequent source of incinerator problems is an undersized "backmix zone." The backmix zone is followed by the "plug flow" zone. This is the part of the furnace chambers where no overfire air is introduced, so the combustion gases travel with little or no back-mixing.

Air Requirements

Having determined the grate area, the combustion-chamber volume, and the MSW charging rate, the most important design parameter remaining to be determined is the air requirement of the incinerator. The stochiometric air requirement of a combustion process represents the amount of air that will provide two oxygen atoms for each carbon atom in the fuel to achieve complete combustion. The amount of air introduced in excess of this stochiometric quantity is called "excess air." In most incinerators, 80% to 100% excess air is required to achieve complete combustion of all organics in the MSW (57). Incinerator operation is optimized by providing as much oxygen as is needed to achieve complete combustion, but not more, because any additional air will reduce thermal efficiency and the incinerator will increase NO_x generation.

The cold air volume required for proper combustion in the incinerator per unit weight of MSW can be calculated as follows (19):

$$\text{Total Cold Air Volume (cu.ft./lb)} = Bo\,(1 - a - M)(S)(1 + e)$$

Where:

    Bo = dry and inert-free (DIF) heating value of the MSW in Btu/lb
    a = inert or ash fraction of the MSW
    M = moisture fraction of the MSW
    S = cubic feet of stochiometric cold air required per BTU of heat release
    e = excess air fraction

If the combustibles are mostly organic and the waste does not contain metals, the value of S is about 0.01. Therefore, a simple rule guiding the incineration of a variety of organic fuels (including coal) is to provide one cubic foot of cold air for each 100 BTU of heat release.

The dry and inert-free (DIF) heating value of MSW is between 8,000 and 10,000 BTU/lb and therefore the stochiometric air requirements ($Bo \times S$) range from 80 to 100 cubic feet, or 6.4 to 8 pounds, of air per pound of DIF waste. The actual total air is usually 150% to 200% of the stochiometric (50 to 100% excess air), or about 10 pounds of air per pound of MSW. The variation between wastes is mostly due to the moisture content of the waste, while the total mass flow through incinerators is mostly a function of the total air flow. The particulate emissions from incinerators tend to increase with heat release and with underfire air flow, while it tends to decrease with increasing fuel particle sizes (Fig. 5.27).

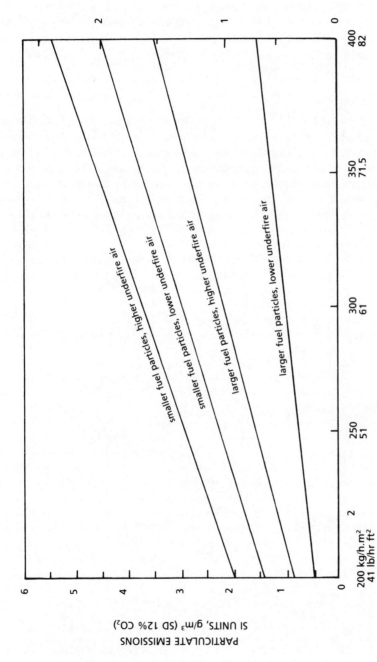

Fig. 5.27 Effects of combustion rate, underfire air, and fuel particle size on particulate emissions generated by combustion of wood wastes (29).

117

The underfire air and overfire air are usually provided by separate blowers. Underfire air is usually more than half of the total (50% to 70%). Flue-gas temperature and smoking are both controlled by modulating the total air flow and the underfire-to-overfire air ratio. For most American grate designs, the required underfire air pressure is about 3 inches of water. The overfire air pressure is adjusted to give high enough entrance velocities at the nozzles to guarantee high turbulence and long enough residence times to assure complete combustion.

Furnace Reaction Model

Mathematical modeling of the incinerator furnace provides a better understanding of the processes taking place. Figure 5.28 describes the gas-phase (II) and the solid-phase (I) zones in a top-charged incinerator (overbed feed). Here the bottom zone (IA) is in a state of combustion, generating the exothermic heat that is needed in the drying/pyrolysis zone (IB) above it. Drying and pyrolysis are endothermic processes and therefore the temperature tends to drop in zone IB. In one test, the temperatures in the overbed flame (IIA) and in the combustion zone (IA) were

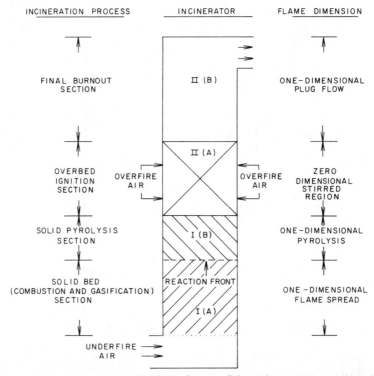

Fig. 5.28 Incinerator for continuous overbed feed of waste. Schematic represents: solid bed, zone I, with overbed zone II. For overbed feed, zone I has subzones, including I(A), the combustion and gasification section on the grate, and I(B), drying and pyrolysis above I(A). Zone II (overbed combustion) has a backmix (stirred) region called subzone II(A), followed by a plug flow burnout region—subzone II(B). With underfeed, zone I would be inverted, with drying and pyrolysis below the combustion subzone, and with reaction front moving down instead of up as shown (19).

measured at around 1800° F (982° C), while in the drying/pyrolysis zone (IB) the temperature was detected as only 1100° F (593° C). In slagging incinerators the combustion-zone temperatures reach 2500° to 3000° F (1371° to 1684° C).

As the moisture content of the MSW increases, the combustion and flame zones can become decoupled. When this occurs, the burning bed can no longer be relied upon as a "flame holder." When the bed is mixed or tumbled, the zones described in Figure 5.28 get partially reversed, as if some of the feed were charged from the bottom ("underfeed"). Such "underfeed" zones will act as "flame holders" and the products from such "bottom feed" zones will leave the bed at a much higher temperature, resulting in a more complete final burnout.

The burning rate Fa (lb/hr/sq.ft.) is related to the rate of underfire-air supply Ga as follows (19):

$$Fa = (f \times Ga)/(1 - a - M)S (1 - Vm)$$

Where:

f = oxygen utilization factor, which is dimensionless and can exceed unity if combustion is incomplete, and therefore some CO is formed

Ga = rate of underfire-air supply per unit area of bed (cu.ft./sq.ft.)

a = ash fraction in the MSW

M = moisture fraction in the MSW

Vm = pyrolysis fraction, representing the fraction of the bed that is lost due to pyrolysis, which is about 0.7 for cellulose-type materials

S = stochiometric ratio of cold air required per pound of MSW (cu.ft./lb)

The equation shows that the burning rate can be increased by increasing the flow rate of underfire-air supply, but this potential for an increased burning rate is limited by the resulting gas velocities, which will carry more and more fly ash into the flue gases.

Modeling the Solid-Bed Zone

In a top-feed bed (Fig. 5.28), zone IA is the char combustion zone. In this zone the reactions occur only between carbon and oxygen molecules, since drying and pyrolysis has already taken place in zone IB above.

Combustion:

$$MSW + O_2 \rightarrow CO_2 + H_2O + Ash + Heat$$

Pyrolysis:

$$MSW + Inert \ Gas \rightarrow H_2 + CO + CO_2 + CH_4 + Charcoal \ Heat$$

The bed in zone IA can be regarded as a porous carbon block, with oxygen diffusing through the channels of the block, producing a mixture of carbon monoxide and carbon dioxide. As CO_2 builds up and oxygen runs out, the CO_2 reacts with the carbon to form CO in the gasification region.

In the top-feed model, the ignition plane is located between zones IA and IB and it propagates upward by radiation. A pyrolysis zone is interposed between the drying zone at the top and the combustion zone at the bottom of the solids. Pyrolysis is an endothermic (heat using) reaction in which long organic molecules disintegrate into H_2, CO, CO_2, CH_4 and various saturated and unsaturated hydrocarbons. Drying is also endothermic. It consumes heat not only to provide the latent heat of vaporization for the moisture but also in the water-gas-reaction, where steam reacts with carbon to form CO and hydrogen. In the drying/pyrolysis zones, heat as well as oxygen is consumed because the by-products of pyrolysis and the water-gas reactions tend to compete for oxygen.

As the moisture content of the MSW increases, more heat is necessary to increase the rate of burning (Fa). With an increase in the firing rate, the temperature in zone IA also rises, which in turn causes the ash to fuse into clinkers, thereby starting the slagging process. The tumbled bed provides some compromises between these limiting effects by breaking up the various distinct zones.

The Overbed Combustion Zone

The gas-phase zone is also separated into two distinct zones. Zone IIA is called the "backmix zone" because this is where the overfire air is introduced to create turbulence and backmixing. Figure 5.29 illustrates an elegant design, where 260 small air nozzles direct the secondary (overfire) air to the region where both the gas temperature and its velocity are at maximum. The volume of zone IIA is only about 5% of the total gas-phase furnace volume, but 60% to 90% of the gas-phase combustion takes place in this zone. The plug-flow zone (IIB) is 95% of the gas-phase volume but handles only the remaining 10% to 40% of the gas-phase combustion. As the supply pressure of the overfire air is increased, the combustion intensity in the backmix zone (IIA) will increase and the flames will be shortened.

Top-feed incinerators require less underfire air than the tumbled or bottom-flow designs, but they are likely to smoke more, even with substantial quantities of excess air, if most of the air is supplied as underfire air. If the same amount of total air is supplied but is distributed evenly between the underfire and overfire nozzles, the resulting flameholding and backmixing effects will eliminate smoking. If the flame is stabilized, overbed turbulence will eliminate odor as well as smoke.

Calculating Heat Generation

Calculating the amount of heat generated through the incineration of MSW is important in determining how much auxiliary fuel is needed for combustion. The moisture content of MSW ranges from 20% to 50% by weight. The combustible content of MSW usually ranges from 25% to 70% by weight and is composed of cellulose, proteins, fats, waxes, rubber, and plastics. As shown in Table 1.19, the heating value of MSW depends on its composition. The average heating value of cellulose-based material is about 8,000 BTU/lb; rubber is about 10,000 BTU/lb, fats

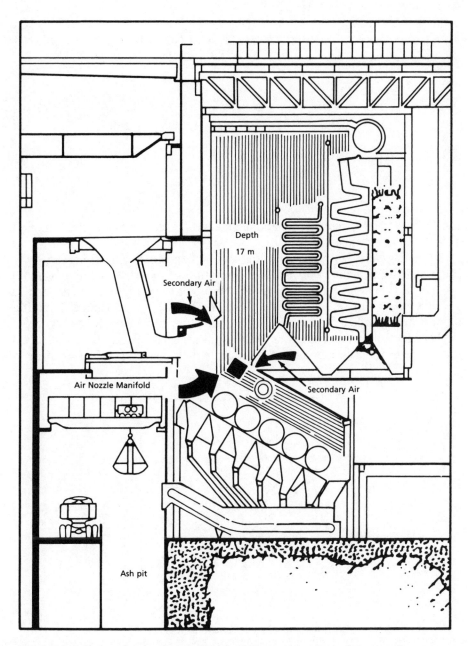

Fig. 5.29 Installation of transverse manifold for 260 secondary air nozzles over roller grate at Wuppertal, Germany (34).

Table 5.6
HEAT CONTENT OF SOLID WASTE AS A FUNCTION OF MOISTURE AND
NONCOMBUSTIBLES (18)

Noncom- bustible, %	10		15		20		25	
Moisture, %	Comb., %	Heat Content[a]	Comb., %	Heat Content[a]	Comb., %	Heat Content[a]	Comb., %	Heat Content[a]
50	40	3400	35	2975	30	2550	25	2125
40	50	4250	45	3825	40	3400	35	2975
30	60	5100	55	4675	50	4250	45	3825
20	70	5950	65	5525	60	5100	55	4675

[a] Btu/lb

about 17,000 BTU/lb, and plastics about 20,000 BTU/lb. The higher the heat content in MSW, the more plastic it contains.

Assuming that the average heating value of the combustible material is 8,500 BTU/lb and the moisture and inert concentration of the MSW is known, the heat content of MSW can be estimated using Table 5.6 or Figure 5.30. If the heating value of the combustibles is other than 8,500 BTU/lb, the numbers in Table 5.6 and Figure 5.30 must be multiplied by the ratio of the actual heating value divided by 8,500.

The elemental composition of MSW is given in Table 1.3. Most of it is carbon and hydrogen with trace amounts of sulfur and other oxidizable elements. A material

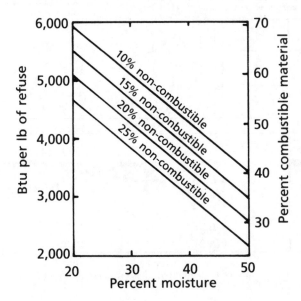

Fig. 5.30 Moisture-heat-content relation with 8,500 BTU/lb combustible material (18).

Table 5.7

MATERIALS BALANCE FOR FURNACE
(IN LB/100 LB OF REFUSE) (18)

Input:

Refuse			
Combustible material			
Cellulose	52.74		
Oils, fats, etc.	5.86	58.6	
Moisture		22.4	
Noncombustible		19.0	100.0
Total air, at 140% excess air			
Oxygen	191.0		
Nitrogen	633.0		824.0
Moisture in air			11.0
Residue quench water			5.0
Total			940.0

Output:

CO_2 (28 × 3.667)			102.7
Air—Oxygen (191 − 80)	111.0		
Nitrogen	633.0		744.0
Moisture			
In refuse		22.4	
From burning cellulose		29.3	
From burning hydrogen		5.4	
In air		11.0	
In residue quench water		5.0	73.1
Noncombustible material			19.0
Unaccounted for			1.2
Total			940.0

balance of burning 100 pounds of MSW is given in Table 5.7. This MSW was assumed to have a heat content of 5,000 BTU/lb, a moisture content of 22.4%, a noncombustible content of 19%, and contain 28 pounds of carbon and 0.6 pounds of hydrogen (18). It was also assumed that 1–3 pounds of combustibles will escape unburned and 140% excess air will be needed to provide sufficient cooling of the refractories. Therefore, the total air provided is assumed to be 2.4 times the stochiometric requirement, or 8.24 pounds of air per pound of MSW.

If a heat-balance calculation is performed for the above example, it would show that the enthalpy of each pound of the exiting gases would be 455 BTUs higher than the enthalpy of the 80° F air that enters (18). Based on this enthalpy rise, the expected flue-gas temperature of 1680° F can be read from Fig. 5.31. This flue-gas temperature is low enough to protect the refractory.

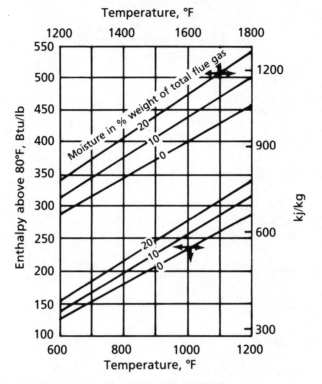

Fig. 5.31 Enthalpy of flue gas above 80°F (18).

Incinerator Accessories

The heart of the incinerator is the furnace and its waste-handling and charging equipment. For the proper operation of an incinerator, various gas and ash-handling devices are needed (Fig. 5.32). On the gas side of the incinerator, the air required to support combustion and provide cooling is "forced" into the furnace by the forced-draft fans. The air passes through ducts and its flow is modulated by dampers. The dampers are used to throttle the air, either manually or automatically, thereby maintaining the desired operating conditions. After combustion, the hot flue gases leave the furnace through refractory-lined ductwork (called breechings) and usually enter some type of gas-cleaning device, such as a wet scrubber (Fig. 5.32). When the pressure drop through the gas-cleaning equipment is high, the flow of hot gases must be "induced" through the use of induced-draft fans. If the pressure drop is low, the chimney effect of a tall stack will suffice to "pull" the hot gases through.

Fans

Forced-draft fans supply the incinerator with both its combustion and cooling air, while induced-draft fans remove the hot combustion gases (when chimney effect

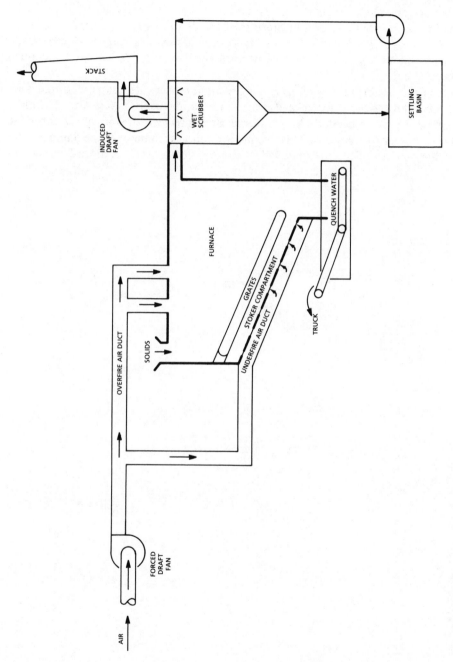

Fig. 5.32 Gas-side incinerator components.

125

alone is insufficient to do so). The fan designs are either axial or radial flow, called propeller or centrifugal, respectively (Fig. 5.33). Axial fans are more likely to be used on incinerator applications because they are lighter, less expensive, and more compact in design. Their characteristic curves are shown in Figure 5.34, where the degrees of rotation refer to the angle of the variable blade pitch that is used to control the fan.

When the underfire and overfire air are required at different pressures, they can be provided by the same forced-draft fan station, using dampers to distribute and regulate the air as needed. Another approach is to furnish separate fans for the underfire and overfire services. This second approach eliminates some dampers and reduces the amount of controls needed, while the first method is more flexible and

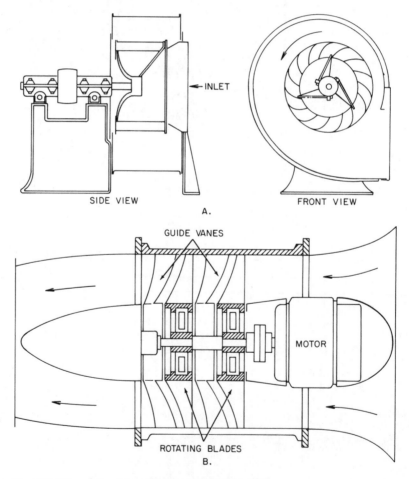

Fig. 5.33 (A) centrifugal fan; (B) two-stage axial fan (20).

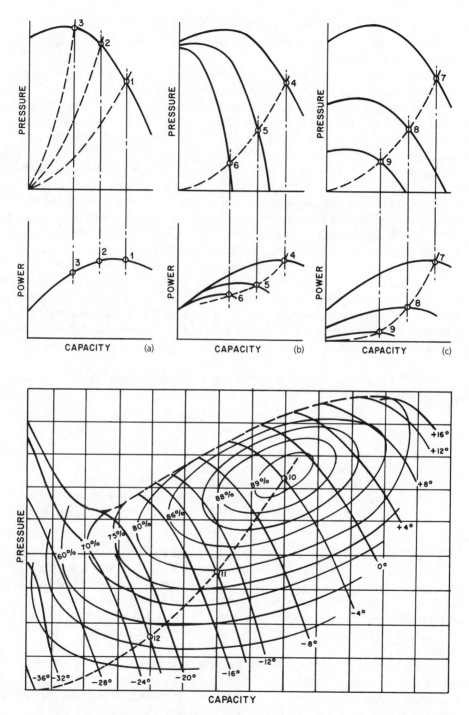

Fig. 5.34 (Above) Fan characteristics with (a) damper control, (b) variable-inlet vane control, (c) variable speed control. (Below) Characteristics of an axial-flow fan with variable pitch control (22).

127

more suited to variable-load conditions which occur as the MSW composition or moisture content changes.

Induced-draft fans need to be protected against excessive temperatures. In some installations cooling is provided by the gas-cleaning equipment, such as a wet scrubber, while in others ambient air must be admitted through barometric dampers to lower the temperature at the fan inlet (Fig. 5.35). The duct where the cooling air is mixed with the flue gases should be sized for a high enough velocity for good mixing, about 40 ft/s (12 m/s).

The selection of the induced-draft fans should be based on a balance between capital and operating costs. For a high-temperature design that can operate at 750° F (400° C), the capital cost will be high, but since less cooling air will need to be transported, the operating costs will be low. For low-temperature fans with mild steel blades, which can operate up to 300° F (150° C), the capital costs will be low, but the operating costs will rise because large quantities of cooling air will need to be transported. Specially designed stainless-steel high-temperature fans are available with water-cooled bearings that can operate up to 1100° F (600° C). For up to 600° F (400° C), conventional high-temperature paddle-wheel fans can be used, while above that temperature the venturi type is recommended. In the venturi design the blower is not in direct contact with the hot gases. A reasonable compromise between first costs and operating costs is to operate the induced-draft fans at about 575° F (300° C).

The rate of a fan's power consumption (KW) can be calculated according to the equation (20):

$$KW = (0.815 \times Q \times H)/(10,000 \times E1 \times E2)$$

Where:

Q = gas flow rate in cubic feet per minute
H = fan discharge head in inches of water
$E1$ = fan efficiency (a typical value is 70%)
$E2$ = efficiency of the electric motor driving the fan (a typical value is 75%)

Ducts

Ducts transport the air and hot combustion gases to and from the incinerator. Cold-air ducts are usually made of riveted and/or soldered galvanized sheet steel. The ducts that transport hot gases are lined with refractory materials and are called breechings. In large incinerators the breechings are made of refractory-lined masonry and their cross-sectional areas are frequently larger than that of the chimney. They are designed to transport the gases at velocities of 1,200 to 2,400 ft/min (6 to 12 m/sec).

Ducts are usually under positive pressure, since they force the cold air into the furnace. The breechings are usually under negative pressure, since they pull the hot gases out of the furnace. Ducts from the forced-air fan must therefore be protected from overpressure (PSH–02 in Fig. 5.36). Breechings leading to the induced-draft

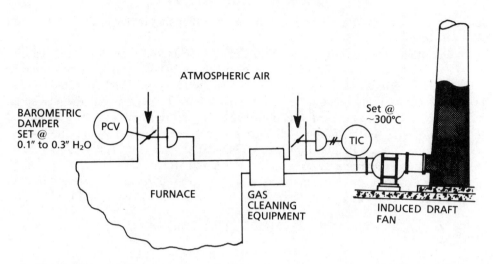

Fig 5.35 Cooling by atmospheric air.

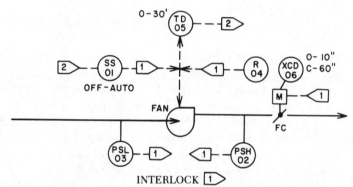

INTERLOCK [1]>

	Conditions				Actions	
SS-01	PSH-02	PSL-03	R-04 (Reset)	Fan	XCD-06 (Damper)	TD-06 (Note #1)
Off	—	—	—	Stop	Close	Time
Auto-On	High	—	—	Stop	Close	Time
		Low	—	Stop	Close	Time
	Not high	Not low	Not reset	Stop	Close	Time
			Reset	Run	Open	Reset to start

— = Any state or condition (don't care)
Note #1: Shut-down delay used in interlock [2]>

Fig. 5.36 Safety interlocks should be provided on all fans (22).

fan, on the other hand, must be protected from collapsing under excessive vacuum (PSL–03 in Fig. 5.36).

The other safety devices shown in Figure 5.36 include the discharge damper XCD–06, the time delay TD–05, and a reset button R–04. The interlock table in Figure 5.36 describes the operation: If the discharge pressure is high, or if the suction pressure is low, or if the shutdown time delay has not been reset, the discharge damper cannot open and the fan cannot start. Therefore the fan system is allowed to start only when both suction and discharge pressures are normal and the fan has been off long enough for the time delay (TD–05) to time out. This delay protects the fan from overheating due to frequent starts and stops.

The pressure drop in ducts is due to wall friction, dynamic losses, and to duct shape or direction. The friction-head loss (H) can be calculated as follows:

$$H = (fL/D)(V/4,004)^2$$

Where:

H = head loss due to friction in inches of water
f = nondimensional friction coefficient
L = length of the duct in feet
D = diameter of the duct in feet
V = velocity of the flowing air in ft/min

Dampers and Screens

Dampers are used to control the flowrate of cold air into the incinerator and the flowrate and/or temperature of the hot gases leaving the furnace. On the cold-air side, all air inlets are provided with manually operated cast-iron shutoff dampers. In retort-type incinerators, the underfire and overfire air openings are provided with rotating circular-spinner shutters (20). On in-line incinerators, the overfire (secondary) air openings are usually rectangular and the top-hinged dampers or the center-hinged butterfly-type plate dampers are used for their manual throttling. The sliding rectangular dampers tend to warp and are not recommended.

In existing incinerators, the use of closed-loop automatic control, and thus automatic modulating dampers, is not common (22). In newer installations, air flows and air distributions are automatically monitored and controlled. The total cold-air flows are usually controlled by automatically adjusting the variable blade pitch of the forced-draft fans (Figs. 5.34 and 5.49) and by using dampers or butterfly valves to distribute this total air accurately between the underfire and overfire ducts or individual air-ports (23).

The furnace draft is controlled by either throttling the chimney effect or modulating the induced-draft fan so that the desired draft (vacuum) is maintained inside the furnace. The furnace draft is usually maintained between 0.1 and 0.3 inches of water vacuum (25 to 75 Pa vacuum) and is controlled by barometric dampers that admit just enough atmospheric air to maintain this level of vacuum. The least expensive barometric dampers are pivoted steel plates that are balanced to open as the draft rises (Figs. 5.37 and 5.38). Furnaces are operated under vacuum for reasons

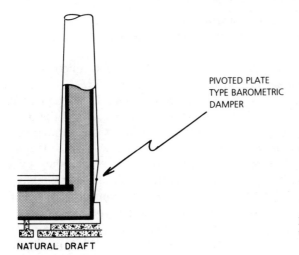

PIVOTED PLATE
TYPE BAROMETRIC
DAMPER

NATURAL DRAFT

Fig. 5.37 Natural draft chimney
with barometric damper control.

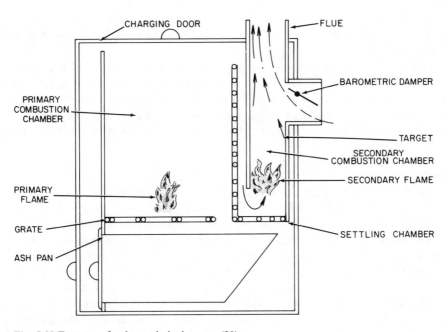

CHARGING DOOR

FLUE

BAROMETRIC DAMPER

PRIMARY
COMBUSTION
CHAMBER

TARGET

SECONDARY
COMBUSTION CHAMBER

SECONDARY FLAME

PRIMARY
FLAME

GRATE

ASH PAN

SETTLING CHAMBER

Fig. 5.38 Features of a domestic incinerator (20).

131

of safety: If the pressure of the surrounding atmosphere is higher than that of the combustion zone, then unburned fuel and smoke cannot leak out and cause accidents or health risks.

Drafts produced by the chimney effect of the stack can be controlled by placing a damper in the breeching between the furnace and stack (see Fig. 5.4). The disadvantage of using a manual damper is that it will require continuous readjustment, particularly during startup, because the chimney effect will increase as the hot-gas temperature rises. In addition, the gas flowrates will increase as the burning rate is increased. Therefore barometric dampers at the furnace (Fig. 5.35) or in the stack (Fig. 5.37) give better vacuum control than manual dampers. Depending on the location of the manual damper, its operating temperature will also vary. Thus the damper might be constructed of temperature-resistant, refractory-lined, or water-cooled steel. The damper can be the sliding gate or guillotine type.

Dampers are also used to protect the induced-draft fans from excessive temperatures. These control loops consist of a temperature controller (TIC in Fig. 5.35) that throttles the outside air damper to admit only as much cold outside air as is required to keep the total mixture at the desired temperature. When a barometric damper is used to maintain the furnace draft and an outside air damper is used to admit cooling air, these two loops are likely to interact. To eliminate interaction and cycling, the two loop responses must not be similar. The usual solution is to make the pressure loop (PCV) fast and the temperature loop (TIC) slower. All dampers must be sized for maximum flow at minimum pressure drop. Maximum flow in cooling dampers is the air needed to cool the maximum amount of hot gases that will be generated when the furnace is fully loaded.

Screens to trap larger particulate matter from leaving the furnace with the flyash are installed either in the stack or in the breeching between the furnace and stack. They are usually made of 4- to 10-mesh (4 to 10 lines per inch) stainless steel. In municipal incinerators, counterbalanced, double-hung screens are used for convenience; one screen can be operated while the other is being cleaned or repaired.

Stacks

Chimneys are vertical flues that transport combustion gases into the atmosphere. Because the hot gases inside the chimney are not as dense as the cold air outside the chimney, the barometric pressure on the outside will be greater than the pressure inside. This difference in the weight of the two gas columns allows the chimney to generate a draft. Once the hot gases reach the top of the chimney, the warm gases are dispersed in the atmosphere. If there are few accessory devices attached to the incinerator, the pressure drop that the flue gases have to overcome is small and the chimney itself will provide the required draft (Fig. 5.4). On incinerators with heat-recovery and air-pollution-control devices, the installation of an induced-draft fan will be required to generate required pressure-drop (Fig. 5.32).

The draft generated by a chimney rises with its height and with a rise in the temperature of the flue gases. Chimneys usually produce drafts in the range of 1–4 inches of water column and produce gas-flow velocities in the range of 25 to 50 ft/s (8 to 16 m/s). Figure 5.39 shows the most economical chimney diameter for various gas flowrates (24). Figure 5.41 shows the stack height as a function of average gas temperature and draft requirement. If the dimensions of the stack and its inlet temperature are known, Figure 5.40 can be used to determine the amount of cooling that takes place as the gases travel up the stack; from this, the stack outlet temperature can be obtained. The average stack temperature is the arithmetic average of the inlet and outlet temperatures, so the *assumed* average stack gas temperature in Figure 5.42 can be checked and corrected using the actual outlet temperature obtained from Figure 5.40.

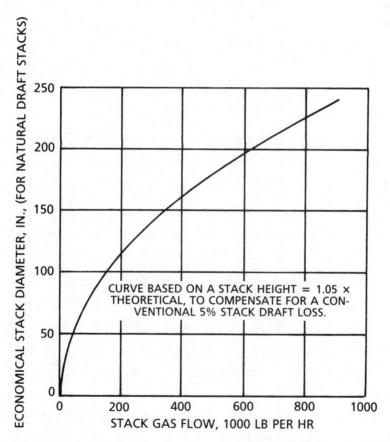

Fig. 5.39 Economical stack diameter for a range of gas flows (24).

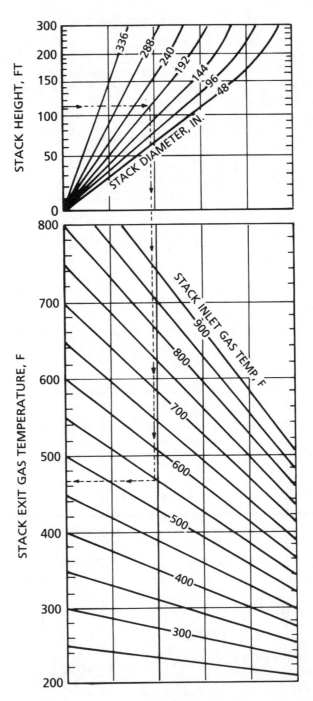

Fig. 5.40 Stack exit-gas temperature (its approximate relationship to stack diameter and height) (24).

134

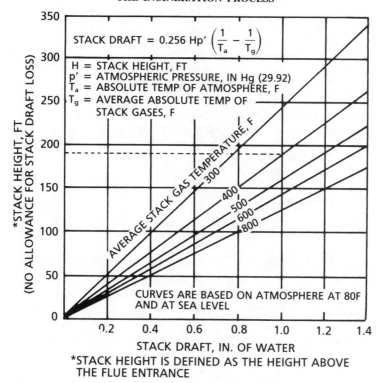

Fig. 5.41 Stack height required for a range of stack drafts and average stack-gas temperatures (24).

Chimney Materials and Accessories

The least expensive stacks are made of steel and instead of being self-supporting are supported by guy-wires. The inner surface of the steel stack is usually lined with a thin layer of bonded fused glass. The main disadvantages of steel chimneys are their high maintenance and unsightly appearance.

The strongest, most economical large chimneys are made of reinforced concrete and frequently are lined with fireclay brick, which can handle flue-gas temperatures up to about 1000° F (540° C). Masonry chimneys also have an inner layer of fireclay brick, surrounded by a 2- to 6-inch (5 to 15 cm) annular space and an outer shell of ordinary brick (20).

Chimney accessories include iron caps at the top, which prevent deterioration caused by moisture and cleanout doors at the bottom for inspection and maintenance purposes. A checklist for performing a chimney inspection is given in Table 5.8. The chimney cap usually has a lead-coated copper lightning arrestor and warning lights, which are required by the Federal Aviation Agency. Ladders and catwalks are provided for maintenance.

Table 5.8
CHIMNEY INSPECTION CHECKLIST (20)

---

1. Is lightning rod system correctly grounded?
2. Are there excessive soot deposits in chimney bottom?
3. Can cracked lining be seen through cleanout door?
4. Is the cleanout door broken, or has it been left open?
5. Is there loose packing where breeching enters chimney?
6. Are there cracks or spalled areas on chimney surface?
7. Are outside metal bands (if any) corroded?
8. Are all clamps on lightning rod cable properly tightened?
9. Is the outside ladder corroded?
10. Are any of the lightning rod points loose or missing?
11. Is the cap (if any) loose or defective?

In advanced installations the flue-gas temperatures and compositions are measured at two elevations: at six diameters above the bottom and at nine diameters below the top of the chimney. As a minimum, one temperature sensor and two capped test openings are provided at each elevation 90° from each other. In larger, more sophisticated installations, on-stream analyzers continuously monitor and control excess oxygen, carbon monoxide, opacity or particulates, oxides of sulfur and nitrogen, chlorine, SO_2 and HCl.

Ash-Handling Drag Conveyors

The residue from the incinerator is discharged either through manual dump grates or by more sophisticated means. In batch furnaces, the ash-collection chambers or hoppers are placed below the grates to collect the residue. The hoppers are sized for several hours of ash collection. Ash hoppers (Fig. 5.42) are provided with water spray or a water bath to minimize fire hazards. In larger-batch installations, dump trucks drive in beneath the hopper, so when the discharge gate is opened, the

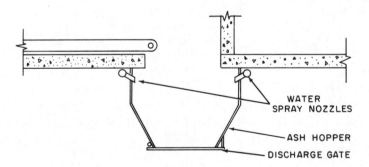

WATER
SPRAY NOZZLES

ASH HOPPER
DISCHARGE GATE

Fig. 5.42 Ash-hopper layout (20).

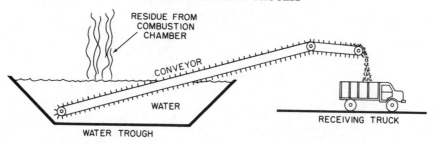

Fig. 5.43 Residue cooling and removal (17).

residue falls into the truck. The temperature of the discharged ash must be monitored and the ash cooled to protect personnel from being burned.

In continuously-fed incinerators, the ash is discharged continuously through a chute into a conveyor trough, which is filled with water to cool the residue before it is hauled away to final disposal (Figs. 5.43, 5.44, and 5.45). The chute is submerged under the quench water to seal the furnace outlet and prevent the entry of atmospheric air. In newer, mass-burning facilities, the use of full-size discharge chutes tends to minimize hangups of large pieces of residue, but even in these systems, stringy wire items such as fencing and trolling wire can become entangled in the residue conveyor and therefore should be removed from the MSW feed.

The residue conveyor (Fig. 5.44) pulls the settled residue from the bottom of the water trough and transports it up to an ash hopper, storage bin, roll-off carrier, or dump truck. Some operators prefer to provide a separate discharge system for the fly ash because in a conventional water trough it tends to float and is hard to

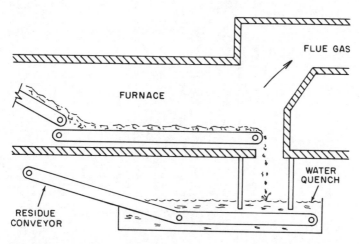

Fig. 5.44 Water-sealed ash conveyor for continuous-feed incinerator (20).

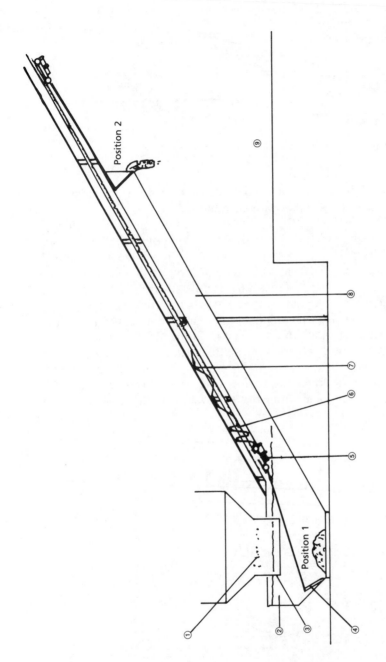

Fig. 5.45 Schematic cross section of residue removal dredge. (Courtesy of Sigoure Freres [75].)

① SLIDE
② TANK
③ SKIRT
④ DREDGE
⑤ MOTORIZED TROLLEY
⑥ RACK
⑦ ELECTRIC CABLE
⑧ INCLINED CHANNEL
⑨ ASH REMOVAL AREA

Position 1

Position 2

remove by bottom-dragging conveyors. Other operators prefer to dump the fly ash into the water trough because it eliminates the cost of a separate ash-conditioning system. The trough can be constructed of steel or concrete and usually has two conveyor troughs so that the residue-discharge system will have a full standby. Having a full spare also allows for switching between the systems to provide even wear and scheduled maintenance.

As the conveyor carries the residue, most of the quench water runs off and returns into the trough. Conveyors should run at velocities not exceeding 5 to 10 ft/min (1.5 to 3 m/min) to guarantee good dewatering and to minimize wear (37). The moisture content of the ash by weight is usually in the range of 25% to 40% or more. Reducing the water content of the ash minimizes transportation costs and water pollution. This can be achieved by maximizing the residence time of the residue on the wet-drag conveyor, by reducing the speed of variable-speed conveyors, or by running constant-speed conveyors at low speeds.

The wet-drag conveyors can operate at up to 45° slopes (37), but some operators prefer lower slopes to protect against bulky items, such as kitchen appliances, from rolling back. Most conveyors operate continuously to minimize shear-pin breakage. Those that operate intermittently achieve better dewatering because of increased retention time.

To achieve even wear, the conveyor chain should be turned over after a few years of operation. It is also advisable to return the conveyors overhead to provide better access for housekeeping purposes. Conveyor systems are subject to severe abrasion and wear because of ground glass, ceramics, concrete, and small metal pieces in the ash. This means pumps, drag chains, sumps, wear strips, and flights must be carefully designed. The conveyor also should have replaceable bottoms and rails; wearing shoes and chains should be made of hard alloy steel, with Brinell hardness of 460 to 512, such as No. 698 Beaumont "Beaucalloy" (37).

Ash-Handling Accessories

The residue-handling system must be carefully designed to minimize the emission of water pollutants. The ash can be acidic or alkaline, so the water pH must be controlled in the range of 6 to 9 pH. The water can also contain high concentrations of BOD, dioxins, heavy metals, and other suspended or dissolved toxic or polluting constituents. For these reasons, the ash-handling system should be operated in the "zero discharge" mode (37). This requires a system of water recirculation and clarification, including properly designed basins, sumps, and easily maintainable pumping stations. To capture all the water that might drain off in the process of ash transfer, catch troughs should be added at the point where the conveyor transfers the ash into the receiver.

In colder regions, with subfreezing temperatures in the winter, the ash-handling system must be protected against freezing. In cold areas, heated trucks transport the ash, and the fly-ash conveyors are insulated to protect them from corrosion and caking. The ash conveyor can unload the wet residue into a temporary storage container or directly into a transport vehicle for removal from the site. In mass-burn systems, the direct discharge into dump trucks is the best and simplest system. In smaller plants, however, the use of twenty-ton dump trucks can create problems because it takes several hours to fill the truck evenly. Instead, operators often use roll-off ash carriers for temporary storage.

Incinerator Instrumentation

The process that takes place in a heat-recovery incinerator (HRI) is similar to the combustion process that occurs in a boiler. The main difference is in the fuel, which is MSW in one case and coal, oil, or gas in the other. Another important difference is in the purpose of these processes: The main task of a boiler is to generate steam, while the main purpose of an HRI is to convert MSW into ash. Therefore steam demand determines the firing rate of a boiler, while the firing rate of an HRI is dependent on the supply of MSW. As a consequence, the fuel flow to the boiler is continuously modulated while that to the incinerator is held relatively constant.

The dynamic characteristics of boilers and incinerators are similar on their air side. The "dead time" of a larger furnace—the time it takes for the furnace outlet temperature to *start* changing, *after* a change in the total airflow—is about 60 seconds. This dead time or transporation lag is largely due to the displacement volume of the furnace, and therefore, the lower the airflow, the longer the dead time tends to be. Since the MSW feed rate is relatively constant, the total airflow rate is adjusted only to reflect the moisture and heating value variations; therefore the incinerator dead time is also relatively constant. Incinerators oscillate at a 2- to 5-minute period and their control modes (integral and derivative) are set to equal their maximum dead time.

Boilers have been around for a long time and thus their controls have evolved to a relatively high level of sophistication (Fig. 5.46). Incinerators, on the other hand, are more recent and the controls on older units are simple and manual. In plants built before 1970, the instruments are pneumatic or analog electronic and are located out in the field without centralization. They usually serve only monitoring functions, while most of the control is done manually by the operators. The only closed-loop controls in these older plants are the temperature and draft controllers. Table 5.9 lists the typical instrumentation found on existing continuous-feed incinerators.

Even in modern incinerator installations that have sophisticated and expensive computerized controls, the control strategies implemented are frequently simple and primitive. It is the exception, not the rule, to see closed feedback loops with feed-forward trimming or on-stream analyzers providing automatic optimization of the process. Most often, even in modern incinerators, where expensive distributed control systems (DCS) with CRT displays are housed in air-conditioned, central control rooms, all these controls do is monitor the operating parameters of the process.

To minimize pollution while maximizing both the burning rate of MSW and the generation of steam, the key incinerator variables must be maintained automatically at their optimum values (or desired setpoints). These variables include the furnace

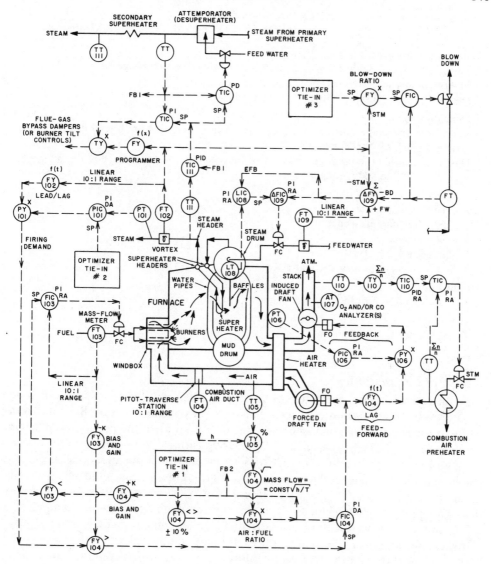

Fig. 5.46 An example of good boiler controls without optimization (22).

temperature and draft, the temperature at the inlet to the induced-draft fan, the completeness of combustion, and the emissions of particulate and gaseous pollutants. The controlled variables are maintained at their optimum values through manipulating the flowrates and flow ratios of underfire, overfire, and cooling air streams and water sprays.

Table 5.9
INSTRUMENT LIST FOR CONTINUOUS-FEED INCINERATOR (20)

Temperature Recorders
   Furnace temperature detected at furnace sidewall near outlet, range 38° to 1250°C.
   Stoker compartment temperatures, range 38° to 1250°C.
   Dust collector inlet temperature, range 38° to 500°C.

Temperature Controllers
   Furnace outlet temperature controlled by regulating total air from forced draft fan; set
      point in 800° to 1000°C range.
   Dust collector inlet temperature controlled by regulating water spray into flue gas; set
      point in 300° to 400°C range.

Draft Gauges
   Forced-draft-fan outlet duct
   Induced-draft-fan inlet duct
   Furnace outlet
   Stoker compartments
   Differential gauge across dust collector

Draft Controller
   Furnace draft control by regulating damper opening

Oxygen Analyzer
   Furnace outlet

Temperature Monitoring

In conventional incinerators without heat recovery, the temperature is monitored in four locations and controlled in two of these locations: (1) The temperature
at the furnace outlet is monitored and controlled to protect the refractory at around
1475° F (800° C). The usual means of lowering furnace temperature is to increase
total airflow at the forced-draft fan. (2) The temperature at the inlet of heat-sensitive
equipment, such as dust collectors or induced-draft fans, is monitored and controlled
at about 575° F (300° C). The temperature controller can modulate the water flow
into the spray chamber (Fig. 5.47) or throttle the opening of the air damper (Fig.
5.35). In the other two locations, the stack and stoker compartment, where damaging
overheating can occur, the temperatures are only monitored.

The most frequently used temperature sensors are thermocouples. Resistance
bulbs are also used, but only at lower temperatures and where temperature differences need to be accurately detected. In the combustion zone, total radiation, infrared, and fiberoptic pyrometers are used in addition to thermocouples to measure
these higher temperatures (25).

For incinerator applications, the most popular sensors are platinum-rhodium
(ISA type R and S) thermocouples, which can be used up to 3100° F (1700° C). At
lower temperatures, up to 2200° F (1200° C), chromel-alumel (ISA type K) thermocouples are also used. Both thermocouple types are suitable for incinerators'

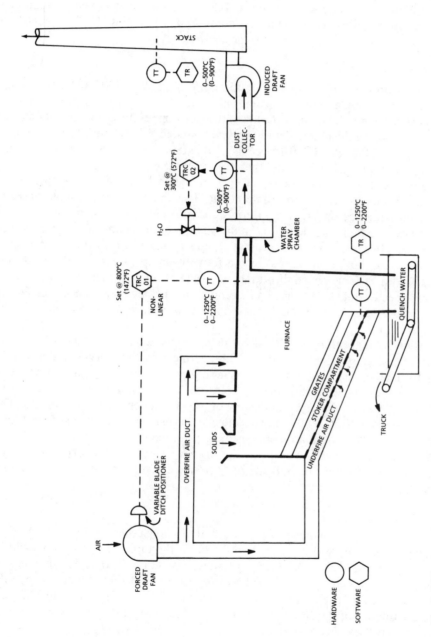

Fig. 5.47 Temperature monitoring and control.

143

oxidizing atmospheres (20), but they should be protected by ceramic thermowells, which are impervious to gases. For up to 1850° F (1000° C), the best protection is provided by fused silica wells, while at temperatures reaching 2750° F (1500° C), refractory porcelain tubes give the best protection.

Temperature Control

The measurement of the furnace temperature controller (TRC–01) is frequently calculated from multiple sensors to determine the average, median, or highest of several measurements. One configuration of furnace temperature control is illustrated in Figure 5.47, where the manipulated variable assigned to TRC–01 is the blade pitch of the forced-draft fan. If TRC–01 detects a rise in temperature, it increases the total cold airflow into the incinerator and thereby lowers the temperature.

In other control configurations, TRC–01 might throttle various dampers (Fig. 5.49) or fan speeds to achieve the same effect. The choice of manipulated variable does have an influence on the incinerator's cost of operation. When the airflow is throttled through damper modulation (on the suction or discharge of the forced-draft fan or in the underfire and overfire ducts), the expensive transportation energy introduced by the fan motor is lost in the form of damper pressure drop. On the other hand, with variable-speed throttling, no waste occurs. Because the capital cost of variable-speed controls is high and because damper controls are expensive to operate, the most cost-effective and best-quality compromise is to use variable blade-pitch control (Fig. 5.34).

The control algorithm for TRC–01 is usually selected as a nonlinear PID (proportional-integral-derivative) algorithm. The nonlinear feature serves to desensitize this loop to small and sudden temperature variations. By providing this neutral zone within which the temperature can vary freely without resulting in any corrective action, the airflow is stabilized and cycling or unstable operation is eliminated. The proportional band of these temperature controllers is usually tuned to less then 100%, while their integral and derivative modes are set to about one minute.

When auxiliary fuels are used in the primary and/or secondary combustion chambers, the burners must be fully instrumented with temperature and flame-failure safety controls (23).

Furnace Draft Measurement

To prevent the outleakage of smoke, fuel, or hot gases, all furnaces operate under a slight vacuum, or draft. Ideally the draft at the top of the furnace should be maintained between 0.05 to 0.1 inches of water vacuum (12 to 25 Pa vacuum). If measured at a point below the top of the furnace, the draft should increase by 0.01 inch of water vacuum per foot of elevation. Therefore, if the pressure tap is 20 feet (6 m) below the top of the furnace, the draft reading should be about 0.3 inches of water vacuum (75 Pa vacuum).

When installing draft gauges, large (1-inch-diameter) pipe should be used to minimize the pressure drop between the furnace and the draft gauge. Because of the turbulence of the hot gases and flames inside the furnace, draft measurements tend to be noisy. The most traditional and simplest way to desensitize the draft measurement is to use a relatively wide range for this measurement. Typical is a "compound" range of +2 inches to −3 inches of water column (+500 to −750 Pa). It is also possible to filter out measurement noise mechanically, by using volume chambers, or electronically, by using noise filters or inverse derivative control modes. The draft controller is usually set at about 0.1 inches of water vacuum (−25 Pa) and tuned as a mostly integral controller, with a wide proportional band, not less than 100%.

On incinerator furnaces the draft gauges also can be useful in signaling operational problems, such as overloaded or burned-out beds and plugged gas passages. For the local indication of draft pressures at various points on the furnace, inclined manometers or mechanical slack diaphragms can be used. The draft gauges should be fully enclosed (for cleanliness) and provided with large, easily readable scales. The draft gauge should be scaled so that the indicator pointer will be at midscale during normal operation.

Furnace Draft Control

Furnace draft can be controlled by barometric dampers (Figs. 5.35, 5.37, and 5.38) or by throttling the flow generated by the induced-draft fan. The suction draft generated by the induced-draft fan can be controlled by damper modulation on the suction or discharge sides of the fan or, more efficiently, by adjusting the blade pitch of the fan.

Figure 5.48 illustrates a sophisticated draft-pressure control system. The furnace draft is controlled in a feedback manner by modulating the blade pitch of the induced-draft fan; the total airflow (modulated to maintain constant furnace temperature, Fig. 5.47) is used as a feedforward addition to this loop. As furnace temperature controls air inflow (Fig. 5.47) and furnace draft controls outflow (Fig. 5.48), the two loops would interact if the feedforward decoupling was not applied. The purpose of FY–02 in Figure 5.48 is to introduce the correct furnace residence time into the feedforward algorithm. As soon as the rate of air entering the furnace changes, the FY–02 lag is initiated, slowly increasing the outflow of the hot gases so that the furnace draft will remain unaffected. In effect, the furnace temperature controller (TRC–01 in Fig. 5.47) will adjust both the forced and induced-draft fan flows equally. The only purpose of the furnace draft pressure controller PIC-01 in Figure 5.48 is to provide some feedback trim to that model, so as to overcome minor measurement or modeling errors.

The feedback draft controller (PIC–01 in Fig. 5.48) can be provided with a narrow neutral band around its setpoint of 0.1 inch of water vacuum (−25 Pa). As long as the draft pressure is within this neutral band of, say, 0.09 to 0.11 inches of water vacuum, the controller will not change its output signal at all. This can have a stabilizing effect when the furnace draft measurement is noisy and also can eliminate the need to keep changing the fan's blade pitch. The draft controller should be

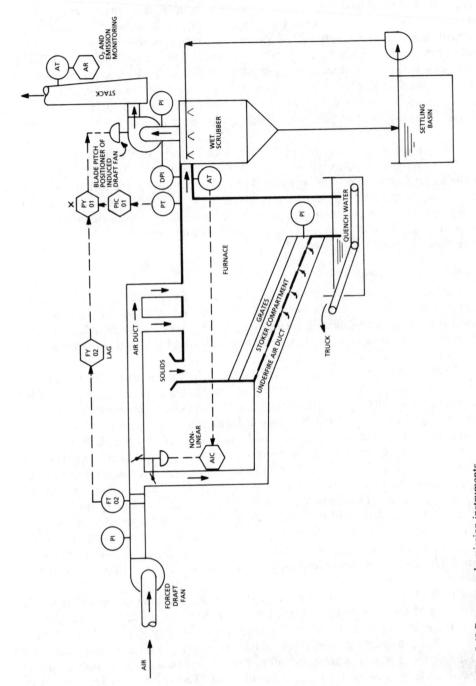

Fig. 5.48 Pressure and emission instruments.

provided with a mostly integral algorithm, having little gain or with wide proportional band (over 100%).

Fan Controls

Both the forced and induced draft fan stations require at least three levels of control instrumentation. The first level is provided to guarantee operational safety. The second level serves to integrate individual fan units into a single "load following" fan station. The purpose of the third level is to find the optimum pressure and automatically maintain it.

SAFETY INTERLOCKS

In Figure 5.36 the safety interlock numbers are given as 1 and 2 inside the "interlock flags." Interlock 1 will stop the fan if PSH–02 detects excessively high pressures at the fan discharge or if PSL–03 senses excessively high vacuums on its suction side. These pressure switches will protect the ductwork or breechings against bursting or collapsing when accidental blockages occur. Once the fan is stopped, it can be restarted only by pressing the reset button (R–04). This gives the operator an opportunity to correct the cause of the abnormal pressure before restarting the fan. When the fan is stopped, its discharge damper (XCD–06) closes automatically, thereby protecting against flow reversal that otherwise could occur in a multifan station. This damper is the fast-opening, slow-closing design. When the fan is on, the damper is always open.

The time-delay relay (TD–05) guarantees that once the fan is started it will continue running for a preset period unless the safety interlock turns it off. This protects the motor from excessive on-off cycling, which can cause overheat damage. Interlock 2 is the fan-cycling interlock described in Figure 5.49.

CYCLING OF MULTIPLE FANS

A fan station is optimized when it transports the required amount of gas at the minimum cost. This minimum cost is obtained by minimizing the number of fans in operation and by maximizing their energy efficiency. The control system in Figure 5.49 fulfills both these goals. Interlock 2 provides the logic for starting or stopping the second fan unit (Fig. 5.49, upper right). As the demand for forced draft air is increased, the speed (or blade pitch) of the first fan is also increased. When this fan has reached its maximum capacity (point A on the fan curve—Fig. 5.49, upper left), PSH–07 activates interlock 2, which in turn starts the second fan.

If the second fan is started while the control signal from PIC–10 is still at its maximum, the operating point on the fan curve will jump to point C and a substantial upset will occur in the fan-discharge pressure. A signal generator (PY–07) is introduced to eliminate this upset. When interlock 2 starts the second fan, it will also actuate FY–07, which will drop its output to x. The value x corresponds to the speed

148

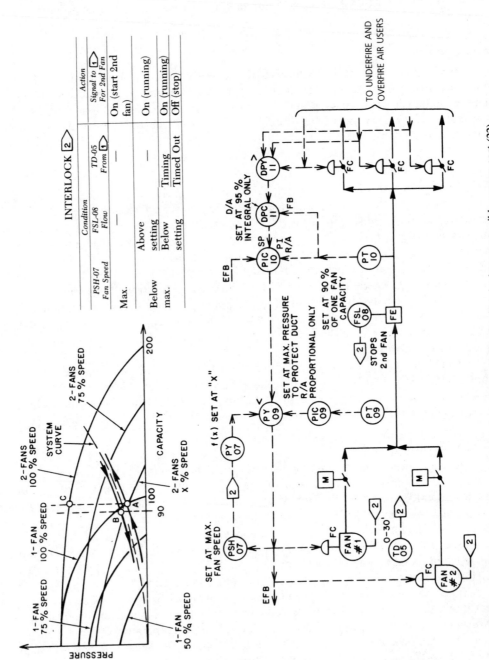

Fig. 5.49 A forced-draft fan station is optimized when it meets demand at the lowest possible energy cost (22).

(or blade-pitch position) of a two-fan operation at point A. The low signal selector PY–09 will immediately select this output x for control and thus prevent the upset. After actuation, the output of PY–07 slowly rises to full scale; shortly after starting the second fan, control smoothly returns to PIC–10.

When both fans are in operation and the demand for air is dropping, FSL–08 will stop the second fan when a single unit can meet the total load. FSL–08 is set at 90% of the capacity of one fan (point B on the system curve). When total airflow drops to that setting, interlock 2 will stop the second fan if the time delay TD–05 has timed out. This shutdown delay protects the fan from excessive cycling. Setting TD–05 for 20 minutes guarantees that the fan cannot be started and stopped more frequently than three times an hour. This control system will cycle any number of parallel fans. If n is the number of fans in operation, then FSL–08 is set for 90% of the capacity of $(n - 1)$ fans and PY–07 is set to generate a value of x, which corresponds to the required speed (or blade position) of $(n + 1)$ fans at point A.

DISCHARGE-PRESSURE OPTIMIZATION

The optimum discharge pressure is the minimum pressure that will still satisfy all air users. If the fan station supplies air to both underfire and overfire dampers, the optimum discharge pressure can be found by observing the opening of the most open damper. If even the most open damper is not fully open, the air-supply pressure can be safely lowered, while if the most open damper is fully open, the supply pressure must be raised. This supply-demand control scheme not only minimizes the use of fan power, it also guarantees that no damper will fully open and thereby lose control. The associated electric power savings can be substantial.

The optimization loop in Figure 5.49 functions as follows: DPY–11 selects the opening of the most open damper and sends that signal to the damper position controller DPC–11 as its measurement. This controller is set at 95% and therefore will reduce the supply-pressure setpoint until the most open damper reaches an opening of 95%. In this process, all other dampers are also opened, thus saving fan power. DPC–11 is an integral-only controller, tuned for an integral time (minutes/repeat) that is ten times that of PIC–10. This provides stable and smooth control even if the dampers are unstable or cycling. The feedback signal (FB) protects DPC–11 from reset windup when PIC–10 is switched from cascade to manual.

PIC–10 is the load-following controller. It compares the optimized setpoint with the actual air-supply pressure and adjusts the speed (or blade pitch) of the forced-draft fan. When the output signal of PIC–10 reaches its maximum, PSH–07 starts another fan. The only purpose of PIC–09 is to protect the fan station from overpressures. Because the setpoint of PIC–09 is much higher than the normal operating values, its output signal will also be high most of the time. Therefore, under normal conditions, the low signal selector PY–09 will select the output of PIC–10 and will disregard PIC–09.

Flow Instruments

The complete automation of incinerators involves more than temperature and pressure instruments. The measurement of solid and gas flows is needed for material balance-control purposes. The detection of flue-gas composition is necessary to evaluate the quality of combustion and to monitor the emission of pollutants.

Material balance of combustion requires that both the fuel and airflows be measured. This is seldom the case when incinerators are involved. The typical receiving station at the incinerator usually includes truck-weighing scales, either manually operated mechanical scales or automatically recording load-cell-type units (25). The latter are more accurate, as well as being maintenance-free and tamper-proof. The continuous weighing of the MSW flow into the furnace can be done at the charging hopper or at the charging grates (Figs. 5.5, 5.17, and 5.21).

While it is possible to detect the "rate of weight change" in the hopper or the mass flow of solids on the grates (gravimetric feeders), this is seldom done, partly because of the high cost and maintenance of such continuous weighing devices and partly because the MSW mass flow rate registered by a gravimetric feeder is of value only if the heating value and composition of the MSW are constant. Therefore, measurement of the solid-waste flow into a mass-burn incinerator is not as important as measuring it when the fuel is RDF (refuse-derived fuel).

In existing incinerators, the air and gas flows are measured only during performance testing by portable point velocity sensors, such as pitot tubes or hot-wire anemometers (25). This practice is undesirable because the required airflows are not constant but vary with MSW feed rate and composition. It would be preferable to match each load condition with a corresponding airflow distribution in the ducts, and this can be done only if those ducts have continuous-flow sensors. Therefore the installation of such airflow transmitters as FT–02 in Figure 5.48 is highly recommended. (For a selection of flow sensors for air, water, and steam services, see Table 5.10.)

Stack Sampling

The traditional method of determining the composition of stack gases is through laboratory analysis of manually obtained grab samples. The manual testing of incinerators is specified in Bulletin T6–71 of the Incinerator Institute of America. These performance tests are difficult and time-consuming to perform reliably. Even when done correctly, they give only one operating point and not a continuous performance curve. (Chapter 8 discusses incinerator performance testing in more detail.)

An important part of performance testing is the collection of representative samples (Fig. 5.50). When collecting particulates, it is not sufficient to draw samples from all points of the cross section (as illustrated in Fig. 5.50); the samples also must be drawn isokinetically. This means that the sample should be drawn at the same velocity at which the flue gases and particles are traveling in the stack. If the sample is drawn at a higher velocity, excess particles will be drawn in; if the sample is drawn

Table 5.10
FLOW SENSOR ERRORS ON INCINERATORS (23)

Flow Streams Measured	Type of Flowmeter	Error at 100% Flow in % of Measurement	Error at 10% Flow in % of Measurement
Steam and Water	Vortex shedding[a]	1	1
	Orifice	1	Useless (10% with two d/p cells)
Air	Area averaging pitot traverse station	0.5	Useless if one d/p cell is used; 5% with two
	Multipoint thermal	2	20 (better with dual range)
	Piezometer ring, orifice segment Venturi section airfoil section	3	Useless

[a] Limited to 750°F (400°C)

at a lower velocity, some of the particles will pass by the sample. Therefore, accurate samples can be obtained only if the sampling velocity is adjusted to match the stack gas velocity at each point of the stack cross section.

Excess-Oxygen Analyzers

Incinerator flue gases are analyzed to evaluate the quality of combustion and to check the performance of the pollution-control devices. This analysis is frequently done on an intermittent and manual basis, using laboratory evaluations of grab samples and less often through the use of continuous onstream analyzers (23), which are expensive and not that reliable. In the last decade, however, progress has been made to improve these analyzers.

Complete combustion with a fuel such as MSW is difficult to achieve. Large quantities of excess air are needed to guarantee that all the combustibles will burn completely. Of the several different types of analyzers, all detect the completeness of combustion. These include the excess oxygen, opacity, and carbon-monoxide analyzers. Table 5.11 gives a list of stack analyzer features and suppliers.

Excess-oxygen analyzers detect the amount of unused oxygen in the flue gases. The higher the concentration of excess oxygen in the stack gases, the less likely it is to find unburned fuel in the incinerator residue. The better excess-oxygen analyzers are configured as probes (Fig. 5.51), so no sampling is needed. The working element of these probes is zirconium oxide, which generates an electronic signal that relates

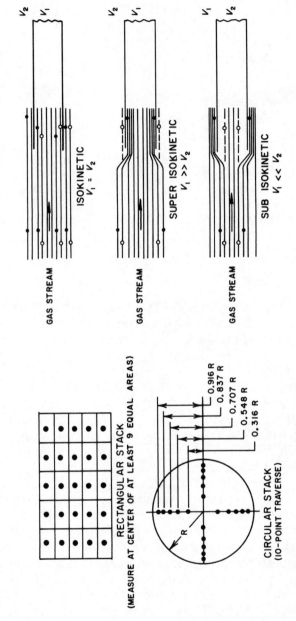

Fig. 5.50 Particle collection and sampling velocity (23).

152

Table 5.11
COMBUSTION STACK GAS ANALYZERS

Company Name	Functions	Gases Sensed	Measuring Method	Sampling Method
Anacon	CM FC	O_2 CO HC TRS SO_2	IR UV ZI MS	EX
Anarad	CM	O_2 CO CO_2 HC H_2O	IR	IN EX
Astro Resources	CM	O_2 HC	ZI	EX
Bacharach	CM	O_2	ZI	IN
Bailey Controls	CM FC PI	O_2 CO HC	CA ZI	EX
Bambeck	CM FC	CO	IR	IN
CEA	PI	CO CO_2	IR	EX
Cleveland Controls	CM	O_2	ZI	IN
CSI	CM	SO_2 H_2S NOX CO_2 HC	UV FL	EX
Datatest	CM FC PI	O_2 CO CO_2 HC OP SO_2	IR UV NE OC ZI	IN EX
Dynatron	CM FC	O_2 CO OP	IR GD OC ZI	IN
Eberline	CM	RD	RA	IN EX
Econics	CM FC	CO CO_2 HC OP	IR OC	IN
Energetics Science	CM	CO	EC	EX
Gas Tech	PI	O_2 CO HC	CA	EX
Honeywell	PI	O_2		EX
Infrared Industries	CM	O_2 CO CO_2 HC	IR CO	EX
Interscan	PI	CO CO_2	IR	EX
Kent	CM	O_2 CO CO_2	IR TC ZI PA	IN
Lear Siegler	CM FC	O_2 CO CO_2 SO_2 NOX	IR UV TC PA	EX
Leeds & Northrup	CM FC	O_2 CO_2 HC	TC NE ZI PA	IN EX
Lynn Products	FC PI	O_2 CO OP	PO	EX
Marlin Mfg	CM	O_2	ZI	EX
Milton Roy	CM	O_2	ZI	IN
Moniteq	CM FC	CO	IR	IN
Monitor Labs	CM PI	CO CO_2 SO_2 NOX	IR CT FT	IN
MSA	CM FC	O_2 CO CO_2 HC	IR TC CA ZI PO PA	EX
Neutronics	PI	O_2	EC	EX
Permutit	CM	CO_2	GD	EX
Princeton Sensors	CM FC	CO HC	IR	IN
Process Analyzers	CM FC	O_2 CO CO_2 HC	GC	EX
SUMX	CM FC PI	O_2 CO	IR ZI PA	EX
Teledyne Analytical	CM PI	O_2 CO CO_2 HC	IR UV TC CA ZI PO	IN EX
Testoterm	PI	CO_2		IN EX
Thermco	PI	O_2 CO_2 HC H_2	TC EC	EX
Western Research	CM	SO_2	UV	EX
Westinghouse	CM FC	O_2 CO HC	IR ZI	IN EX

Key to abbreviations

CA	catalytic bead	GD	gas density	OP	opacity
CM	continuous monitoring	H_2	hydrogen	PA	paramagnetic
CO	carbon monoxide	H_2O	water	PI	periodic inspection
CO	coulometry	H_2S	Hydrogen sulfide	PO	polarography
CO_2	carbon dioxide	HC	hydrocarbons & combustibles	RA	radiation
CT	chemiluminescent	IN	in situ	RD	radioactivity
EC	electrochemical	IR	infrared absorption	SO_2	sulfur dioxide
EX	extractive	MS	mass spectrometry	TC	thermal conductivity
FC	feedback control	NE	nephelometry	TRS	total reduced sulfur
FL	flame photometry	NOX	oxides of nitrogen	UV	ultraviolet absorption
FT	fluorescent	O_2	oxygen	ZI	zirconium oxide
GC	gas chromatography	OC	opacimetry		

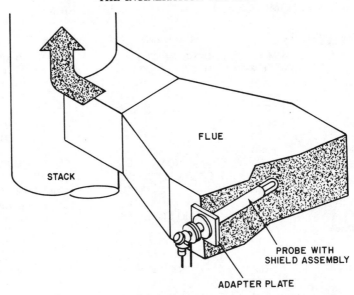

STACK

FLUE

PROBE WITH
SHIELD ASSEMBLY

ADAPTER PLATE

Fig. 5.51 The probe-type oxygen analyzer should be installed close to the combustion zone, but at a point where the temperature is below the limit for the zirconium oxide detector (23).

to the oxygen concentration in the stack. The excess-oxygen probe can be used for monitoring and trend-recording purposes, or it can be integrated into an automatic, closed-loop control system.

When used in a closed loop configuration, the excess-oxygen control loop usually trims the furnace temperature controls in Figure 5.47 or adjusts the ratio of underfire/overfire air, as does the AIC in front of the wet scrubber in Figure 5.48. While the system shown in Figures 5.47 and 5.48 is preferable because it assigns separate manipulated variables (total airflow and underfire/overfire ratio) to the two controlled variables (furnace temperature and excess air), some prefer to manipulate total air only. In that case the furnace temperature controls of Figure 5.47 are converted either into a cascade (Fig. 5.52) or a selective (Fig. 5.53) control config-uration. Neither of these solutions is perfect, but both are better than having no closed-loop excess-oxygen control at all. The problems with Figure 5.52 include its limited dynamic performance (the AIC must be faster than the TRC) and the AIC's lack of an independent setpoint. Therefore the system will not control excess oxygen at a single value but will allow the TIC to vary these settings (within the permitted limits) to keep furnace temperature constant. The control system in Figure 5.53 is better because the dynamics of the two loops are not in series, but the conflict still exists insofar as only one of the setpoints can be met and only one of the controllers—

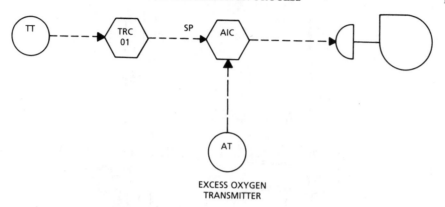

Fig. 5.52 Cascade control configuration.

the one that requires more air—will control the total airflow into the incinerator. This system thus will err in the direction of more excess air.

Continuous On-line Analyzers

For incinerators being built today, continuous on-line analyzers are essential. Manual stack sampling is now done more to perform a calibration check than to operate the plant. The continuous, in situ analyzers illustrated in Figure 5.54 look through the whole cross-sectional area of the stack and provide simultaneous readings for particulates (fly ash and unburned fuel), oxides of carbon, sulfur, nitrogen, and other pollutants such as chlorine. Some of these readings serve to limit and

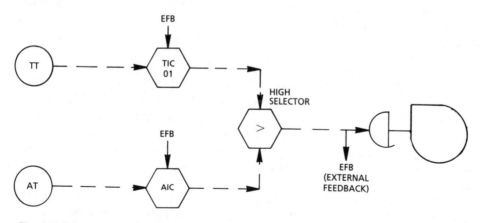

Fig. 5.53 Selective control configuration.

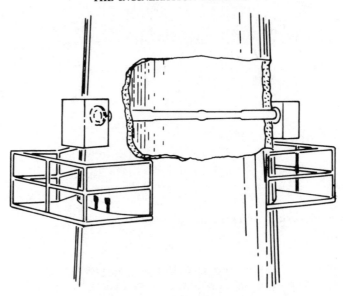

Fig. 5.54 Continuous on-line stack monitor (23).

control emissions, while others, such as CO, can also be used to control the quality of combustion by modifying the air-fuel ratio of the furnace.

CO gives a good indication of the completeness of combustion because the desired CO concentration in the flue gas is unaffected by the type of fuel used. This makes CO control superior to excess-oxygen control because the requirement for excess oxygen *does* vary with fuel composition. In CO analysis, the combustion process is likely to be optimized when the CO concentration in the flue gas is between 100 and 200 ppm (parts per million), regardless of fuel composition. Because the composition of MSW is highly variable, this is an important consideration in favor of CO control.

Another possible advantage of CO control has to do with the reported correlation between dioxin formation and CO concentration in the flue gas (Fig. 5.55). If this theory is proved, then by monitoring the quality of combustion through the detection of CO concentration in the flue gas (and by maintaining high stack temperatures), one can indirectly monitor the concentration of dioxins. Since present methods of dioxin measurement involve such maintenance-prone and expensive methods as gas chromatography and mass spectroscopy, CO-based indirect dioxin controls would be welcome.

MSW can contain many toxic substances, so the plant should be monitored for the presence of toxic gases. Table 5.12 lists the capabilities of a number of toxic-gas detectors.

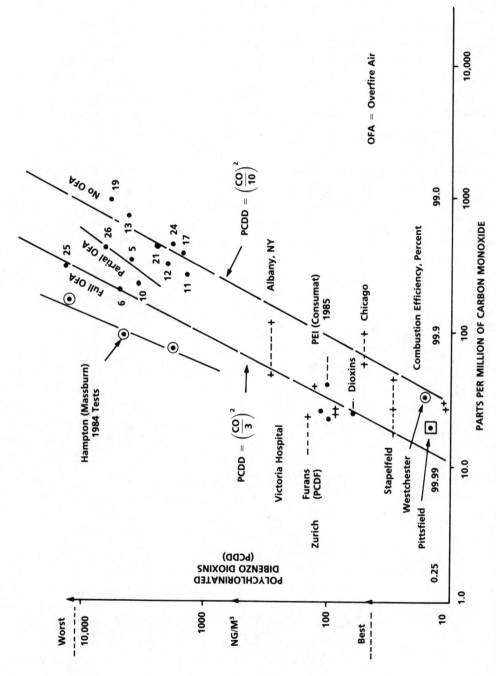

Fig. 5.55 Total PCDD emissions and predicted cancer-risk levels (38).

157

Table 5.12
TOXIC GAS DETECTION

I. Applications	Cl₂	HCl	HF	F₂	Br₂	ClO₂	I₂	NH₃	HCN	CO₂
Measurement range	0–3 PPM 0–9.9 PPM 0–99 PPM	0–9.9 PPM 0–15 PPM 0–99 PPM	0–9.9 PPM 0–99 PPM	0–9.9 PPM 0–99 PPM	0–9.9 PPM 0–99 PPM	Consult Factory	Consult Factory	0.6–75 PPM 0.8–99 PPM	0.2–30 PPM 0.8–99 PPM	850–10,000 PPM
Response time (when exposed to 5 times the TLV)	<6 Sec	<6 Sec	<10 Sec	<6 Sec	<10 Sec	Consult Factory	Consult Factory	<15 Sec	<20 Sec	<10 Sec
Temperature range	14° F to 122° F (−10° C To 50° C)									
Relative humidity range	20% to 90% RH									
Power requirements	15 to 30 Vdc at 0.05 amps									
Transmission link	4-20 mA via Two Wires									
Sensor life	3 Years	3 Years	3 Years	3 Years	3 Years	3 Years	3 Years	3 Years	3 Years	3 Years
Linear or log	Linear	Linear	Linear	Linear	Linear	Linear	Linear	Log	Log	Log
Electrical classification	Intrinsically Safe For Operation In NEC Class 1, Division 1, Groups A, B, C And D When Used With Specified Zener Barriers. Explosion Proof for NEC Class 1, Division 1, Groups B, C, And D									
Oxygen required	No	No	No	No	No	No	No	No	No	No
Arctic sensor available	YES	YES	YES	YES	YES	YES	YES	YES	YES	YES
Sensor simulator available	YES	YES	YES	YES	YES	—	—	YES	YES	YES

II. Applications	SO_2	Amines	H_2S-(A)	CO	NO_2	H_2	AsH_3	PH_3	SiH_4	H_2S-(B)
Measurement range	0–9.9 PPM 0–99 PPM	Consult Factory	0.2–30 PPM 0.8–99 PPM	0–99 PPM 0–500 PPM	0–9.9 PPM	50–300 PPM 0–4% by Vol. 0–9.9% by Vol. 0–99% LEL	0–.99 PPM	0–.99 PPM	0–9.9 PPM	0–99 PPM
Response time (when exposed to 5 times the TLV)	<10 Sec	Consult Factory	<10 Sec	<10 Sec	<10 Sec	<60 Sec to 90% of Concentration	<10 Sec	<10 Sec	<10 Sec	<10 Sec
Temperature range	14° F to 122° F −10° C to 50° C			10° F to 105° F −12° C to 40° C		30° F to 105° F	55° F to 80° F 15° C to 25° C	30° F to 105° F 0° C to 40° C	30° F to 105° F 0° C to 40° C	10° F to 105° F −12° C to 40° C
Relative humidity range	20% to 90%					5% to 100%	30% to 70%	20% to 90%	20% to 90%	25% to 90%
Power requirements	15 to 30 Vdc at 0.05 amps									
Transmission link	4-20 mA via Two Wires									
Sensor life	3 Years	3 Years	3 Years	1 Year	1 Year	3 Years	1 Year	1 Year	1 Year	1 Year
Linear or log	Linear	Log	Log	Linear	Linear	Linear	Linear	Linear	Linear	Linear
Electrical classification	Intrinsically Safe For Operation In NEC Class 1, Division 1, Groups A, B, C And D When Used With Specified Zener Barriers. Explosion Proof for NEC Class 1, Division 1, Groups B, C, And D									
Oxygen required	No	No	No	1%	1%	No	1%	1%	1%	1%
Arctic sensor available	YES	YES	YES	NO	NO	NO	NO	NO	NO	NO
Sensor simulator available	YES	NO	YES	NO	NO	YES	NO	NO	NO	NO

Continuous Emission Monitoring

The Clean Air Act of 1969 applied to only 250 million BTU/hr or larger steam-generating power plants. In recent years the requirements have been expanded to cover other types of sources, such as steam-generating incinerators. In 1987 the EPA issued requirements for emission control and monitoring for all new or reconstructed steam-generating units with heat inputs exceeding 29 megawatts (100 million BTU/hr). In 1988 a bill in Congress (H.R. 4902) proposed to set standards for the emissions of some ten pollutants from MSW incineration facilities. All of this points to an eventual need for developing continuous emission monitoring (CEM) packages for incinerators.

Currently, most incinerators are required to monitor CO, SO_2, NO_x, and opacity (visible emissions). The EPA does not yet require the monitoring of HCl, but various state agencies do. The capabilities of a typical CEM unit therefore will include the measurement of CO, CO_2, SO_2, NO_x, total hydrocarbons, opacity, and HCl. Such monitoring packages are already available from analyzer suppliers and are likely to become more widely used as the number of incinerator installations increases in the United States.

On-Site Incinerators

This section will describe some of the smaller incinerator units used on site in domestic, commercial, and industrial applications, and some of the more specialized designs, such as fluid-bed and multiple-hearth incinerators.

On-site incineration is a simple and convenient means of reducing the waste transportation problem since it reduces the volume of disposable waste. On-site incinerators tend to be smaller and their fuel more predictable in composition than the MSW burned in municipal incinerators. For these reasons it is appropriate to discuss separately the considerations that affect the location, selection, and operating practices of these on-site units.

The on-site incinerator should be located close to the larger sources of waste and to expected waste-collection routes. On-site incinerators are constructed of 12-gauge steel casing with high-temperature (over 1000° F) insulation and high-quality refractory lining. Indoor installations are preferred, but even when the incinerator is situated outdoors, the charging and cleanout operations still should be shielded from the weather. Incinerator rooms should be designed for two-hour fire resistance and should comply with the National Fire Protection Association (NFPA) recommendations contained in bulletin NFPA No. 82.

The Incinerator Institute of America (IIA) separates incinerators into nine classes according to their use and size (see Table 5.13) and provides minimum construction and performance standards for each class. Similar classifications and construction standards have also been instituted by the NFPA in its standard "Incinerators and Rubbish Handling." The IIA has also classified the wastes into seven types (Tables 1.16 and 1.17). Local and state codes must be complied with when selecting an incinerator.

Table 5.13
INDUSTRIAL WASTE TYPES (16)

Type of Wastes	Description
1	Mixed solid combustible materials, such as paper and wood
2	Pumpable, high heating value, moderately low ash, such as heavy ends, tank bottoms
3	Wet, semisolids, such as refuse and water treatment sludge
4	Uniform, solid burnables, such as off-spec or waste polymers
5	Pumpable, high ash, low heating value materials, such as acid or caustic sludges, or sulfonates
6	Difficult or hazardous materials, such as explosive, pyrophoric, toxic, radioactive or pesticide residues
7	Other materials to be described in detail

Table 5.14
CLASSIFICATION OF INCINERATORS (14)

Class I—Portable, packaged, completely assembled, direct feed incinerators having not over 5 ft³ storage capacity, or 25 lb/hr burning rate, suitable for type 2 waste.

Class IA—Portable, packaged or job assembled, direct fed incinerators 5 ft³ to 15 ft³ primary chamber volume; or a burning rate of 25 lb/hr up to, but not including, 100 lb/hr of type 0, type 1 or type 2 waste; or a burning rate of 25 lb/hr up to, but not including, 75 lb/hr of type 3 waste.

Class II—Flue-fed, single chamber incinerators with more than 2 ft² burning area for type 2 waste. This type of incinerator is served by one vertical flue functioning both as a chute for charging waste and to carry the products of combustion to atmosphere. This type of incinerator has been installed in apartment houses or multiple dwellings.

Class IIA—Chute-fed multiple chamber incinerators for apartment buildings with more than 2 ft² burning area, suitable for type 1 or type 2 waste. (Not recommended for industrial installations.) This type of incinerator is served by a vertical chute for charging wastes from two or more floors above the incinerator and a separate flue for carrying the products of combustion to atmosphere.

Class III—Direct fed incinerators with a burning rate of 100 lb/hr and over, suitable for type 0, type 1 or type 2 waste.

Class IV—Direct fed incinerators with a burning rate of 75 lb/hr or over, suitable for type 3 waste.

Class V—Municipal incinerators suitable for type 0, type 1, type 2 or type 3 wastes, or a combination of all four wastes, and are rated in tons per hour or tons per 24 hours.

Class VI—Crematory and pathological incinerators, suitable for type 4 waste.

Class VII—Incinerators designed for specific by-product wastes, type 5 or type 6.

Note: For waste type numbers, see Tables 1.16 and 1.17.

Table 5.15
INCINERATOR CAPACITY CHART (14)

Classification	Building Types	Quantities of Waste Produced
Industrial buildings	Factories	Survey must be made
	Warehouses	2 lb/100 ft²/day
Commercial buildings	Office buildings	1 lb/100 ft²/day
	Department stores	4 lb/100 ft²/day
	Shopping centers	Study or survey required
	Supermarkets	9 lb/100 ft²/day
	Restaurants	2 lb/meal/day
	Drug stores	5 lb/100 ft²/day
	Banks	Study of plans or survey required
Residential	Private homes	5 lb basic & 1 lb/bedroom
	Apartment buildings	4 lb/sleeping room/day
Schools	Grade schools	10 lb/room & ½ lb/pupil/day
	High schools	8 lb/room & ½ lb/pupil/day
	Universities	Survey required
Institutions	Hospitals	15 lb/bed/day
	Nurses' or interns' homes	3 lb/person/day
	Homes for aged	3 lb/person/day
	Rest homes	3 lb/person/day
Hotels, etc.	Hotels—1st class	3 lb/room and 2 lb/meal/day
	Hotels—medium class	1½ lb/room & 1 lb/meal/day
	Motels	2 lb/room/day
	Trailer camps	6 to 10 lb/trailer/day
Miscellaneous	Veterinary hospitals	Study of plans or survey required
	Industrial plants	
	Municipalities	

Incinerator Selection

The first step in selecting an incinerator is to record the volume, weight, and class(es) of waste collected for a period of at least two weeks. The survey should be checked against typical waste-production rates, such as those in Table 5.15. If no waste survey is available, Table 5.15 can serve as the basis for an approximate estimate. The maximum daily operation can be estimated as three hours for apartment buildings, four hours for schools, six hours for commercial buildings, hotels, and other institutions, and seven hours per shift of industrial installations (14).

The results of the waste survey will help determine whether to install a continuous or batch-type incinerator. Batch-type units consist of a single combustion chamber (Fig. 5.38). If the batch furnace has no grate, the ash accumulation will reduce the rate of burning. The batch incinerator is sized according to the weight of each type of waste per batch at so many batches per day. The continuous incinerator consists of two chambers, one for charge storage and the other an evacuated chamber for combustion. The charge chamber can be loaded at any time. Sizing is based on pounds per hour burning rate required.

The nature and characteristics of the waste are usually summarized in a form such as that in Table 5.16. Most incinerator manufacturers (Table 5.17) offer standard, pre-engineered packages for waste types 0, 1, 2, 3, and 4 (Tables 1.16 and 1.17). Waste types 5 and 6 almost always require unique designs because the physical, chemical, and thermal characteristics of these wastes are highly variable. Type 6 wastes tend to have low heating values but still contain materials that can cause intense combustion. Plastics and synthetic rubber tend to decompose at high temperatures and form complex organic molecules that require auxiliary heat and high turbulence before they are fully oxidized. In extreme cases, three combustion chambers are necessary; flue gases from the secondary combustion chamber must be recycled back into the primary chamber in order to complete the combustion process.

Incinerator Charging

Incinerators can be charged manually or automatically; they also can be charged directly (see Fig. 5.4) or from charging rooms (Figs. 5.56 and 5.57). Direct incinerators are the least expensive but are limited in their hourly capacities to 500 pounds, while indirect incinerators can operate at capacities up to 1,000 lb/hr. A manually charged incinerator (Fig. 5.56) is fed through a bell-covered chute from the floor above the furnace. This labor-saving design also guarantees good combustion efficiency and protection against "flashbacks." The separate charging room is also convenient to sort the waste for recycling (14). Incinerators can also be fed from the same floor

Table 5.16
WASTE ANALYSIS SHEET (27)

% Ash _____ % Sediment _____ % Water _____

Waste material soluble in water? _____ Water content well mixed, emulsified? _____

If there are solids in the liquid, what is their size range? _____

Conradson carbon _____ Corrosion (copper strip) _____

Is the material corrosive to carbon steel? _____ Corrosive to brass? _____

What alloy is recommended for carrying the fluid? _____

Distillation data (if applicable) 10% at _____ ° F; 90% at _____ ° F; end point _____ ° F.

Flash point _____ ° F; fire point _____ ° F; pour point _____ ° F.

Viscosity _____ SSF at 122° F or _____ SSU at 100° F.

pH _____; acid number _____; base number _____

Heating value _____ Btu/gallon. Specific gravity (H_2O = 1.0) _____

Will the material burn readily? _____

Toxic? (explain) _____

Table 5.17
PARTIAL LIST OF INCINERATOR SUPPLIES (26)

Air Preheater Co., Inc., American Schack Co., Inc., Aqua-Chem, Inc., BSP Corp., Div. of Envirotech Systems, Inc., Bartlett-Snow, Beloit-Passavant Corp., Best Combustion Engrg. Co., Bethlehem Corp., Brule Pollution Control Systems, C-E Raymond, Carborundum Co., Pollution Control Div., Carver-Greenfield Corp., Coen Co., Combustion Equipment Assoc. Inc., Copeland Systems Inc., Dally Engineering-Valve Co., Dorr-Oliver Inc., Dravo Corp., Environmental Services Inc., Envirotech, First Machinery Corp., Foster Wheeler Corp., Fuller Co., Garver-Davis, Inc., Haveg Industries Inc., Holden, A. F., Co., Hubbell, Roth & Clark, Inc., Intercontek, Inc., International Pollution Control, Inc., Ishikawajima-Harima Heavy Industries Co., Ltd., Kennedy Van Saun Corp., Klenz-Aire, Inc., Koch & Sons, Inc., Koch Engrg. Co., Inc., Kubota, Ltd., Chuo-Ku, Tokyo, Japan, Lawler Co., Leavesley Industries, Lurgi Gesellschaft fuer Waerm & ChemotecHnik mbH, 6 Frankfurt (Main) Germany, Maxon Premix Burner Co., Inc., Melsheimer, T., Co., Inc., Midland-Ross Corp., RPC Div., Mid-South Mfg. Corp., Mine & Smelter Supply Co., MSI Industries, Mitsubishi Heavy Industries, Ltd., Tokyo, Japan, Monsanto Biodize Systems, Inc., Monsanto Enviro-Chem Systems Inc., Nichols Engrg. & Research Corp., North American Mfg. Co., Oxy-Catalyst, Inc., P.D. Proces Engrg. Ltd., Hayes, Middlesex, England, Peabody Engrg. Corp., Picklands Mather & Co., Prenco Div., Plibrico Co., Prenco Mfg. Co., Pyro Industries, Inc., Recon Systems Inc., Renneburg & Sons Co., Rollins-Purle, Inc., Ross Engrg. Div., Midland-Ross Corp., Rotodyne Mfg. Corp., Rust Engrg. Co., Div. of Litton Industries, Sargent, Inc., Surface Combustion Div., Midland-Ross Corp., Swenson, Div. of Whiting Corp., Tailor & Co., Inc., Takuma Boiler Mfg. Co., Ltd., Osaka, Japan, Thermal Research & Engrg. Co., Torrax Systems, Inc., Vulcan Iron Works, Inc., Walker Process Equip., Div. of Chicago Bridge & Iron Co., Westinghouse Water Quality Control Div., Infilco, Zink, John, Co., Zurn Industries, Inc.

where the furnace is located (Fig. 5.57). This arrangement also permits sorting and is labor-efficient, although the radiant heat can be uncomfortable for the operator.

In high-rise buildings, installation of a waste chute eliminates the labor involved in charging the incinerator (Fig. 5.58). The chute automatically directs the solid waste into a top-charged, mechanical incinerator. The charging rate can be regulated by rotary star feeders or by charging gates that open at 15- to 30-minute intervals. Both offer protection against momentary overloading.

Hydraulic plungers or rams offer a more controlled method of automatic charging. The movement of the reciprocating plunger forces the refuse from the bottom of the charge hopper into the furnace (see Fig. 5.9). This is the most common method of automatic charging for capacities exceeding 500 pounds per hour.

Incinerators that burn sawdust or shredded waste are frequently charged by screw feeders or pneumatic conveyors (Fig. 5.10). Screw feeders are at least 6 inches

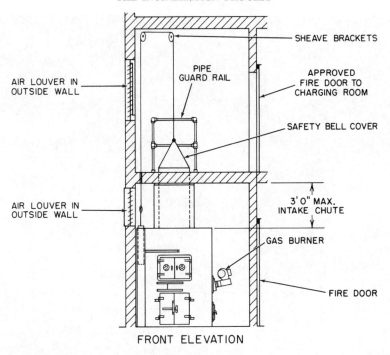

Fig. 5.56 Top-charging incinerator (14).

(15 cm) in diameter and are designed with variable pitch to minimize the compression (and therefore blocking) the shredded waste. Container charging, which is being used in a few isolated cases, has the advantage of protecting against exposure to flashback caused by aerosol cans or the sudden combustion of highly flammable substances.

Incinerator Accessories

For smaller incinerators, chimneys provide sufficient draft to discharge flue gases at a high enough point where no nuisance is caused by the emissions. A fully loaded chimney should provide at least 0.25 inches of water draft (-62 Pa). Table 5.18 lists the diameters and heights of chimneys according to the weight rate of wastes burned in a continuously charged multiple-chamber incinerator. The table assumes that the incinerator uses no dilution air and that the breechings between the furnace and chimney are of minimum length (14). The lining thicknesses shown are for outdoor chimneys; chimneys inside buildings need additional insulation.

Fig. 5.57 Incinerator floor charging showing a guillotine power-operated door and the combustion controls (14).

For proper incinerator operation, the cold-air supply to the furnace should not be restricted. In most designs the furnace receives its air supply from the incinerator room. The air supply should be sized for about 15 pounds of air per pound of MSW burned. If the air supply is insufficient, the mechanical ventilation system of the building can cause smoking due to downdrafts. When a chimney's natural draft is insufficient, fans are installed to generate the required draft. In on-site incinerators

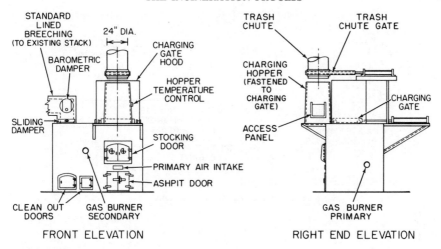

STANDARD LINED BREECHING (TO EXISTING STACK) · 24" DIA. · BAROMETRIC DAMPER · SLIDING DAMPER · CLEAN OUT DOORS · GAS BURNER SECONDARY · CHARGING GATE HOOD · HOPPER TEMPERATURE CONTROL · STOCKING DOOR · PRIMARY AIR INTAKE · ASHPIT DOOR

FRONT ELEVATION

TRASH CHUTE · TRASH CHUTE GATE · CHARGING HOPPER (FASTENED TO CHARGING GATE) · CHARGING GATE · ACCESS PANEL · GAS BURNER PRIMARY

RIGHT END ELEVATION

Fig. 5.58 Incinerator with automatic charging system (14).

the forced-draft air is usually introduced "underfire." The introduction of overfire air to improve combustion efficiency is not widely used in on-site units.

When the waste is wet or its heating value is low, auxiliary fuels are needed to support combustion. In continuously charged incinerators, the primary burner is sized for 1,500 BTU per pound of Type 3 waste or for 3,000 BTU per pound of Type 4 waste (see Table 1.16). The heat capacity of the secondary burner is also 3,000 BTU per pound of waste. When the incinerator is fully loaded, the secondary burner runs only for short periods at a time.

Incinerator Controls

On-site incinerators are frequently operated automatically from ignition to burn-down (Fig. 5.59). The cycle is started by the microswitch on the charging door,

Table 5.18
CHIMNEY SELECTION AND SPECIFICATION
CHART (14)

Incinerator Capacity All Types Waste, lb/hr	Chimney Size		Lining Thickness	Steel Casing Thickness
	Inside Diameter	Height above Grate		
100–150	12"	26'	2"	10 ga.
175–250	15"	26'	2"	10 ga.
275–350	18"	32'	2"	10 ga.
400–550	21"	37'	2½"	10 ga.
600–750	24"	39'	3"	¼"
800–1400	30"	44'	3"	¼"
1500–2000	36"	49'	3½"	¼"

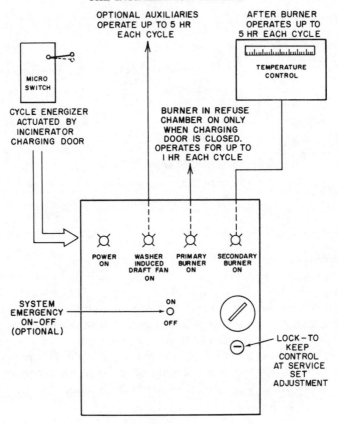

Fig. 5.59 Incinerator control system (14).

which automatically starts the secondary burner, the water flow to the scrubber, and the induced-draft fan. When the door is closed, the primary burner is started and stays on for an adjustable time of up to an hour, or until the door is reopened. Unless interrupted by a high-temperature switch, the secondary burner stays on for up to five hours. In order to provide each charge with the same preset burn-down protection, the secondary burner timer is reset every time the charging door opens. Both burners are provided with overtemperature and flame-failure safety controls (23). The induced-draft fan and scrubber water flow are also controlled by a separate cycle timer to guarantee airflow and scrubbing action during burning. In some installations the charging sequence is also automated.

Domestic and Multiple-Dwelling Incinerators

Domestic incinerators are sized to handle a few pounds of solid wastes per person per day. In single dwellings a typical incinerator would have about 40,000

BTU/hr of auxiliary heat capacity. Because domestic incinerators are much less efficient than their municipal counterparts, the amount of auxiliary fuel used is high. The domestic incinerator in Figure 5.38 has two combustion chambers. The main purpose of the secondary chamber is to eliminate smoke and odor. As a result, the pollutant emissions from domestic incinerators are not excessive (Table 5.19).

In multiple dwellings the main purpose of incineration is to reduce the volume of the MSW prior to disposal. The refuse from a dwelling of 500 residents producing 2,000 lb/day of MSW at a density of 4 lb/cu.ft. will fill 100 trash cans. If incinerated on site, the residue will fit into 10 trash cans. Incinerators in multiple dwellings can either be chute-fed or flue-fed. In the chute-fed design, the wastes are discharged into the chute and then into the incinerator feed hopper in the basement (see Fig. 5.58). In the flue-fed design (Fig. 5.60), the chimney also serves as the charging chute for the waste, which falls onto grates above an ash pit inside a boxlike furnace. The main purpose of the charging door is to ignite the waste, while the purpose of the underfire and overfire air-ports is to manually set the airflows for smokeless burning. The walls of the incinerator consist of two brick layers with an air space in between. The inner layer is made of 4.5 inches of firebrick and the 9-inch outer layer is made of regular brick.

Flue-fed apartment house incinerators have a draft-control damper in the stack, right above the furnace. This damper is pivoted and counterweighted to close when a chute door opens to charge refuse into the furnace. As a result, draft at the furnace remains relatively constant. In order to withstand flame impingement, the draft-control damper should be made of 20-gauge 302 stainless steel.

Miscellaneous On-Site Incinerators

Some incinerator designs have been specially developed for on-site industrial applications. The outstanding design feature of the retort incinerator (Fig. 5.61) is the multiple chambers connected by lateral and vertical breechings; the combustion gases must pass through several U-turns for maximum mixing. The in-line design (Fig. 5.62) also emphasizes good flue-gas mixing. Here the combustion gases are mixed by passing through 90° turns in the vertical plane only. Both designs are available in mobile styles for use in such temporary applications as land clearance

Table 5.19
INCINERATOR EMISSIONS—TYPICAL VALUES [20]

Pollutant	New Domestic Wastes Incinerator	Municipal Incinerator with Scrubber
Particulates, grain/SCF	0.01–0.20	0.03–0.40
Carbon monoxide, ppm	200–1000	<1000
Ammonia, ppm	<5	—
Nitrogen oxides, ppm	2–5	24–58
Aldehydes, ppm	25–40	1–9

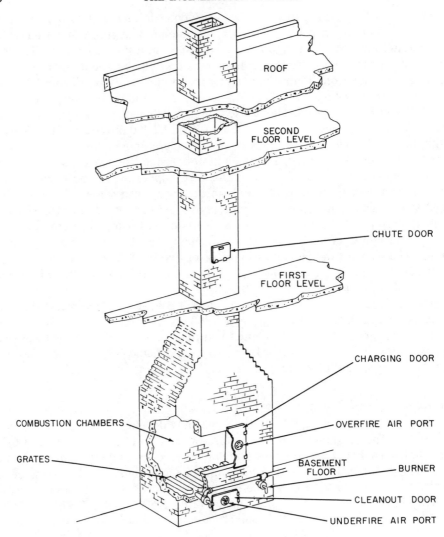

ROOF

SECOND
FLOOR LEVEL

CHUTE DOOR

FIRST
FLOOR LEVEL

CHARGING DOOR

COMBUSTION CHAMBERS

OVERFIRE AIR PORT

GRATES

BASEMENT
FLOOR

BURNER

CLEANOUT DOOR

UNDERFIRE AIR PORT

Fig. 5.60 Flue-fed incinerator (20).

or housing construction. The retort design is for smaller waste-burning capacities (under 800 lb/hr), while the in-line design is for higher burning rates.

Rotary incinerators for burning solid or liquid wastes can be continuous or batch and can be charged manually or by automatic rams. Their capacities range from 100 to 4,000 lb/hr. For burning wastes that contain chlorinated organics, the incinerator chamber must be lined with acid-resistant brick and the combustion gases must be sent through absorption towers to remove the acidic gases from the flue gas.

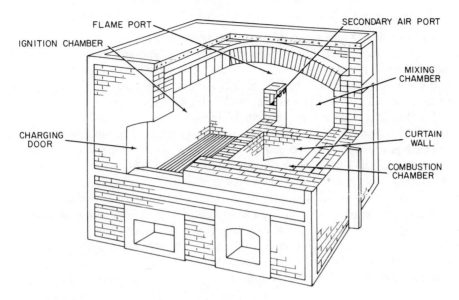

Fig. 5.61 Retort type multiple-chamber incinerator (20).

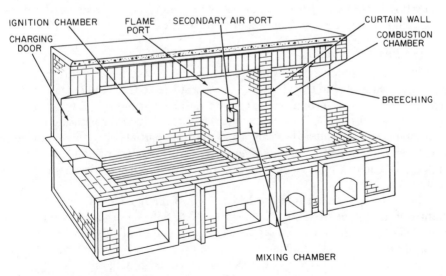

Fig. 5.62 In-line multiple-chamber incinerator (20).

171

Multiple-Hearth Incinerators

Multiple-hearth incinerators were discussed in Chapter 4 in connection with sewage-sludge incineration. In addition to that application, they can be used to recalcinate lime sludge (to burn $CaCo_3$ into CaO and CO_2), to regenerate activated carbon or diatomaceous earth, and to make charcoal briquettes from pits, shells, and lumber waste, as well as in other reclaiming and roasting operations.

The furnace of a multiple-hearth incinerator is a refractory-lined steel shell divided by horizontal brick arches into several hearths or compartments (see Fig. 4.6). The liquid or solid waste can be charged by screw or belt conveyors to the top, where it falls by gravity from one hearth to the next. The waste is moved inward, then outward, on alternate hearths by the teeth on the rotating rabble arms. The hollow shaft and the arms are cooled by recycled air, which rises as the waste moves downward. The residue is discharged at the bottom. As a wet waste stream travels from the top of the furnace to the bottom, it passes through zones of drying, combustion and deodorization, and cooling. If the wet waste contains 75% moisture, the hourly incinerator capacity can be calculated as 7 to 12 pounds of waste per square foot of furnace area. The incineration of most sludge materials requires auxiliary fuels.

Because the outlet-gas temperatures from multiple-hearth incinerators tend to be low, odor can be a problem. This can be alleviated somewhat by not evaporating the odor-producing substances before the waste enters the combustion zone (26). Another factor that can help alleviate odor is the "thermal jump" phenomena, the sudden temperature rise that occurs in the drying zone when the moisture content of the sludge drops to 48%.

The amount of ash produced depends on the percentage of inerts in the waste. Residue that is sterile and free of putrescible or odorous materials can be disposed of in landfills or recycled as filler material for brick, concrete, or roadfills.

Fluidized-Bed Incinerators

Fluidized-bed incinerators are used to burn pulp and paper industry wastes, to reduce the volume of industrial sludges, or to oxidize sodium and calcium salts. The furnace consists of a vertical cylinder with an air-distribution plate at the bottom (Fig. 4.11); the furnace is filled with sand. As air is forced through the openings of the distribution plate, the bed expands and the sand particles become fluidized (they are no longer in contact). Fluidized beds obey Archimedes' principle of buoyancy: In a fluidized bed of sand, which behaves similarly to a viscous fluid, wood will float and stones will sink.

The waste is injected into the heated bed of fluidized sand, operating at temperatures in the range of 1300° to 1500° F (740° to 850° C). The temperature cannot exceed 2000° F (1128° C) because sand starts to soften at this temperature. Good control of air velocity is important because a drop in velocity can cause the bed to collapse, while a rise in velocity can result in sand entrainment of the flue-gas stream. The incinerator operates with about 25% excess air (28).

Fluidized-bed incinerators are expensive to operate because of the high power requirements of fluidization and the specialized collection devices required to remove the entrained particles from the flue gas. Some advantages of this method are that the heated sand particles store large quantities of readily available heat and the movement of these particles prevents the formation of hot spots or stratification. The turbulence in the fluidized bed also guarantees high heat-transfer rates, further contributing to rapid combustion and uniform bed temperatures. The temperature variations throughout the bed tend to be less than 10° F (6° C). Another advantage is that this incinerator produces complete combustion, which eliminates odor problems and reduces the need for pollution-control equipment and equipment maintenance.

Chapter 6
Air Pollution
and the Health Effects
of Incineration

□
■ ■

THIS CHAPTER discusses the pollution and health effects of MSW incineration, providing in the first part of the chapter some general data on the types and quantities of incinerator emissions and on the acceptable limits set by various regulating agencies. The role of the EPA and of the state and local agencies will also be discussed, together with an example of how the permitting process works. In the second part of the chapter, the release of toxic and cancer-causing substances will be discussed, with particular emphasis on the emission of heavy metals and dioxin. The discussion will include the various approaches to determining "how much is safe" and the different cancer risk assessments.

The Effects of Air Pollution

The Clean Air Act of 1970 set maximum emission limits for lead, sulfur dioxide, nitrogen oxides, carbon monoxide, ozone, and particulate matter. Except for ozone and particulates, which were reported to have increased in the late 1980s, the concentration of most of these pollutants has abated in the last decade. Figure 6.1 gives the quantities of air pollutants released in each state in 1988. Tables 6.1 and 6.2 give yearly quantities of air pollutants released; Table 6.3 summarizes air-quality standards as amended by Public Law 91-604 of the Clean Air Act.

The American Conference of Governmental Industrial Hygienists has established threshold limit values (TLV) for the allowable atmospheric concentrations of many airborne substances. The TLV values represent those airborne concentrations that are believed to be safe even if workers are exposed to them repeatedly day after

174

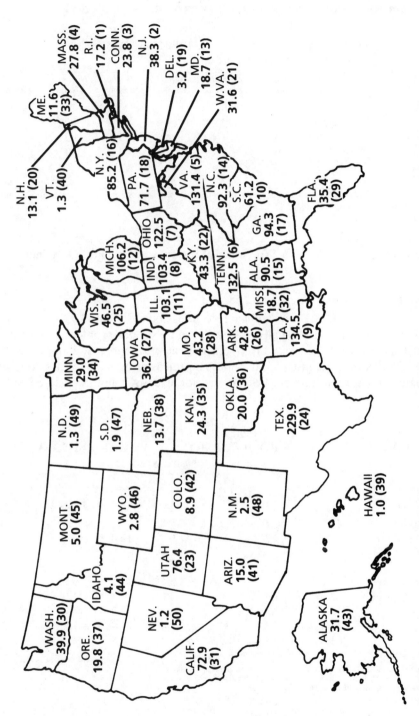

Fig. 6.1 Poisoning the air, state by state. Figures are millions of pounds of toxic industrial pollutants released into the air in each state. Number in parentheses is state's ranking in terms of pounds per square mile. (U.S. Environmental Protection Agency; The *New York Times*, March 23, 1989.)

175

Table 6.1

NATIONAL AIR POLLUTANT EMISSIONS, 1965 (60)

(millions of tons annually)

	Totals	% of Totals	Carbon Monoxide	Sulfur Oxides	Hydro-carbons	Nitrogen Oxides	Particles
Automobiles	86	60	66	1	12	6	1
Industry	23	17	2	9	4	2	6
Electric power plants	20	14	1	12	1	3	3
Space heating	8	6	2	3	1	1	1
Refuse disposal	5	3	1	1	1	1	1
Totals in millions	142		72	26	19	13	12

Source: "The Sources of Air Pollution and Their Control," Public Health Service Publication No. 1548, Government Printing Office, Washington, D.C., 1966.

day. Table 6.4 lists the TLV values for some of the substances emitted by incinerators (49).

Incinerator Emissions and Their Limits

The concentration of odorous and gaseous emissions (SO_x and NO_x) generated by continuous incinerators are usually low and pose no serious problems. The emissions of HCl and HF gases generated from burning plastics represent a more serious hazard (66). The largest quantity of air pollutants released by incinerators consists of particulates and carbon dioxide. Carbon dioxide usually is not considered to be

Table 6.2

ESTIMATED EMISSIONS OF AIR POLLUTANTS BY WEIGHT, NATIONWIDE, 1969 (61) (in millions of tons per year)

	Pollutant						
Pollution Source	CO	Partic-ulates	SO_2	HC	NO_2	Total	Change, % 1968–1969[a]
Transportation	111.5	0.8	1.1	19.8	11.2	144.4	−1.0
Fuel combustion in stationary sources	1.8	7.2	24.4	0.9	10.0	44.3	+2.5
Industrial processes	12.0	14.4	7.5	5.5	0.2	39.6	+7.3
Solid waste disposal	7.9	1.4	0.2	2.0	0.4	11.9	−1.0
Miscellaneous	18.2	11.4	0.2	9.2	2.0	41.0	+18.5
Total	151.4	35.2	33.4	37.4	23.8	281.2	+3.2
Change, % (1968–1969)	+1.3	+10.7	+5.7	+1.1	+4.8		

Source: The Mitre Corp. MTR-6013. Based on Environmental Protection Agency data.
[a] Computed by the 1969 method from the difference between 1969 estimates and 1968 estimates. The new method results in higher values for 1968 than those computed by EPA for 1968.

Table 6.3
AMBIENT AIR-QUALITY STANDARDS (66)

Pollutant	Primary	Secondary, if not same as primary
Particulates, $\mu g/m^{3b}$		
annual geometric mean	50	
max. 24-h conc.[a]	150	
Sulfur dioxide, $\mu g/m^3$		
annual arith. aver.	80(0.03 ppm)	
max. 24-h conc.[a]	365(0.14 ppm)	
max. 3-h conc.[a]		1,300(0.5 ppm)
Carbon monoxide, mg/m^3		
max. 8-h conc.[a]	10(9 ppm)	
max. 1-h conc.[a]	40(35 ppm)	
Ozone, $\mu g/m^3$		
1-h max.[a]	1.5(0.002 ppm)	
Lead particulates, $\mu g/m^{3b}$		
Calendar qtr, av.		
Nitrogen dioxide, $\mu g/m^3$		
annual arith. aver.	100(0.05 ppm)	

Note: All values subject to periodic revision.
[a] Not to be exceeded more than once a year.
[b] Clean Air Act as amended under Public Law 91-604.

an air pollutant, although it is the main cause of the general warming of the earth—the greenhouse effect. The range of typical municipal incinerator emissions is given in Table 5.18, and the composition of unabated MSW-based flue gas is given in Table 6.5.

The limits and standards set for incinerator emissions vary around the world. Table 6.6 gives the standards of the Canadian Ministry of Environment. In 1964 the West German Ministry of the Interior issued technical guidelines, referred to as TAL limits, for the protection of the air quality (Table 6.7). West Germany tightened the TAL limits for refuse incinerators in 1974 and again in 1984.

The early limits established by the Clean Air Act of 1970 were applied only to large steam generators, but in 1987 they were extended to smaller steam generators as well, including HRIs with steam-generating capacities of 100,000 lb/hr or more. In 1988 legislation (H.R. 4902) was introduced to set national standards to regulate incinerator emissions. Pollution-control regulations vary. Some are clear and establish performance standards through specific design requirements, while others leave many issues to personal judgment. The EPA standards of performance for new stationary sources also apply to incinerators.

The new source emission standard for municipal incinerators with capacities exceeding 45 TPD is 0.08 grains of particulates per standard cubic foot (0.18 gram

Table 6.4
THRESHOLD LIMIT VALUES (49)

Substance	ppm	mg/M^3
Ammonia	50	35
Antimony & compounds (as Sb)	—	0.5
ANTU (alpha naphthyl thiourea)	—	0.3
Arsenic & Compounds (as As)	—	0.5
Cadmium oxide fume (as Cd)	—	0.1
Calcium oxide	—	5
Carbon dioxide	5,000	9,000
Carbon monoxide	50	55
Chlorinated diphenyl oxide	—	0.5
Chlorine	1	3
Chromium, sol. chromic, chromous salts as Cr	—	0.5
Fluoride (as F)	—	2.5
Hydrogen chloride	5	7
Iron oxide fume	—	10
Lead	—	0.2
Mercury-Skin	—	0.1
Mercury (organic compounds)-Skin	—	0.01
Nickel, metal and soluble cmpds, as Ni	—	1
Nitric acid	2	5
Nitric oxide	25	30
Nitrogen dioxide	5	9
Phosphorus (yellow)	—	0.1
Sulfur dioxide	5	13
Sulfur hexafluoride	1,000	6,000
Sulfuric acid	—	1
Zinc oxide fume	—	5

Table 6.5
UNABATED FLUE-GAS
COMPOSITION (53)

CO_2	9.2% dry volume
O_2	10.2% dry volume
N_2	80.6% dry volume
H_2O	16.0%
SO_2	150 ppm, wet volume
HCl	700 ppm, wet volume
NO_x	250 ppm, wet volume
CO	100 ppm, wet volume
Particulate	1.5 (0.16) gr/SDCF (g/Nm³)

Table 6.6
INCINERATOR AIR EMISSIONS USING ELECTROSTATIC
PRECIPITATOR (PRESENT), BAGHOUSE FILTER (NEW), AND
COMPARING RESULTS WITH CANADIAN MINISTRY OF
ENVIRONMENT STANDARD (56)

| | Present | New Emissions Estimate | | |
	Emission Rate from Stack (10^3 $\mu g/m^3$)	Emission Rate from Stack (10^3 $\mu g/m^3$)	Half hour Average Maximum Ground Level Concentration ($\mu g/m^3$)	Ministry of Environment Standard (10^3 $\mu g/m^3$)
Particulate	475	14.9	2.65	100
Calcium	7.2	0.21	0.04	20
Iron	5.4	0.19	0.03	10
Lead	3.6	0.10	0.02	10
Mercury	0.2	0.004	0.001	5
Zinc	5.4	0.17	0.03	100
Hydrogen Chloride	138	138	24.8	100
Sulfur Dioxide	270	270	48.5	830
Nitrogen Oxides (as NO₂)	48	48	8.7	500
	(ppm)	(ppm)		(ppm)
Unburned Hydrocarbons (as methane)	100	50		100
Carbon Monoxide	1000	800		2000

Table 6.7
COMPARISON OF EMISSIONS LIMITS FOR
DIFFERENT VERSIONS OF TA LUFT (53)

Pollutant(s)	1964 mg/m^{3a}	1974 mg/m^{3b}	1984 mg/m^{3b}
Solid Phase			
Total particulate matter	150	100	50.
Class I particulates (Cd, Hg, Tl)	NA	20	0.2
Class II particulates (As, Cr, Co, Ni, Se, Te)	NA	50	1.0
Class III particulates (Sb, Pb, F, Cu, Mn, V)	NA	75	5.0
Gaseous Phase			
Carbon monoxide (CO)	NA	1,000	100.
Fluorides (Fl⁻)	NA	5	5.
Hydrogen chloride (Cl⁻)	NA	100	50.
Sulfur dioxide (SO₂)	NA	NA	200.

Note: 800°C or 1,472°F and 0.3 sec. min. furnace conditions. NA denotes "not applicable" or "not available."
ᵃ Wet, corrected to 7% vol. CO_2, 1,013 mb and 32°F or 0°C.
ᵇ Wet, corrected to 11% vol. O_2, 1,013 mb and 32°F or 0°C.

per standard cubic meter) corrected to 12% of CO_2. The 12% CO_2 reference corresponds to about 50% excess air (18) and protects against cheating by using air dilution to reduce particulate loading. Some local regions place much stricter limits on particulate emissions; in the Los Angeles area, for example, the limit is 0.01 grains per standard cubic foot (57). Gas and vapor pollutants are controlled according to the capabilities of the "best available technology" (BAT). BAT is around 80% removal for NO_x, 90% to 95% for SO_2 and 95% or more for HCl (53); and 97% to 99% of lead and mercury (202).

Actual Incinerator Performance

The quantity and quality of emissions depend on the nature of the refuse. If the inerts are removed from the MSW prior to burning (NRT in Table 6.8), pollution emissions are reduced. Refuse-derived fuels (RFD) in general tend to be less polluting because the quality of combustion improves as inerts are removed and particle sizes are made smaller and more uniform.

The actual performance of HRIs has been evaluated by the U.S. Army (Table 6.9). The equipment of some thirty manufacturers was grouped into four categories according to their operating principles: modular starved-air units, rotary kilns, excess-air grate designs, and fluid-bed combustion systems. Other studies have evaluated the effects of different types of pollution-control devices (discussed in Chapter 7) on RFD-burning incinerators. Table 6.6 shows the estimated effect of replacing electrostatic precipitators ("Present") with a "New" system consisting of a spray

Table 6.8

REDUCTION IN POLLUTANT EMISSIONS BY BURNING
INERT FREE (NRT) FUEL INSTEAD OF INTERMIXED
MSW (54)

Regulated Pollutants	NRT Fuel (lb/ton)	MSW Fuel (lb/ton)	Ratio: NRT/MSW
Sulfur dioxide (SO_2)	2.77	2.80	0.99
Particulates (Part)	29.4	42.5	0.69
Nitrogen oxides (NO_x)	1.28	2.20	0.58
Lead (Pb)	0.13	0.27	0.48
Carbon monoxide (CO)	1.66	4.50	0.37
Complex hydrocarbons (NMHC)	0.06	0.23	0.25
Non-Regulated Pollutants			
Hydrogen fluoride (HF)	0.008	0.031	0.25
Sulfates (SO_4)	0.27	0.97	0.28
Ammonia (NH_3)	0.033	0.121	0.27
Cadmium (Cd)	0.0064	0.0236	0.27
Chromium (Cr)	0.0029	0.0079	0.36
Arsenic (As)	0.0021	0.0032	0.66
Hydrogen chloride (HCl)	6.2	5.3	1.17

Note: Data in pounds of pollutant per ton of MSW processed. Measurements are prior to emissions control equipment and are not direct air releases.

Table 6.9

RELATIVE PERFORMANCE OF FOUR DIFFERENT HEAT RECOVERY INCINERATOR (HRI) DESIGNS

Unit Types	Starved	Rotary	Grate	FBC
Number of units sold each year	5–15	3–10	8	3
Percent industrial versus municipal	94	100	85	97
Company contracts to operate units?	Yes	Yes	Yes	Yes
Number of units currently in service	1–2000	1–20	18–1455	34
Expected life of unit (years)	10–15	10–30	20–40	20
Average equipment availability (%)	90	88	90–95	90
Size range of units (TPD)	2–100	2–320	1–1250	10–400
Steam generation range (lb/hr)	1K–50K	720–72K	6K–250K	3K–250K
Expected thermal efficiency (%)	40–70	50–75	30–70	60–85
Fans routinely provided?	Yes	Yes	Yes	Yes
Pre-heating combustion air?	Avail.	Yes	Yes	Avail.
Type of waste fuel (MSW, Indust.).	MSW & 1	MSW & 1	MSW & 1	1 & RDF
Pollution control devices supplied	None	Baghouse ESP & scrubber	Baghouse ESP & scrubber	Baghouse multiclone
Expected uncontrolled emissions				
Particulate (gr/dscf)	0.13–0.08	0.5–0.03	1.1–0.41	1–3
Nitrogen oxides (ppm)	—	—	<35	100–130
Other measured pollutants (Cl)	—	—	Varies	—
Opacity (%)	—	—	20	10–20
Expected controlled emissions				
Particulate (gr/dscf)	0.13–0.08	.03–.005	0.05–0.01	0.03
Nitrogen oxides (ppm)	—	—	<35	100–130
Other measured pollutants (Cl)	—	—	Varies	—
Opacity (%)	—	—	0–3	10
Ash water solids content			<50%	
Other pollutants in the ash water	0	—	—	0
Add'l pollution control devices	Scrubber	Varies	Scrubber	None

FBC—Fluid Bed Combustion; ESP—Electrostatic Precipitator

Table 6.10

KIEL APC SYSTEMS PERFORMANCE, CONTROL OF GASEOUS POLLUTANTS (53)

Gaseous Pollutant		Unit of Measurement	Processing Line #2[a] 5-11/12-76	Processing Line #3[b] With Choke 6-22-81	Without Choke 6-23-81
HCl (measured as Cl⁻)					
Scrubber Inlet	Max.	mg/Nm³ wet @ 11% O₂	N.A.	1,220	1,520
	Avg.		1,170	1,150	1,280
	Min.		N.A.	1,065	1,140
Scrubber Outlet	Max.	mg/Nm³ wet @ 11% O₂	44	57	176
	Avg.		24[c]	47	138
	Min.		N.A.	38	109
Removal Efficiency	Avg.	%	98.0 +1.2/−1.6	95.9 +1.0/−1.3	89.2 +3.6/−4.6
Sample Size	n	#	24	6	6
HF (measured as F⁻)					
Scrubber Inlet	Max.	mg/Nm³ wet @ 11% O₂	N.A.	12.2	14.3
	Avg.		9.3	10.6	11.7
	Min.		N.A.	9.1	10.8

182

Scrubber Outlet	Max.		N.A.	2.0	3.5
	Avg.	mg/Nm³ wet @ 11% O₂	0.4[c]	1.6	1.9
	Min.		N.A.	1.1	1.1
Removal Efficiency	Avg.	%	96.0 +1.7/−2.3	84.9 +6.0/−6.9	83.8 +8.5/−16.2
Sample Size	n	#	24	6	6
SO₂					
Scrubber Inlet	Max.		N.A.	170	190
	Avg.	mg/Nm³ wet @ 11% O₂	550	150	157
	Min.		N.A.	120	130
Scrubber Outlet	Max.		350	31	38
	Avg.	mg/Nm³ wet @ 11% O₂	250	27	33
	Min.		N.A.	24.5	28
Removal Efficiency	Avg.	%	55.0 +3.5/−9.1	82.0 +3.6/−7.8	79.0 +6.3/−8.2
Sample Size	n	#	24	6	6

[a]Source: Personal communication with plant operator, 8-5-76
Specific Lime Rate: 4.71 kg Lime/MT Refuse
Specific Water Rate: 1.32 MT Water/MT Refuse
Specific Steaming Rate: 2.10 MT Steam/MT Refuse
Specific Steam Consumption: 0.21 MT Steam/MT Refuse

[b]Source: Acceptance Test Report #123 UM 00310 TuV Norddeutschland 8-19-81
Specific Lime Rate: 3.70 kg Lime/MT Refuse
Specific Water Rate: 0.63 MT Water/MT Refuse
Specific Steaming Rate: 2.27 MT Steam/MT Refuse
Specific Steam Consumption: 0

[c]Dilution with hot air bleed does not affect these results because of correction to 11% O₂.
N.A. = Not Available

cooler followed by a baghouse filter. The RFD fuel is shredded and ferrous metals are removed prior to burning. In the past, electrostatic precipitators and wet scrubbers were used to meet EPA emission-control requirements for particulates. Precipitators have been found to be more reliable and predictable, while wet scrubbers had problems with corrosion and with the formation of "white plumes" (water vapor). Because of a growing concern about the discharges of acidic gases, future incinerator designs probably will be provided with dry scrubbers using lime powder and baghouse filters.

Table 6.10 provides data on the effectiveness of acid-gas scrubbing systems at the Kiel station in West Germany, which met TAL 74 emission requirements (see Table 6.7). In this system, electrostatic precipitation is followed by wet scrubbing. The chimney draft is increased (to minimize plume formation) by reheating the scrubber outlet gases by the gases entering the scrubber. To protect against corrosion, the heat-recovery exchanger in this installation was made of glass. By the mid-1980s, twenty-seven of West Germany's forty-six municipal incinerators were equipped with acid-gas scrubbers and were believed to operate within the TAL 74 guidelines.

In addition to the type of fuel and pollution-control system used, the combustion temperature also has an effect on the nature and quantity of emissions. The emissions from the stack in Figure 6.2 are given in Table 6.11 at four different operating temperatures in the secondary combustion chamber. The primary combustion temperature is approximately the same (676° to 703° C) for all four tests, but the temperature in the secondary combustion chamber varies as follows: "Low" = 780° C, "Long" = 890° C, "Normal" = 910° C, "High" = 1080° C.

In the Würzburg plant in West Germany, one of the most modern installations in Europe, a dry scrubber is followed by baghouse filters. The air-emission testing at this plant (44) showed that combustion was almost complete (35 ppm of CO at 7% O_2, less than 1 ppm of hydrocarbons and no vinyl chloride in the flue gases), and the particulate loading was one-twentieth of the EPA limit (0.004 grains per dry standard cubic foot at 12% CO_2). The actual test results for particulates, heavy metals, and vinyl chloride are listed in Tables 6.12, 6.13, and 6.14.

Air Pollutants Emitted by Incinerators

The most common air pollutants emitted by incinerators are particulates, acidic emissions (sulfur dioxide and hydrogen chloride), and nitrogen oxides. The characteristics, emission limits, and preferred emission-control methods will be outlined for each. The more toxic emissions of heavy metals and dioxins will be discussed later.

Particulates

The federal source-emission standard limits the concentration of particulates in incinerator emissions to 0.08 grains per standard cubic foot, corrected to 12% CO_2. In some localities this limit is as low as 0.01 grains per standard cubic foot (57). (The particulate loading is tied to a given concentration of CO_2 to prevent cheating by dilution with air or steam.) In metric units, the federal standard corre-

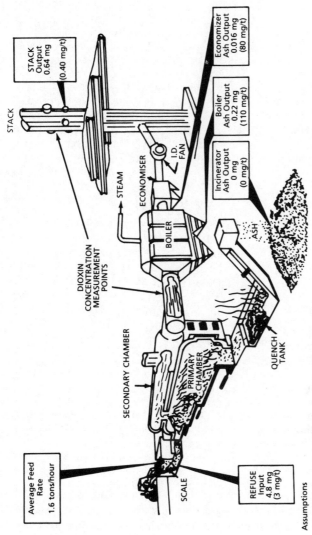

STACK

STACK
Output
0.64 mg
(0.40 mg/t)

STEAM

ECONOMISER

I.D.
FAN

Economizer
Ash Output
0.016 mg
(80 mg/t)

Boiler
Ash Output
0.22 mg
(110 mg/t)

Incinerator
Ash Output
0 mg
(0 mg/t)

BOILER

DIOXIN
CONCENTRATION
MEASUREMENT
POINTS

ASH

SECONDARY CHAMBER

PRIMARY
CHAMBER

QUENCH
TANK

Average Feed
Rate
1.6 tons/hour

SCALE

REFUSE
Input
4.8 mg
(3 mg/t)

Assumptions

1. 90% Reduction of Refuse on a weight basis for Total Ash.
2. Incinerator Bottom Ash is approx. 98.5% of Total Ash.
3. Boiler Ash is approx. 1.25% of Total Ash.
4. Economizer Ash is approx. 0.13% of Total Ash.

Refuse Input and stack outputs are based on actual
measurements.

Fig. 6.2 The equipment used in the test reported in Table 6.11. Dioxin mass input/output is based on 12 test-run average (48).

185

Table 6.11

EMISSIONS MEASURED AS COMBUSTION TEMPERATURES
CHANGED (For Incinerator Shown in Figure 6.2) (48)

Contaminant	Stack Concentration (at 12% CO_2) Dry			
	Normal	Long	High	Low
TSP (mg/Nm³)	208	230	247	167
HCl (mg/Nm³)	1,085[a]	1,070	1,165	950
PCDD (ng/Nm³)	107	107	62	123
PCDF (ng/Nm³)	143	156	95	98
PCB (ng/Nm³)	801	58	0	126[a]
PAH (ng/Nm³)	7,005	8,010	6,653	12,490[a]
Chlorophenol (ng/Nm³)	4,346	3,773	2,706	6,591[a]
Chlorobenzene (ng/Nm³)	4,321	3,161	3,968	4,884[a]
Cadmium (mg/Nm³)	0.9	0.8	0.8	0.6
Lead (mg/Nm³)	13.5	15.2	15.2	8.4
Chromium (mg/Nm³)	0.04	0.03	0.1	0.03
Nickel (mg/Nm³)	0.2	0.3	0.5	0.5
Mercury (mg/Nm³)	0.7	0.5	0.9	0.5
Antimony (mg/Nm³)	0.6	2.6	0.5	0.5
	Stack Emissions per ton of Feed (as fired)			
TSP (g/ton)	843	874	977	682
HCl (g/ton)	4,400	4,130	4,480	3,930
PCDD (mg/ton)	428	400	228	516
PCDF (mg/ton)	570	574	340	411
PCB (mg/ton)	3,413	245	0	574[a]
PAH (mg/ton)	29,305	33,201	26,956	54,514[a]
Chlorophenol (mg/ton)	18,403	15,042	10,814	28,973[a]
Chlorobenzene (mg/ton)	18,014	12,807	16,061	22,045[a]
Cadmium (g/ton)	3.8	3.0	3.2	2.6
Lead (g/ton)	54.8	57.8	60.0	34.2
Chromium (g/ton)	0.2	0.1	0.4	0.1
Nickel (g/ton)	1.0	1.0	2.2	1.9
Mercury (g/ton)	2.8	2.0	3.6	2.2
Antimony (g/ton)	2.3	9.6	2.1	1.9

[a]Based on average of two tests only.

sponds to 180 milligrams per cubic meter. Canada's standard is 100 mg/m³ (see Table 6.6), West Germany's is 50 mg/m³, and the standards in the Los Angeles area correspond to a limit of 23 mg/m₃ (57). Actual emission concentrations reported from West Germany (44) were lower than any of these limits (9 mg/m³).

Emissions of particulates include fly ash and unburnt fuel; smoke can be either liquid or solid matter in the exhaust gases, which hinders visibility. Smoke is distinguished from a steam plume by its persistence. Smoke is eliminated by providing sufficient air, turbulence, and residence time to complete the combustion process

Table 6.12

PARTICULATE EMISSION DATA FROM WÜRZBURG INCINERATOR (44)

Date	Grains/DSCF	Grains/DSCF (12% CO_2, dry)	Total Condensible Particulate (mg)	Sulfate ($H_2SO_4 \cdot 2H_2O$, mg)	Chloride (HCl, mg)	Other by Difference (mg)
12/03/85	0.0049	0.0067	12.7	12.8	1.0	—
12/04/85	0.0040	0.0071	26.3	18.2	0.0	8.1
12/05/85	0.0022	0.0035	23.0	15.0	1.0	7.0
Average	0.004	0.006				

Table 6.13
HEAVY METAL EMISSIONS IN WÜRZBURG FLUE GAS
(44)

Compound	Grains/DSCF at 68°F	μg/Nm³ at 0°C, dry	Weight (% of total particulate)
As	$<1.9 \times 10^{-9}$	$<4.7 \times 10^{-3}$	<0.00025
Cd	1.9×10^{-6}	4.7	0.25
Cr	1.7×10^{-7}	0.42	0.023
Ni	7.7×10^{-8}	0.19	0.01
Mn	5.0×10^{-7}	1.2	0.066
Pb	3.8×10^{-6}	9.3	0.51
Sb	$<1.9 \times 10^{-7}$	<0.47	<0.025

Table 6.14
VINYL CHLORIDE EMISSIONS AT WÜRZBURG (44)

Method	Date	Time	Concentration ppb	μg/Nm³
U.S. EPA Method 18	12/04/85	1035	<20	
	12/05/85	0945	<20	
U.S. EPA VOST	12/03/85	1352–1630	<0.12	
	12/04/85	1008–1028	<0.26	
	12/04/85	1056–1205	<0.15	
	12/05/85	0917–0937	<0.43	
	12/05/85	1005–1025	<0.24	
	12/05/85	1045–1105	<0.49	
VOST Average			<0.31	<0.86

Note: Sorbent tubes on 12/03/85 were combined for initial GC/MS analysis to achieve greater accuracy (six tubes total). As the sequential analysis continued, individual sets of tubes were independently analyzed (two tubes per set).

Table 6.15
PARTICULATE EMISSIONS AND PARTICLE-SIZE DISTRIBUTION FROM REFUSE INCINERATORS WITHOUT POLLUTION CONTROLS (20)

Type of Incinerator Unit	Particulate (lb/ton of refuse)	Particle Size Distribution, %				
		>44μ	20–44 μ	10–20 μ	5–10 μ	<5 μ
Municipal (multiple chamber)	17	40	20	15	10	15
Commercial (multiple chamber)	3	40	20	15	10	15
Commercial (single chamber)	10	40	20	15	10	15
Flue-fed	28	40	20	15	10	15
Domestic (gas-fired)	15	40	20	15	10	15

either in the main furnace or in the secondary combustion chamber. Table 6.15 provides data on particulate sizes and emissions from incinerators without pollution-control devices.

Federal standards for particulate emission limits can be met by using wet scrubbers or electrostatic precipitators, but only if their efficiencies are 99% or more. The most common pollution-control devices are electrostatic precipitators, which are provided in combination with mechanical collectors. The use of wet scrubbers, which have a long history in the steel and paper industries, is relatively new in municipal incinerator applications. The best results have been reported by using dry scrubbers followed by baghouse filters (44).

Acidic Emissions

The main concerns with acidic emissions are corrosion and acid rain. If maintained for five hours, an SO_2 concentration of only 0.3 ppm will damage the leaves of sensitive vegetation. The odor threshold for sulfur oxides is also fairly low, around 0.2 ppm. In contrast, the sulfur dioxide concentration in the Antarctic air is around 1 ppb; the EPA standards for ambient air allow an average of 0.03 ppm (80 micrograms per cubic meter) and a maximum of 0.14 ppm (365 micrograms per cubic meter) (59). The actual sulfur dioxide concentration varies from city to city (Fig. 6.3) and depends on the sulfur content of the fuels used (Table 6.16). The sulfur oxide content of incinerator emissions is usually estimated by assuming that all the sulfur in the MSW will be converted to gaseous SO_2.

The limits of acid-gas emissions from incinerators are based on BAT. Without pollution controls, the unabated flue-gas composition from municipal incinerators is around 150 ppm for SO_2 and 700 ppm for HCl (see Table 6.5). In comparison, the emission limits in Germany are 200 mg/m³ (75 ppm) for SO_2 and 50 mg/m³ (33 ppm) for hydrogen chloride (see Table 6.7). In the Los Angeles area the limit for SO_2 concentration in incinerator flue gases is 550 ppm at 3% oxygen on a dry basis (57). Therefore SO_2 is usually not a problem in meeting the emission standards for municipal incinerators. On the other hand, HCl and HF are becoming more of a problem as the plastic content of the MSW increases. (As will be discussed later in the chapter, HCl also plays a role in the formation of dioxins.)

Nitrogen Oxides

All combustion processes produce nitric oxide (NO), a general anesthetic that is slowly converted to nitrogen dioxide in the atmosphere. The incineration of each ton of MSW produces about 2 pounds of nitrogen dioxide. The term NO_x refers to the combined total of all forms of nitrogen oxides. The NO_x concentration of unabated incinerator flue gases is around 250 ppm (53), and it is even higher from the combustion of other fuels (Table 6.17). Table 6.18 gives the estimated annual global emissions of the various nitrogen compounds.

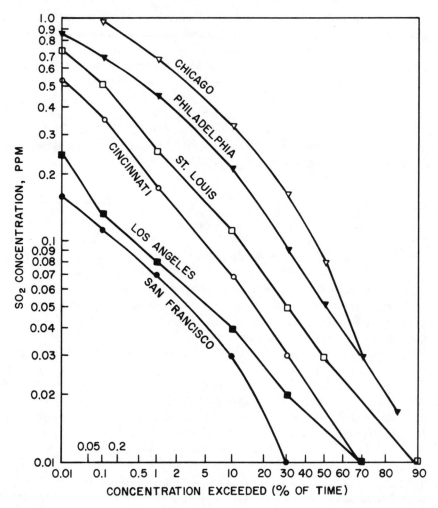

Fig. 6.3 Frequency distribution of sulfur dioxide. Levels in selected American cities; 1-hour averaging time (59).

Table 6.16
ESTIMATED SULFUR DIOXIDE EMISSIONS FROM
STEAM GENERATORS (52)

Type of Fuel	Emissions (lb/MBTU)	Sulfur Dioxide Content in Exhaust Flue Gas (ppm)	Annual Uncontrolled Emissions, 10^6 tons/year
Coal	5.0	2600	17.9
Fuel Oil	3.0	1500	1.8
Natural Gas	Nil	Nil	Nil

190

Table 6.17

ESTIMATED NITROGEN OXIDES EMISSIONS FROM
STEAM GENERATORS (52)

Type of Fuel	Emissions (lb/MBTU)	NO_x Content in Exhaust Flue Gas ppm	Annual Uncontrolled Emissions, 10^6 tons/year
Coal	1.4	1000	4.4
Fuel oil	0.7	530	0.7
Natural Gas	0.4	330	0.8

The NO_x concentration in the unpolluted atmosphere is between 2 and 8 ppb, while in urban air it can reach 1 ppm. NO_x contributes to the chemical destruction of the ozone layer in the stratosphere, which protects the earth from harmful ultraviolet radiation. Nitrogen dioxide is also responsible for triggering a series of chemical reactions that produce photochemical smog. Photochemical smog results from the action of sunlight with oxides of nitrogen and vaporized organics, which produces ozone and a variety of noxious products. NO_2 has an orange-brown color and is lethal to most animal species at concentrations exceeding 100 ppm.

National air-quality standards require that the annual arithmetic mean concentration of NO_2 in the atmosphere not exceed 0.05 ppm (100 micrograms per cubic meter). Normal, uncontrolled NO_x concentration in flue-gas emissions of existing MSW incinerators range from 150 to 300 ppm (dry corrected to 3% O_2); in Los Angeles emission regulations call for 225 ppm (57). NO_x-reduction techniques (discussed in Chapter 7) include combustion modification, catalytic decomposition, adsorption, and ammonia injection.

Table 6.18

ESTIMATED ANNUAL GLOBAL EMISSIONS OF NITROGEN COMPOUNDS
(60)

Compound	Source	Source Magnitude (tons/yr)	Estimated Emissions (millions of tons annually)	Emissions as Nitrogen (millions of tons annually)
NO_2	Coal combustion	$3,074 \times 10^6$	26.9	8.2
	Petroleum refining	$11,317 \times 10^6$ (bbl)	0.7	0.2
	Gasoline combustion	379×10^6	7.5	2.3
	Other oil combustion	894×10^6	14.1	4.3
	Natural gas combustion	20.56×10^{12} (ft^3)	2.1	0.6
	Other combustion	1290×10^6	1.6	0.5
	Total NO_2		52.9	16.1
NH_3	Combustion		4.2	3.5
NO_2	Biological action		500	150
NH_3	Biological action		5900	4900
N_2O	Biological action		650	410

The Permitting Process: An Example

To obtain a construction permit and later an operating permit for a new incinerator, the owner must satisfy the requirements of various federal, state, and local regulatory agencies. The regulatory structure, emission standards, and permitting process vary from region to region. Los Angeles will be used as an example to illustrate some of the steps involved in obtaining a permit (57).

In the mid-1980s, Los Angeles County produced 40,000 tons of MSW per day, of which 98% was landfilled. By the year 2000, the county hopes to recycle over 10% and incinerate 60%. This would more than double the lifespan of its landfills, because only 29% noncombustible and 15% ash of the MSW (a total of 44%) would require landfilling. It was toward this goal that the city of Commerce commissioned a $50 million, 300-tons-per-day HRI in 1983, with a proposed start-up date of 1987 (Fig. 6.4). The unit was designed to produce 115,000 lb/hr (52,000 kg/hr) of steam, burning MSW with a heating value of 6,350 BTU/lb (14,540 kJ/kg). The superheated steam 750° F/650 PSIG (400° C/45 atm.) would generate an estimated net electrical power output of 10,050 kW.

The air-pollution-control equipment at the plant would consist of an ammonia injection system for NO_x control and a spray dryer followed by a fabric filter for acid-gas and particulate removal (Fig. 6.5). Three governmental entities were involved in the permitting process: the EPA, the South Coast Air Quality Management District (SCAQMD), and the California Air Resources Board (CARB).

The Role of the EPA

In 1987 the EPA issued emission requirements for all new, modified, or reconstructed steam-generating units with a heat-input capacity greater than 29 mega-

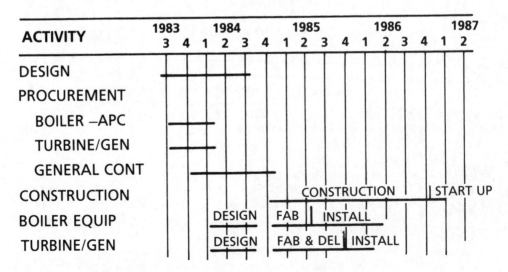

Fig. 6.4 Design and construction schedule (57).

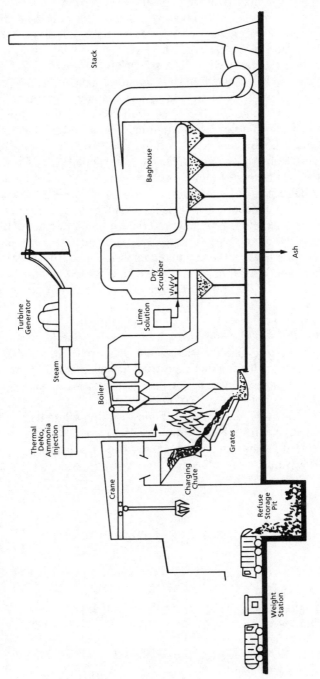

Fig. 6.5 Commerce refuse-to-energy project schematic (57).

193

watts (100,000 lb/hr of steam) for which construction started after June 1986. (Since the construction at Commerce started before this date, this regulation did not apply.) Before a new incinerator is allowed to operate, the EPA must issue a Prevention of Significant Deterioration (PSD) permit. The PSD rule is designed to protect the air from degradation in those areas where it is still relatively clean. In official language, the term "clean" is replaced by the term "nonattainment," meaning that the particular pollutant concentration in that area has not yet "attained the maximum allowable limit;" in other words, it has not yet exceeded the national air-quality standard. By the same terminology, an "attainment area" refers to a polluted region. Therefore, because Los Angeles is an "attainment" (or unclassifiable area under section 107 of the Clean Air Act) for CO, NO_x, particulates, and ozone, meaning that the concentrations of these pollutants have already exceeded the limits set (see Table 6.3), the PSD permit *did not* apply.

The lead and SO_2 concentration in the Los Angeles area does not yet exceed the limits set by the ambient air standard (they are in "attainment"), so for these pollutants the EPA *would* perform a PSD review, *if the emission rates are significant*. For the Commerce incinerator, the estimated emission quantities were calculated to be less than significant; therefore the EPA exempted the plant from PSD review. Consequently, it seems that no pollutant emissions from the Commerce incinerator were evaluated by the EPA (57).

The Role of State and Local Agencies

The South Coast Air Quality Management District (SCAQMD) is the local regulatory agency responsible for issuing construction and operating permits. SCAQMD regulations consist of two separate parts: Part 1 describes the New Source Review (NSR) Rules and Part 2 outlines the Prohibitory Rules.

The NSR requires a preconstruction review for all pollutants that are in "nonattainment" (their concentration is below ambient air-quality-standard limits) and that would be emitted in daily quantities greater than the maximum limits listed in Table 6.19. If these values are exceeded, the facility is supposed to use the "best

Table 6.19
POLLUTANT EMISSION THRESHOLDS (57)

Pollutant	Net Cumulative Emission	
	(kg/d)	(lb/day)
Carbon monoxide	249	550
Sulfur dioxide	68	150
Nitrogen oxides	45	100
Particulate matter	63	150
Reactive organic gases (hydrocarbons)	34	75
Lead compounds	1.4	3

Table 6.20
SOUTH COAST AIR-QUALITY MANAGEMENT DISTRICT PROHIBITORY RULES (57)

Pollutant	Threshold
Total suspended particulate	0.01 gr/SCF[a,b] or 11 lb/hr
Nitrogen oxides	225 ppm[b]
Sulfur dioxide	550 ppm[b]
Carbon monoxide	2000 ppm[b]

[a]Includes condensible fraction.
[b]Calculated at 3% oxygen on a dry basis averaged over a minimum of 15 min.

attainable control technology'' (BACT). The BACT requirement was waived for the Commerce incinerator because the plant produces less than 50 MW of electricity; therefore state law expects only a "good-faith effort.''

The Prohibitory Rules of SCAQMD set limits on the allowable emission concentrations from incinerators (Table 6.20). These limits are greater than the expected uncontrolled concentrations of CO and SO_2, but they are somewhat below the expected uncontrolled concentration of NO_x and are much lower than the expected uncontrolled concentration of particulates. The limit for total suspended particulates is 0.01 grain/scf (0.023 gram/scm), which in the Commerce plant corresponds to an emission rate of 11 lb/hr (5 kg/hr). Based on the allowable emission limits, the pollution-control equipment was selected and the anticipated removal efficiencies (Table 6.21) were applied to the uncontrolled emissions. The resulting emission rates in pounds-per-hour units were incorporated into the SCAQMD construction permit. Once testing proves that the plant is capable of complying with these emission rates, an incinerator will receive a permit to operate. The third governmental entity, the California Air Resources Board (CARB), ensures that local districts adhere to the national regulations. As such, it presented no additional requirements for the Commerce plant.

In summary, the total permitting process in connection with the Commerce incinerator resulted in the active involvement of only one agency (SCAQMD), which issued both the construction and operating permits based on the requirements set in the Prohibitory Rules.

Table 6.21
ESTIMATED REMOVAL EFFICIENCIES
(57)

Pollution Control Equipment	Pollutant	Removal Efficiency (Percent)
Fabric filter	Particulates	99.7
Spray dryer	Sulfur dioxide	80
	Hydrogen chloride	90
Thermal DeNO$_x$	Oxides of nitrogen	20–50

Emissions of Toxic and Cancer-Causing Substances

The most toxic substances released by incinerators include heavy metals such as mercury, which is mostly released into the atmosphere, lead and cadmium, which leave with the bottom ash, and chemicals called furans and dioxins, which can enter both the atmosphere and the collected fly ash and have been implicated in birth defects and in causing cancer (179). The health concerns associated with incinerator emissions increase as the weight percentage of plastics rises in the MSW (see Table 1.6) and as the practice of high-temperature "mass burning" spreads. In 1990 the average weight percentage of plastics in the MSW of the United States was about 8% (199). The burning of plastics can contribute to the formation of dioxins, while mass-burning can increase the emission of toxic heavy metals.

In mass-burning plants, unsorted MSW is burned at temperatures around 2500° F (1370° C) for periods of up to 45 minutes. In such "slagging" incinerators, car batteries that contain lead and cadmium, dry-cell batteries containing zinc and magnesium, button-cell batteries and consumer goods that contain mercury and other heavy metals will all melt or vaporize. As a consequence, they either become a potential source of air pollution, or, if removed from the flue gases, will contaminate the collected fly ash and scrubbing fluids. The increase in furnace temperatures tends to reduce the dioxin and tends to increase the metal concentration in the stack gases (48).

The Dioxin Controversy

"Dioxin" is a common term for a large number of chlorinated chemicals, including some 200 polychlorinated dibenzo-p-dioxins (PCDDs) and polychlorinated dibenzo-furans (PCDFs). The EPA classifies dioxins as *possible* human carcinogens. It is known that dioxins cause cancer in laboratory animals, but their effects on humans are still being debated. The EPA relies on chemical industry data in gauging the effect of dioxin on human health. BASF and Monsanto have published studies on the consequences of exposure to 2,3,7,8,-TCDD by their workers. A copy of one of these studies is available from BASF (196). Greenpeace, an environmentalist group, objects to this practice and has petitioned the EPA to revoke all dioxin guidelines that are based on industry studies. They also have petitioned that EPA identify all dioxin sources (paper bleaching, incineration) and eliminate all of them by the year 2000.

The danger caused by man-made carcinogens in general and the effects of dioxin in particular are both controversial issues. Bruce Ames, director of the Environmental Health Sciences Center at Berkeley, in the January 1991 issue of *Journal* argues that, with the exception of lung cancer, the mortality rates of other forms of cancer have been falling in the United States since 1950. He also argues that animal tests at near-toxic doses cannot predict the cancer risk in humans and that 99.9% of carcinogens in the environment and in our food supply are not synthetic in origin, but natural, usually produced by plants as natural pesticides to defend themselves from fungi and insects. He even goes so far as to claim that cabbage and broccoli contain a chemical that behaves just like dioxin. While Ames is not alone in denying that modern technology harms public health, he is in the minority, and this minority will not be able to reverse the trend toward reduced emissions of synthetic toxins and carcinogens.

Dioxins are by-products of pesticide and wood-preservative manufacturing, vehicular sources (leaded gasoline contains ethylene dichloride), and of such combustion processes as incineration.

Dioxin is present in the total environment, including the food and water supplies, the soil, and the atmosphere. In North America the total dioxin intake through ingestion and inhalation is about 54 pg/day (38). How much of this intake is through inhalation has not been clearly established, but Massachusetts has established a guideline of 1.1 pg/day of PCDD intake, which corresponds to inhaling 20 m³ of air per day. Dioxins are concentrated through the food chain and are deposited in human fat tissues. In some cases a total of 1.0 ppb of dioxins has been found in mother's milk (1).

A key ingredient in the formation of dioxins is chlorine. Figure 6.6 shows the

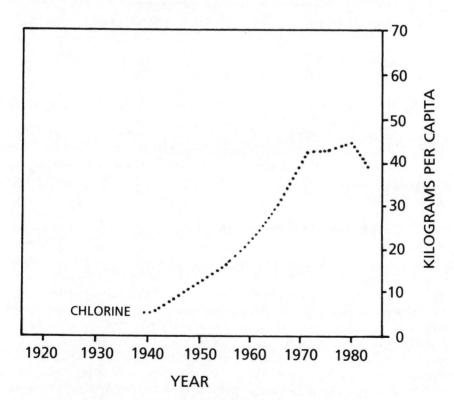

Fig. 6.6 Chlorine production in the United States (39).

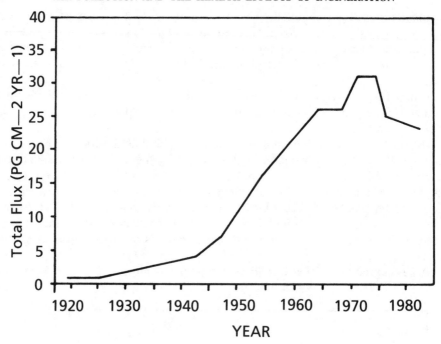

Fig. 6.7 Flux of PCDD and PCDF to Siskiwit Lake (40).

trends in chlorine production in the United States (39); Figure 6.7 shows the dioxin concentration in the sediments of Lake Siskiwit. The correlation between the two curves seems to indicate that an increase in the use of chlorine results in an increase in the dioxin concentration in the environment, even in areas where the only dioxin source is the atmosphere.

Dioxin Emissions from Incineration

Supporters of incineration argue that dioxin emissions can be controlled by maintaining proper furnace conditions. Modern plants have increased afterburner temperatures from 1000°–1500° F required by the 1973 EPA standard (AP–40) to 1600°–1800° F or more and thereby dioxin emissions have been lowered. They also argue that improved combustion controls and better monitoring can further reduce these emissions. They point out that modern HRIs generate less dioxin emission than the older ones, which provided no energy recovery. (Unconfirmed reports from Wheelabrator Technologies indicated that a 2,250 TPD HRI in Bridgeport, Connecticut, operated at dioxin emission levels of 0.01 to 0.07 nanograms per cubic meter.) Supporters of incineration also point to a World Health Organization report, which says that "well operated . . . incinerators contribute only a small fraction of the apparent overall daily intake of PCDD and PCDF even for people living in areas where the emission levels are the highest" (41).

Opponents of incineration argue that the level of dioxins in the environment is already too high, that these toxic chemicals bioaccumulate in the human body, that most dioxins in the atmosphere are received from incinerators, that furnace temperature and combustion efficiency have little effect on dioxin emissions, that the only acceptable level of dioxin emission is zero discharge, and that incinerators never will be capable of providing that.

The reasons for such diagonally opposite views can be explained partly by commercial interests and partly by limited factual, reliable data. The review of the available data (Figure 6.8) shows that the emissions of dioxins from incinerators range from as low as 0.01 ng/m^3 to as high as 10,000 ng/m^3 (38). (A nanogram is 10^{-9} gram.) This extremely high ratio between emission concentrations can be explained by the differences in the chlorine content of the wastes burned and by the differences in the design of the incinerator used, but there is little question that *all incinerators* emit dioxins (Tables 6.22, 6.23, and 6.24).

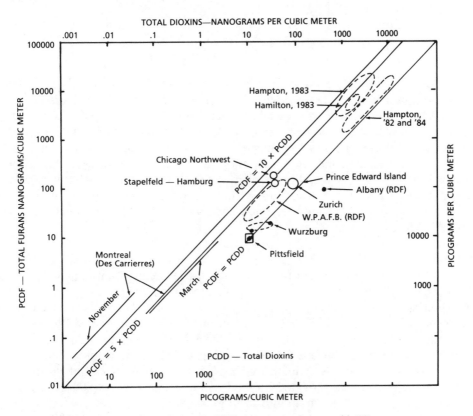

Fig. 6.8 Dioxin versus furan emissions in HRI plants around the world (38).

Table 6.22
CONCENTRATIONS OF FURANS AND
DIOXINS IN SWEDISH INCINERATORS (in
ng/nm³ at 10% CO_2) (87)

Plant Fuel	Malmö RDF	Linköping RDF
CO content, mg/nm³ dg 10% CO_2	184	71
Rec. furnace temperature, °C	805	901
FE, %	99,5	99,94
Dioxins		
Total TCDD	0,3	0,6
of which 2,3,7,8-TCDD	0,02	0,03
Total PnCDD	0,1	0,1
of which 1,2,3,7,8-PnCDD	<0,03	<0,04
Furans		
Total TCDF	4	4,5
of which 2,3,7,8-TCDF	0,1	0,2
Total PnCDF	6	4
1,2,3,7,8/1,2,2,4,8-PnCDF	0,3	0,3
2,3,4,7,8-PnCDF	0,9	0,3

The lowest dioxin-emission reports to date (0.01 to 0.07 ng/m³) come from Wheelabrator Technologies and the Bridgeport incinerator. This incinerator is not typical, however, because it uses fabric filters to collect the fly ash. Fabric filters are used in only about 2% of the operating incinerators in the country. Other examples of low levels of dioxin emissions include a new mass-burning incinerator in Zurich, Switzerland, where the emission was reported to be 50 ng/m³ in 1981 (36). Similar results were achieved in Würzburg, Germany, in 1985, as detailed in Table 6.25 (44). The average dioxin concentration of HRIs is reported to be between 400 and 500 ng/m³ (31), although some industry representatives feel this can be reduced to about 150 ng/m³ or even lower (43).

How Much Dioxin Is Safe?

In the early 1980s, when the United States incinerated only about 5% of its MSW, the total dioxin emission from incinerators (24 kg/yr) already exceeded the total emission from all the automobiles in the country (45). At a dioxin concentration in the incinerator flue gas of 250 ng/m³, each ton of incinerated MSW adds one milligram of dioxin to the atmosphere. If the United States incinerated a similar portion of its MSW as Europe (30–50%) or Japan (70%), this would amount to the yearly burning of about 80 million tons of MSW. Even if burned at the reduced emission concentrations provided by modern plants, this would still substantially increase the emissions of dioxin into the atmosphere.

Table 6.23

DIOXIN AND FURAN EMISSIONS FROM THE STACKS OF SOLID-WASTE-FIRED PLANTS (43)

(Data are nanograms per standard cubic meter of flue gas)

Facility	Dioxin Emissions[b]	Dioxin and Furan Emissions[a]			Heat Recovery
		Total	Toxic Equivalents		
			Swiss	EPA	
Westchester County, New York	7	30			Yes
Würzburg, Germany	12–42	22–100			Yes
Stapelfeld, Germany	31				Yes
Chicago, (N.W.) Illinois	42	192	17.3	1.03	Yes
Eskjo, Sweden	73	541	45.3	2.42	Yes
Hamburg (S.M.), Germany	101				Yes
P.E.I., Canada	107	222	12.1	1.00	Yes
Zurich, Switzerland	113	202	11.0	1.07	Yes
Hamburg (B.S.), Germany	128				Yes
Como, Italy	280	683–2,382	39.2	4.28 (19.8)	Yes
(RDF plant), Denmark	316				Yes
Albany, New York	316	196–458	15.2 (33.9)	2.37 (4.8)	Yes
Plant 1, Italy	475	535	24.8	4.03	No
Plant 6, Italy	566	639	52.9	7.21	No
Plant 5, Italy	1,020	1,503	93.9	8.74	No
Plant 2, Belgium		1,756	75.7	6.40	?
Florence, Italy		1,830			No
Plant 1, Belgium		2,193	129.0	12.90	?
Zaanstadt, Holland	1,294	2,455	150.0	16.1	No
Valmedrera, Italy	1,568				No
Tsushima, Japan		4,524	400.0	69.0	No
Hamilton, Canada	3,680	12,500	778.0	91.9	Yes
Hampton, Virginia	4,250	10,230	860.0	82.2	Yes
Plant 4, Italy	4,339	7,487	330.0	28.1	No
Toronto, Canada	5,086	1,900			No
Plant 3, Italy	7,491	8,952	704.0	83.7	No
Plant 2, Italy	48,808	56,460	3,934.0	490.0	No

[a]Compiled by Barry Commoner (Risk Assessment of the Health Effects of PCDD and PCDF Emissions from the Proposed Hennepin County Trash-Burning Incinerator), Centre for the Biology of Natural Systems, Queen's College, Flushing N.Y., January 17, 1986.
[b]Compiled by Kemp and Kay Jones (Roy F. Weston).

The amount of dioxin generated partly depends on the chlorine content of the incinerated waste. Chlorine has been found to be higher in urban waste (0.89 wt%) than in rural waste (0.45 wt%) (46). Also, urban MSW contains more plastics than does rural MSW. Although bleached paper and similar products also contribute to the total chlorine content in the incinerator fuel, there is a direct relationship between the amount of plastics in the waste and the probable dioxin emissions.

Table 6.24
MUNICIPAL WASTE COMBUSTORS: DIOXIN AND FURAN EMISSIONS SUMMARY (42)

Plant Name	Tons/Day	Emission Factor (2378 TCDD EQ mg/ton)	Stack Emission Conc. (2378 TCDD EQ ng/m³)	Total Measured (CDD/CDF mg/nm³)	Reference
Type 1—Mass Burn/Water Wall					
Chicago	1452	0.060	10.717	189	(Haile et al., 1983)
Zurich/Josephstrasse	343	0.072	11.867	145	(Swiss EPA, 1982)
Hampton[a]	227	4.624	738.037	6828	(Scott Env. Tech., 1985)
Hamburg/Stapelfeld	415	0.034	5.673	115	(Nottrodt et al., 1984)
MVA-I Borsigstrasse	480	0.111	18.621	243	(Nottrodt et al., 1984)
MVA-II Stellinger Moor	1173	0.105	15.796	304	(Nottrodt et al., 1984)
Westchester[b]	2250	0.010	1.437	68	(NYS-DEC, 1986a)
Umer[b]	200	0.020	4.755	396	(Marklund et al., 1985)
Iserlohn[a]	768	0.000	1.784	37	(Broker and Gliwa, 1985)
Würzburg[b]	592	0.002	0.449	39	(Ogden Martin, 1986)
Malmo	410	0.013	0.683	7	(Lindfors, 1985)
Avg-Borsigstrasse[a]	1180	0.129	21.520	322	(Nottrodt et al., 1984)
Quebec[a]	931	0.000	20.548	175	(Environment Can., 1981)
Montreal[a]	1200	0.000	0.025	1	(Environment Can., 1984)
Hogorlen[b]	397	0.0053	0.879	53	(Bergstrom, 1986)
Tulsa	680	N/A	0.701	34	(Ogden Projects, 1986)
Linkoping	120	0.0033	0.500	9	(Lindfors, 1985)
Average =		0.039	6.006	133	
Type 2—RDF/Water Wall					
Albany	555	0.027	3.955	53	(NYS-DEC, 1985)
Hamilton[a]	480	11.278	1637.549	8538	(Envirocon, 1984: Ozvacic)
Akron	545	0.194	57.305	632	(Howes et al., 1985)
Dayton	64	0.256	21.402	342	(Higgins, 1982)
Occidental[b]	2179	0.151	19.630	21	(NYS-DEC, 1986b)
Average =		0.1571	25.573	262	

Little Rock	91	0.020	1.695	21	(Higgins, 1982)
Dyersburg	100	0.074	8.253	84	(Howes et al., 1985)
Prince Edwards Island	136	0.055	14.262	175	(Environment Can., 1985)
Lake Cowichan	24	0.289	45.494	601	(Environment Can., 1981)
Cattaraugus County[b]	108	0.033	8.040	—	(NYS-DEC, 1986)
Average =		0.094	15.549	220	

TYPE 4—MASS BURN/REFRACTORY/WASTE HEAT BOILER

Mayport	44	0.034	2.723	25	(Higgins, 1982)
Pittsfield	200	0.086	16.779	176	(MRI, 1986)
Average =		0.060	9.751	101	

TYPE 5—MASS BURN/REFRACTORY

Zaanstrad	206	1.424	177.944	1840	(Olie et al., 1982)
Philadelphia[b]	681	0.481	61.587	1927	(MRI, 1985)
Toronto	576	1.036	130.316	1350	(Clement et al., 1985)
Tsushima[a]	300	3.153	679.040	2264	(Cooper Eng., 1984)
Como Italy	96	2.525	299.951	2006	(Gizzi et al., 1982)
Beveren	94	0.410	231.887	1787	(Janssen et al., 1982)
Milan I	600	0.005	0.355	101	(Cavallaro et al., 1982)
Milan-II[a]	600	0.000	30.680	306	(Cavallaro et al., 1982)
Brasschaart	50	1.940	63.986	1170	(DeFre, 1985)
Harelbeke	109	0.511	242.480	1787	(DeFre, 1985)
Average =		1.041	151.063	1496	

TYPE 6—RDF/FLUID BED

Eskjo	187	NA	0.572	136	(Rappe et al., 1983)

[a]Data not included in calculating the average for technology type.
[b]TCDD equivalents were calculated from isomer specific data
[c]Refers to facilities that have corrected stack emissions in ng/m3 at 12% CO2 and 60 degrees F.
Emission factors & concentrations based on measured flue-gas flow rate.
Capacity (tonnes/day) is the reported plant design capacity.

Table 6.25

FLUE-GAS CONCENTRATIONS OF PCDD AND PCDF AND THE 2,3,7,8-TCDD TOXIC EQUIVALENTS USING EADON EQUIVALENCE FACTORS (44)

	Flue Gas Concentration ng/Nm³ at 12% CO_2				Multiplying Factor for Eadon Toxic Equivalents	Würzburg Average Toxic Equivalents	Suggested Swedish Environmental Protection Board Guidelines for 2,3,7,8-TCDD Toxic Equivalents, 1985[a]
	12/3/85	12/4/85	12/5/85	Average			
Dioxins							
2,3,7,8-TCDD	0.03	0.012	0.012	0.018	1.0	0.018	<0.06
Total TCDD	3.05	1.55	1.13	1.91	—	0.000	
1,2,3,7,8-PeCDD	0.31	0.17	0.12	0.20	1.0	0.200	
Total PeCDD	3.93	2.03	1.67	2.54	—	0.000	
1,2,3,4,7,8-HxCDD	0.12	0.06	0.06	0.08	0.03	0.002	
1,2,3,6,7,8-HxCDD	0.25	0.17	0.15	0.19	0.03	0.006	
1,2,3,7,8,9-HxCDD	0.18	0.09	0.09	0.12	0.03	0.004	
Total HxCDD	4.96	2.46	2.12	3.18	—	0.000	
1,2,3,4,6,7,8-HpCDD	4.00	1.59	1.02	2.20	—	0.000	
Total HpCDD	7.53	2.84	2.54	4.30	—	0.000	
Total OCDD	22.98	3.24	4.28	10.17	—	0.000	
Total PCDD (C_4–C_8)	42.45	12.11	11.72	22.10	—	0.230	

Furans

2,3,7,8-TCDF	0.53	0.14	0.09	0.25	0.33	0.082
Total TCDF	23.25	3.50	2.04	9.60	—	0.000
1,2,3,7,8- and 1,2,3,4,8-PeCDF	1.64	0.57	0.32	0.84	0.33	0.277[b]
2,3,4,7,8-PeCDF	1.23	0.47	0.17	0.62	0.33	0.204
Total PeCDF	17.47	6.53	3.78	9.26	—	0.000
1,2,3,4,7,8- and 1,2,3,4,7,9-HxCDF	0.70	0.24	0.33	0.42	0.01	0.004[b]
1,2,3,7,8,9-HxCDF	0.15	0.05	0.03	0.08	—	0.000
1,2,3,6,7,8-HxCDF	0.83	0.32	0.32	0.49	0.01	0.005
2,3,4,6,7,8-HxCDF	1.15	0.47	0.24	0.62	0.01	0.006
Total HxCDF	11.77	3.72	2.63	6.04	—	0.000
1,2,3,4,6,7,8-HpCDF	2.77	1.55	0.81	1.71	—	0.000
1,2,3,4,7,8,9-HpCDF	0.11	0.05	0.03	0.06	—	0.000
Total HpCDF	3.33	1.88	1.02	2.08	—	0.000
Total OCDF	1.75	0.42	0.48	0.88	—	0.000
Total PCDF (C$_4$–C$_8$)	57.56	16.04	9.95	27.85	—	0.578
Total PCDD and PCDF (C$_4$–C$_8$)	100.01	28.15	21.67	49.95	—	0.808	<1.2

[a]Adjusted from 0.05 ng/Nm^3 for 2,3,7,8-TCDD and 1.0 ng/Nm^3 for TCDD toxic equivalents at 10% CO$_2$.
[b]Worst case—assuming that all of the analysis can be attributed to the 2,3,7,8-isomers.

205

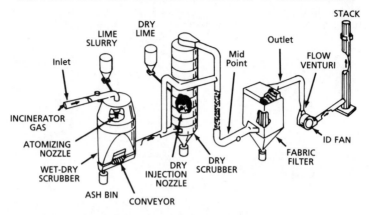

Fig. 6.9 The National Incinerator Testing and Evaluation Program (NI-TEP) air pollution control technology (47).

Another important factor that affects the amount of dioxin emissions is the type of pollution-control device used to collect the particulates. Since the chlorinated organics tend to collect on the surfaces of fly-ash particles, there is a direct correlation between the efficient removal of particulates and the amount of dioxins discharged into the air. (The dioxins removed with the fly ash can also enter the environment when the rainwater percolates through the fly ash in landfills.) In any case, the emission of dioxins into the air can be reduced by reducing the particulate emissions.

Some of the best results in lowering particulate emissions were reported in West Germany, where a combination of dry-lime scrubbing followed by fabric filters resulted in less than 0.007 gr/dscf (grain per dry standard cubic feet) of particulates in the stack gases. While fabric filters seem to give the best collection efficiency of fine particles, only two MSW incinerators in the United States (at Framingham,

Table 6.26
PCDD CONCENTRATION (ng/Nm³ @ 8% O₂) IN FLUE GAS AND
EFFICIENCY OF REMOVAL (47)

Operating Condition	Dry System				Wet-Dry Systems	
	110°C	125°C	140°C	>200°C	140°C	140°C +Recycle
Inlet (ng/Nm³)	580	1400	1300	1030	1100	1300
Mid Point (ng/Nm³)	310	570	540	1140	840	1270
Outlet (ng/Nm³)	0.2	ND	ND	6.1	ND	0.4
Efficiency (%)						
Inlet/Midpoint	47	60	57	(11)	24	2
Overall	>99.9	>99.9	>99.9	>99.4	>99.9	>99.9

( ) denotes negatives.

Massachussetts, and in Bridgeport) use bag filters (42). The performance of the Framingham unit was tested in 1981 and the outlet grain loading was reported to be over 0.03 gr/dscf (42).

Tests were also performed in Quebec City, Canada, to evaluate the dioxin-removal efficiency of dry and wet scrubbers and bag filters at various temperatures. Figure 6.9 illustrates the main components of the pilot plant; Table 6.29 shows the results of the test (47). At temperatures below 150° C, the wet scrubbers removed about 25% of the PCDDs, the dry scrubbers removed about 50%, and the fabric filters removed 99.9% (47). As the temperature was increased above 200° C, the dry scrubber, instead of removing PCDDs, became a *dioxin generator*, and the dioxin concentration increased as the flue gases passed through the dry scrubber.

CAN INCINERATORS SYNTHESIZE DIOXINS?

Tests conducted by the Canadian government at Prince Edward Island showed that the concentration of metals in stack gases tends to rise with furnace temperature, while dioxin concentration tends to be reduced as combustion temperatures rise (48). Figure 6.2 shows the equipment used at Prince Edward Island, the quantities of dioxin found in each of the streams entering and leaving the process, and the points where dioxin concentrations were measured in the flue-gas stream. Table 6.27 lists the concentrations of dioxins and other organic compounds at the boiler inlet and in the stack under three different sets of operating conditions. In all these cases, there are more dioxins present in the stack than at the boiler inlet!

The behavior of dioxins in the incinerator process is not fully understood. Conventional wisdom would suggest that all organics, including dioxins, should be destroyed by incineration, while the above test data contradict this suggestion. Tests in Germany also found that dioxins were formed in the incineration process as PCDD was synthesized on the surface of the fly-ash particles at low temperatures (2). One report attempted to explain the hypothesis that dioxins are synthesized when wood and plastics are burned together (3).

Table 6.27

CONCENTRATIONS OF ORGANIC COMPOUNDS AT BOILER INLET
IN COMPARISON TO STACK (48)

	Concentration (ng/Nm^3 @ 12% CO_2)					
	Long Cycle		High Temperature		Low Temperature	
Compound	Boiler Inlet	Stack	Boiler Inlet	Stack	Boiler Inlet	Stack
PCDD	0	107	1	62	14	123
PCDF	0	156	0	95	10	98
PCB	11[a]	58	8[a]	0	61	126[b]
PAH	10 475	8 010	12 226	6 653	29 481	12 490[b]
Chlorophenol	486	3 773	479	2 706	2 941	6 591[b]
Chlorobenzene	6 595	3 161	4 895	3 968	8 906	4 884[b]

[a]Only one non-zero value (two readings were below detection limit).
[b]Based on average of 2 tests only.

It has been suggested that the derivatives of lignin, a wood constituent, will combine with the chlorine released from such substances as PVC and other chlorinated plastics to form dioxin. It was also suggested, as shown in Fig. 6.10, that while high combustion temperatures do destroy dioxins, when the gas has cooled down to 200°–300° C, dioxins reform on the surfaces of the fly-ash particles. This theory could explain the test results shown in Fig. 6.11 and Table 6.27 (4).

The subject of dioxin formation is controversial. Incinerator manufacturers claim that dioxin emissions can be controlled by maintaining proper furnace conditions, while some tests seem to show that dioxins are formed downstream of the furnace, after the gases have cooled down. If this hypothesis is correct, then fly-ash collectors, which also trap the dioxins that form on the surface of the cooling fly ash, should be located at a point where the flue gases have already cooled below the synthesis temperature of 200°–300° C (see Table 6.27). Even if the particulate collectors were relocated, this would only protect against the formation of added dioxins; it would not destroy the dioxins that already exist at that point.

Cancer-Risk Assessment

As cancer research continues, more and more substances are found to be cancer-causing in animals or potentially cancer-causing in humans. "Cancer risk

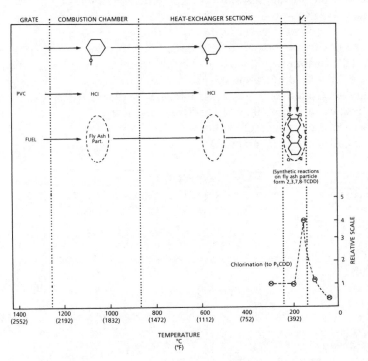

Fig. 6.10 The hypothesis that dioxins are synthesized in the MSW incineration process (3).

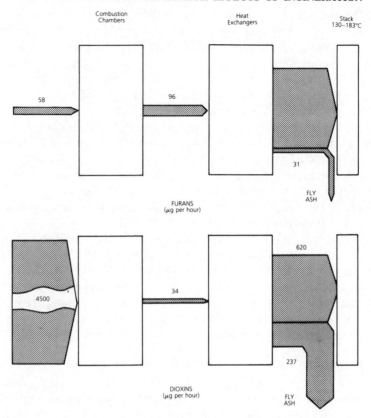

Fig. 6.11 The concentration of furans and dioxins in flue gas increases as the gas is cooled in the heat exchanger (4).

assessment numbers'' have been introduced to express quantitatively the risk associated with the exposure to a particular concentration of a substance for a particular period of time. These numbers reflect either the estimated additional yearly cancer cases in the United States (annual national cancer incidence = ANCI) or the estimated additional cancer cases that are likely to occur in a population of one million if that group is exposed to the particular carcinogen for seventy years (maximum lifetime risk = MLR).

While there is no objective basis for considering a particular cancer risk acceptable or unacceptable, it is likely that social concern will rise as the risk-assessment numbers increase. Section 112 of the Clean Air Act and Section 4f of the Toxic Substances Act assign responsibility to the EPA for initiating regulatory proceedings to control the emissions of airborne carcinogens. The EPA uses the MLR and ANCI numbers as indicators of cancer risk.

The EPA has initiated regulatory proceedings on some carcinogens, such as benzene, that have an ANCI number of 245 and an MLR number of 154 (3). The agency has also begun similar proceedings on other substances, such as methylene chloride, that have an ANCI number of only 2.4 and an MLR number of only 0.83 (3, 10). The ANCI number for the emissions of existing incinerators is between 2 and 40 (10) and the MLR number is between 100 and 1,000 (3, 10). It is further estimated that the average amount of dioxin that already exists in the population corresponds to an MLR number range of 330 to 1,400 (3). According to the Connecticut Department of Environmental Protection, exposure to an ambient dioxin concentration of 0.001 ng/m^3 will result in a cancer risk corresponding to an MLR number of 1.0. The lowest dioxin concentration in any of the incinerator emissions reported to date was in the range of 0.01 to 0.07 ng/m^3. According to a New York study, a person living downwind of an incinerator emitting dioxins at a concentration of 0.07 ng/m^3 will be exposed to a risk of MLR = 1.0 (154).

As of this writing, the EPA has only invoked Section 111 of the Clean Air Act in connection with incinerators. This Section calls only for the use of the "best available technology." Because of mounting public pressure (3) and because the EPA believes that the emissions from existing incinerators represent a significant hazard to public health (10), the agency may also invoke Section 112 of the Clean Air Act in connection with incinerators. This would mean establishing emission standards for incinerators and ambient air-quality standards for dioxins.

Heavy Metals

Mass-burning, high-temperature incinerators operate like metal smelters, evaporating some metals like mercury and concentrating other heavy metals onto the surfaces of the ash particles. In 1990, Detroit's incinerator had to be shut down for several weeks because mercury emissions exceeded the limits set by state law (195). The concentration of heavy metals tends to be lower in the "bottom ash" (the residue that remains on the grate) and higher in the fly ash, which is collected in the electrostatic precipitator or other pollution-control equipment. When scrubbers are used and lime is added, the alkalinity of the ash increases, which in turn can make the heavy metals more soluble in rainwater, which percolates through the ash in a landfill. (The addition of lime changes the metal compounds from the carbonate to the hydroxide form, which in some cases is more soluble in water.) Table 6.28 summarizes the health effects and concentration limits of some waterborne poisonous substances, including heavy metals.

The health effects of heavy metals are well established. Mercury can cause brain damage and can damage the central nervous system. Lead can cause brain damage in children and high blood pressure in adults. Cadmium is a carcinogen. These heavy metals and others, such as zinc, are found in the gas and/or ash discharges from incinerators. (Table 6.29 shows the concentrations of trace metals in the ambient air.) Since most heavy metals bind with the fly ash, emissions into the atmosphere can be substantially reduced by improving the efficiency of particulate

Table 6.28

SELECTED WATERBORNE POISONOUS AGENTS (Toxic Chemicals) AND THEIR PROPERTIES (8)

Name	Formula	Physiological Effect on Man	Maximum Allowable Concentration (mg/l)
Arsenic	As	100 mg causes severe poisoning	0.1
Barium	Ba	A muscle stimulant that can affect heart	1.0
Cadmium	Cd	Cumulative amounts may lead to heart disease	0.01
Chromium	Cr^{6+}	Carcinogenic when inhaled	0.05
Cyanide	CN	Rapid fatal poison; limit has safety factor about 100	0.2
Lead	Pb	Serious, cumulative and acute body poison	0.05
Mercury	Hg	Damages brain and central nervous system	0.005
Nitrate	NO_3^{1-}	Can cause methemoglobinemia in infants	45
Selenium	Se	Can cause loss of hair and dermal changes	0.01
Silver	Ag	Large amounts cause irreversible skin grayness	0.05

collection prior to the discharge of flue gases. Thus, replacing low-efficiency particulate collectors, such as electrostatic precipitators, with high-efficiency ones, such as baghouse filters (see Table 6.6), can substantially lower the heavy metal emissions into the atmosphere.

Table 6.29

SELECTED TRACE METALS FOUND IN AIR (9)

Metal	Approximate Concentration Range, U.S. Cities ($\mu g/m^3$)
Arsenic	0–0.75 (av. 0.02)
Beryllium	0–0.008 (av. < 0.0005)
Cadmium	0–0.3 (av. 0.002)
Chromium	0.0–0.35 (av. 0.015)
Lead	0.1–5 (higher in traffic)
Manganese	0.01–10.0 (av. 0.10)
Nickel	0.01–0.20 (av. 0.032)
Vanadium	0.0–1.4

Chapter 7
Pollution-Control Equipment

□
■ ■

POLLUTION-CONTROL EQUIPMENT for incinerators includes both wet and dry devices. The wet designs include spray towers or chambers and a large variety of scrubbers: centrifugal, cyclonic, packed, floating bed, impingement and orifice types, venturi, and mechanical. Among dry collectors are cyclones and other centrifugal separators, baghouse and other filters, electrostatic precipitators, dry scrubbers, and other dry absorption processes. In addition, this chapter will discuss the various NO_x-control techniques, including furnace modification, fluid-gas recirculation, ammonia injection, and catalytic techniques.

The Capabilities of Pollution Control Devices

Pollution-control equipment can be grouped according to applications, removal efficiencies and pollutant concentration in the emitted gas. Table 7.1 provides a general orientation for gas-cleaning devices, showing the typical applications and optimum particle sizes, concentrations, and gas-temperature ranges for each. Table 6.21 lists the estimated removal efficiencies for some pollution-control equipment on different services. Figure 7.1 shows the requirements of the various regulations for allowable dust content in the cleaned gas and the capabilities of particulate-cleaning devices. Table 6.15 gives the particle-size distribution in the gas discharges of incinerators without pollution-control devices.

The traditional gas-cleaning devices are cyclone separators, impingement baffles, and spray chambers, which remove the larger particulates in the fly ash. The collection efficiencies of dry-collector devices are listed in Table 7.2. In existing incinerators, spray chambers are frequently followed by electrostatic precipitators,

212

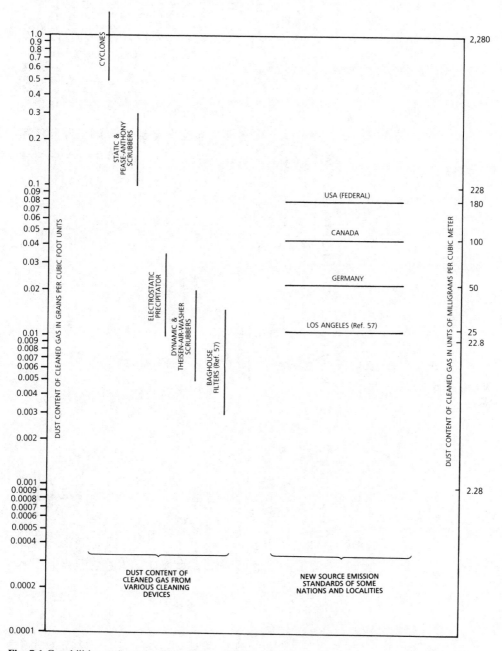

Fig. 7.1 Capabilities and requirements for gas-cleaning pollution-control equipment. Substantial reduction will occur in U.S. figures when the new standards of performance for incinerators take effect (202).

213

Table 7.1

CHARACTERISTICS OF AIR- AND GAS-CLEANING DEVICES (66)

Name of device		Device most suitable for	Removable contaminants	Optimum size particle μm	Limits of gas temperature °F	Opt. conc. ppm by wt[a]	Lim. gas temp. °K
General class	Specific type						
Odor adsorbers	Shallow bed	← Atmospheric air cleaning →	Malodors, gases	(Molecular)	0–100	<1.9	256–311
Air washers	Spray chamber			>20	40–700	<9.5	278–644
	Wet cell			>5	40–700		
Electro. precip., low-voltage	Two-stage, plate		Lints,	<1	0–250	<1.9	256–394
	Two-stage, filter		dusts,	<1	0–180		256–356
Air filters, viscous-coated	Throwaway	Atmospheric air cleaning	pollens,	>5	0–180	<3.81	256–394
	Washable		tobacco	>5	0–250	<3.81	256–356
Air filters, dry-fiber	5–10 μm		smoke	>3	0–180	<1.9	256–356
	2–5 μm			>0.5	0–180	<1.9	256–356
Absolute filters	Paper		Special[b]	<1	0–1800	<1.9	256–1256
Industrial filters	Cloth bag	← Stack gas cleaning →		>0.3	0–180[c]	>90	256–356
	Cloth envelope			>0.3	0–180[c]	>190	256–356
Electro. precip., high-voltage	Single-stage, plate			<2	0–700	>190	256–644
	Single-stage, pipe			<2	0–700	>190	256–644
Dry inertial collectors	Settling chamber		Dusts, fumes,	>50	0–700	>9,520	256–644
	Baffled chamber			>50	0–700	>9,520	256–644
	Skimming chamber			>20	0–700	>1,905	256–644

		smokes, mists ← Stack gas cleaning →			
	Cyclone	>10	0–700	>1,905	256–644
	Multiple-cyclone	>5	0–700	>1,905	256–644
	Impingement	>10	0–700	>1,905	256–644
	Dynamic	>10	0–700	>1,905	256–644
Scrubbers[d]	Cyclone	>10	40–700	>1,905	256–644
	Impingement	>5	40–700	>1,905	256–644
	Dynamic	>10	40–700	>1,905	256–644
	Fog	<2	40–700	>190	256–644
	Pebble bed	>5	40–700	>190	256–644
	Multidynamic	<1	40–700	>190	256–644
	Venturi	<2	40–700	>190	256–644
	Submerged nozzle	>2	40–700	>190	256–644
	Jet	<5	40–700	>190	256–644
		Gases, vapors, malodors			
Incinerators	Direct	Any	2000	Combustible	1367
Afterburners	Catalytic	(Molecular)	1000	any	811
Gas absorbers	Spray tower	(Molecular)	40–100	>1.9	278–311
	Packed column	(Molecular)	40–100	>1.9	278–311
	Fiber cell	(Molecular)	40–100	>1.9	278–311
Gas adsorbers	Deep bed	(Molecular)	0–100	>1.9	256–311

Source: Jorgensen, "Fan Engineering," Buffalo Forge Co.
[a]Based on std. air @ 0.075 lb/ft³ (1.2 kg/m³).
[b]Bacteria, radioactive, or highly toxic fumes.
[c]500°F for glass (553 K), 450°F for Teflon (505 K), 275°F for Dacron (408 K), and 240°F for Orlon (389 K).
[d]Reheating of scrubbed stack is necessary to avoid plumes.

Table 7.2
DRY COLLECTOR CHARACTERISTICS (50)

Types	Space Required	Maximum Capacity ACFM	Efficiency and Micron Size to which it Applies	Pressure Drop, inches H_2O	Temperature Limit, °F	Cost	Typical Applications
Gravity chamber	large	available space is the only limit	over 50 μ	0.2–0.5	700–1000	low	Precollectors for other devices
Cyclone	large to medium	50,000	50–80% on 10 μ	1–6	700–1000	low	Woodworking, buffing fibers, pneumatic conveyors
Rotary dust separator	medium	18,000	90% on 2 μ	to 8	to 500	moderate	Tar separation, cement dust handling
Multiple cyclone	small	100,000	90% on 5 μ	3–6	700–1000	moderate	Precollectors for electrostatic precipitation
Inertial collectors	small	75,000	90% on 2 μ	3–6	to 700	moderate	Pulverized coal handling
Dynamic precipitators	small	25,000	90% on 2 μ	none; it is a fan	to 500	moderate	Metalworking; coal bunkers

which have a proven record of success. While they require a high initial investment, their operating costs are low and they can remove small particles at efficiencies of 90% to 99.5%. Fly-ash collection with 90% efficiency using electrostatic precipitators was first reported in 1954 in Bern, Switzerland. Because of corrosion considerations, the operating temperatures were limited to 500° F (260° C) (31).

Wet scrubbers of the packed column, baffle-and-spray, and venturi varieties are also effective in fly-ash removal applications. Their removal efficiencies depend on dust-particle sizes (Table 7.3) and scrubber design (Fig. 7.2). Wet scrubbers cause water-treatment problems: the scrubber-fluid circulating equipment needs to be designed for highly corrosive service. Also, the gases must be reheated before discharging them through the stack.

Using dry scrubbers with pulverized lime, followed by fabric filters, allows unreacted lime to collect on the bags and provides a secondary surface for acid neutralization. Dry scrubbers followed by baghouse filters can remove 80% to 90% of SO_2 and 90% to 95% of HCl in the flue gas. One disadvantage of using lime is that it increases the alkalinity of the ash, which can make the metals in the incinerator ash more soluble when leached by rainwater, if their hydroxil is more soluble than their carbonate form. Some of the lowest particulate emissions (0.0025 to 0.0074 gr/dscf) were reported from installations using a combination of dry scrubbers and baghouse filters (44). Baghouse filters are limited in the operating temperatures they can tolerate (Table 7.4), and even the silicone-coated glass fibers cannot operate at temperatures exceeding 575° F (300° C).

Wet Collectors

Spray chambers and scrubbers bring the dust-laden gas in contact with the wash water, which then scrubs out dust particles. Wet scrubbers have been in use for over a century and more recently have been found effective in fly-ash removal.

Table 7.3
EFFICIENCY OF SCRUBBERS AT VARIOUS
PARTICLE SIZES (20)

Type of Scrubber	Percentage Efficiency at		
	50 μ	5 μ	1 μ
Jet-impingement scrubber	98	83	40
Irrigated cyclone	100	87	42
Self-induced spray scrubber	100	94	48
Spray tower	99	94	55
Fluid bed scrubber	99+	98	58
Irrigated target scrubber	100	97	80
Disintegrator	100	98	91
Low energy venturi scrubber	100	99+	96
Medium energy venturi scrubber	100	99+	97
High energy venturi scrubber	100	99+	98

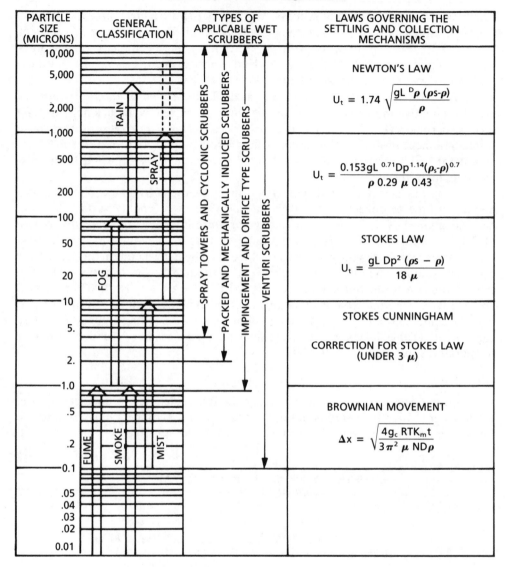

Fig. 7.2 Scrubber comparison chart (20).

More than a hundred different scrubbers are available from dozens of suppliers. Scrubber selection is usually based on inlet dust loading, particle-size distribution, gas-flow rate, and on the collection efficiency required. Performance and reliability tend to be weighed more heavily than capital or operating costs. If they can provide the required collection efficiency, dry collectors are preferred as they offer a predictable performance.

Table 7.4
CHARACTERISTICS OF FILTER FABRICS (50)

Fabric	Maximum Operating Temperature °F	Specific Gravity	Physical Resistance					Chemical Resistance					Relative Cost
			Dry Heat	Moist Heat	Abrasion	Shaking	Flexing	Mineral Acids	Organic Acids	Alkalies	Oxidizing Agents	Solvents	
Cotton	180	1.6	G	G	F	G	G	P	G	F	F	E	Low
Dacron	275	1.4	G	F	G	E	E	G	G	F	G	E	Moderate
Orlon	250	1.2	G	G	G	G	E	G	G	F	G	E	Moderate
Nylon	250	1.1	G	G	E	E	E	P	F	G	F	E	Moderate
Dynel	180	1.3	F	F	F	P–F	G	G	G	G	G	G	Moderate
Polypropylene	225	0.9	G	G	E	E	G	E	E	E	G	G	Low
Creslan	275	1.2	G	G	G	G	E	G	G	F	G	E	Moderate
Vycron	300	1.4	G	F	G	E	E	G	G	G	G	E	Moderate
Nomex	450	1.4	E	E	E	E	E	P–F	E	G	G	E	High
Teflon	500	2.3	E	E	P–F	G	G	E	E	E	E	E	High
Wool	200	1.3	F	F	G	F	G	F	F	P	P	F	Moderate
Glass	550	2.5	E	E	P	P	F	E	E	G	E	E	High

E-Excellent G-Good F-Fair P-Poor

The main advantage of wet scrubbers is their ability to remove simultaneously both gaseous and particulate pollutants. Their disadvantages include the high maintenance and operating costs, corrosion, abrasion, scale deposition, unpredictable performance, and problems with the disposal of the resulting sludge. One problem associated with wet collectors is the loss of "plume buoyancy." To achieve good dilution, the gaseous emissions should leave the stack at high velocities. The higher the intial plume momentum, or buoyancy, the higher the efflux velocity. Plume buoyancy also depends on atmospheric conditions and flue-gas temperature. The higher the stack-gas temperature, the more the plume will lift convectively. The loss in plume buoyancy is caused by the cooling effect of scrubbing, which can be partially corrected by using reheaters. To render the stack discharges essentially invisible, the scrubber outlet gases are subcooled to condense the water vapor and thereby also minimize the acid formation in the wet gases. The corrosion problem can be handled only by selecting the right construction materials and linings, such as stainless steel (304, 316, and 316L) and fiberglass-reinforced plastics (FRP).

The pollution-control system on an incinerator usually consists of more than just a scrubber. One common equipment configuration is a spray chamber or a dry cyclone followed by an electrostatic precipitator or wet scrubber. In some newer installations, dry scrubbers are followed by baghouse filters (Fig. 6.5). The efficiency of wet scrubbers depends on their design, which determines the total energy expended in the gas-liquid contacting process, and on the size of the flue-gas particulates (see Table 7.3). The finer the particle size, the more energy (pressure drop) is usually required to attain the same collection efficiency. Figure 7.2 compares several scrubber designs as to the laws governing their collection mechanisms and their applicability for removing various dust-particle sizes. Table 6.15 gives the particulate-size distribution in incinerator emissions. For incinerator applications, the most frequently used wet scrubbers are the three-stage impingement-type designs.

The Mechanics of Wet Collection

Wet collectors develop an interface between the emitted gas and the scrubbing liquid. The more intimate this interface, the more efficient the removal of the pollutants from the gas. The spray across the flow path of the gas stream impinges on the fine dust particles, increasing their effective size and thus making them easier to collect. Collection efficiency depends on the number and size of the droplets and on the energy imparted to them. Thus scrubber efficiency is a function of the total energy expended in the gas-liquid contacting process.

The dust particles, which, because of their inertia, flow in a nearly straight path, tend to collide with the liquid droplets. Because the droplets are larger and heavier than the dust particles, they are easier to collect by gravitational or inertial forces. The power that is dissipated in fluid turbulence is expressed as power per unit of volumetric gas flow rate (HP/scfm) and is related to the pressure drop through the scrubber. It is this "contacting power" that determines the efficiency of the scrubber. In Figure 7.3 a number of scrubber designs are plotted showing the par-

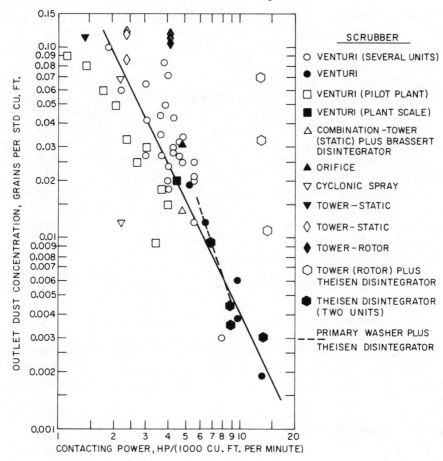

Fig. 7.3 Performance of various scrubbers on blast furnace dust and fume (20).

ticulate concentrations at their outlets as a function of their contacting powers. This plot shows the "rule of contacting power," meaning that scrubbing efficiency depends mainly on the contacting power and not on the geometry, size, or the manner in which the power is applied to the contacting process in the scrubber.

The First Stages of Fly-Ash Collection

The flue gas leaving the combustion zone contains substantial quantities of fly ash. Secondary combustion chambers (Fig. 5.4) and expansion chambers (Figs. 5.21 and 5.23) provide added detention time and reduce the gas velocity so that the gas will carry only the finest ash particles. If the primary combustion chamber is large enough, the secondary combustion chamber is sometimes omitted or used for preliminary gas cleaning.

From the primary or, more often, from the secondary combustion chamber, the gases pass to the gas-cleaning or "spray chamber" area, which serves the dual purpose of fly-ash removal and gas cooling. In this area the gases are blown through a thin film of water, which wets the particulates. As they impinge the water droplets, they drop out (Fig. 5.5). The effectiveness of spray chambers in providing a predictable degree of cooling is limited to installations where both the furnace gas flows and temperatures are maintained at a constant rate. If the gas-cleaning area is of a baffled design, the baffles will send the gases down and around a 180° turn over a water surface (Fig. 5.9), where the larger particles are removed before the gases pass through a series of water sprays and wet baffles for further removal of the finer ash particles.

If the flue gas is cooled by the evaporation of water in a spray chamber, the heat-balance calculation outlined in Figures 5.30 and 5.31 also can be used to determine the amount of water required. If, for example, the flue gases are to be cooled in the spray chamber from 1680° F to 600° F, then 2.06 pounds of spray water will need to be added per pound of refuse burned (18).

The following sections will describe a number of wet scrubber designs, starting with the simplest variety—spray towers and spray chambers. Almost all these scrubber designs have been used on flue-gas applications but not necessarily on incinerator flue gases. It seems appropriate to cover them because as the number of incinerator applications increases, it is likely that the use of wet scrubbers will also rise. Thus designs that have not been used to date could be used in the future. On existing incinerators, the most widely used wet collectors are spray chambers (Fig. 7.4), cross-flow-type packed scrubbers (Fig. 7.5), and direct-contact impingement scrubbers (Fig. 7.6).

Spray Towers and Chambers

One of the simplest wet scrubber designs is a countercurrent spray chamber (Fig. 7.4). The scrubbing water is introduced by spray nozzles and the wetted particles settle and are collected at the bottom of the chamber. Per 1,000 acf (actual cubic feet) of flue gas, 5 to 10 gallons (0.7 to 1.3 liters per acm), of scrubbing liquid is needed. The pressure drop on the gas side of the countercurrent spray chamber is usually less than 1″ H$_2$O (250 Pa) and its dust-collection efficiency is under 80% for all but the coarsest dust particles (over 10 microns in diameter).

Spray towers are also low-efficiency collectors of coarse (over 10 microns) particles. Figure 7.6 illustrates a countercurrent spray tower, in which the liquid droplets are produced by spray nozzles or atomizers and the flue gas enters the tower through a perforated distribution plate at the base of the tower. To avoid entrainment and carry-over of water droplets, the velocity of the spray-water droplets must exceed the gas velocity by 2 to 5 feet per second (1–2 m/s). The higher this velocity difference, the better the collection efficiency.

A special scrubber-design variation is the membrane-type spray chamber (Fig. 7.7). Here the scrubbing water is sprayed concurrently at high pressure (100 to 200

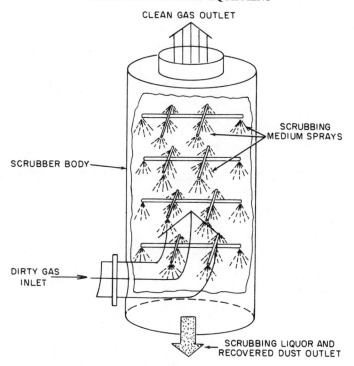

Fig. 7.4 Spray-chamber scrubber (20).

PSIG or 0.7 to 1.4 MPa) against closely spaced, parallel vertical rods, which form an irregular reflecting surface, or membrane. The spray nozzles are arranged so that the atomized water droplets rebound near the membrane surface, causing highly efficient scrubbing action. Added scrubbing takes place as the gas accelerates through the narrow passages between the rods. This design requires 1" H_2O (250 Pa) in gas-side pressure drop and 3 to 6 gallons of scrubbing water per 1,000 acfm (0.4 to 0.8

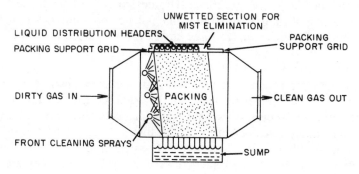

Fig. 7.5 Crossflow scrubber (20).

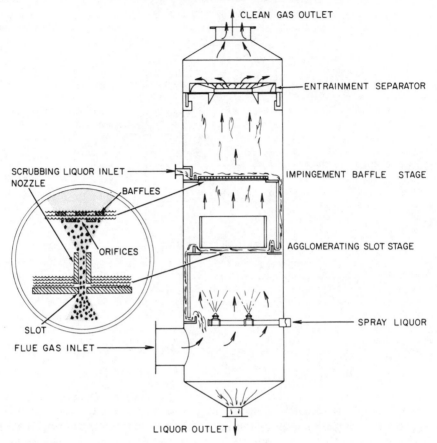

Fig. 7.6 Peabody direct-contact scrubber. (Courtesy of Peabody Engineering Corp. [20].)

liters per acfm) per stage of scrubbing. The design was found to be simple and reliable on fly-ash applications, providing an efficiency of about 99% (20). To obtain higher collection efficiencies, up to three stages can be placed in series.

Centrifugal and Cyclonic Scrubbers

Cyclonic and centrifugal scrubbers are more efficient than spray towers and chambers because the centrifugal force of the spinning gas stream increases the velocity difference between the gas and water droplets (Fig. 7.8). The flue gas is introduced tangentially into the scrubber and forms a rising vortex into which the finely atomized scrubbing water is sprayed. The centrifugal force accelerates both the dust particles and the liquid droplets toward the wall, where they are collected.

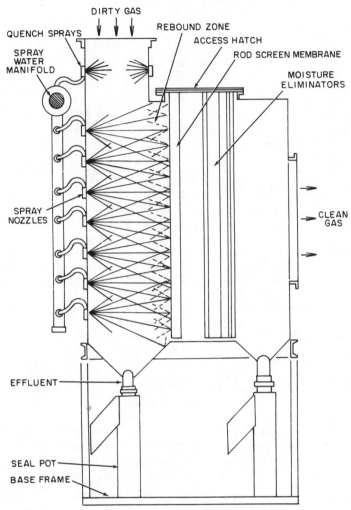

Fig. 7.7 Elbair scrubber. (Courtesy of Krebs Engineers [20].)

The spinning motion of the gas also can be initiated by mechanical baffles or impellers, or by injecting the scrubbing water tangentially.

The tangential velocities of the flue gas at the entrance range from 100 to 120 ft/sec (30 to 36 m/sec) and the scrubbing water is introduced at pressures from 50 to 400 PSIG (0.35 to 2.8 MPa). The finer the dust particles (down to one micron), the higher the required injection pressure of the scrubbing water. Gas-side pressure drops range from 1″ to 8″ H$_2$O (0.25 to 2 kPa) for the different cyclonic scrubber

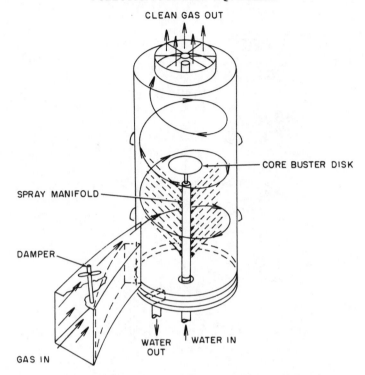

Fig. 7.8 Pease-Anthony cyclonic scrubber. (Courtesy of Chemical Construction Corp. [20].)

designs. The design illustrated in Figure 7.8, falls in the lower part of this range. The scrubbing-water requirement ranges from 4 to 10 gallons per 1000 acf (0.5 to 1.3 acm) of flue gas handled. For particle sizes in excess of 2 microns, the collection efficiencies exceed 90%. The efficiencies of the multiwash designs (Fig. 7.9) can reach 99% on particle sizes exceeding 2 microns and 90% for particles over one micron.

Packed and Floating-Bed Scrubbers

The main application of packed scrubbers is in the absorption of gaseous pollutants, but they are also used for the simultaneous collection of dust and acid-mist entrainment. The wetted packing provides an impingement surface for the deposit of dust, which is then washed away by the scrubbing water. The packing can be held in place (fixed bed), movable (floating bed), or covered with liquor (flooded bed). For particle sizes in excess of 2 microns, the collection efficiency is 80% to 95% for the fixed bed, 95% to 99% for the floating bed, and 99% for the flooded bed. In the fixed-bed design, the slower the gas flow, the higher the collection efficiency, while in the floating-bed design the velocity must be high enough to cause floating

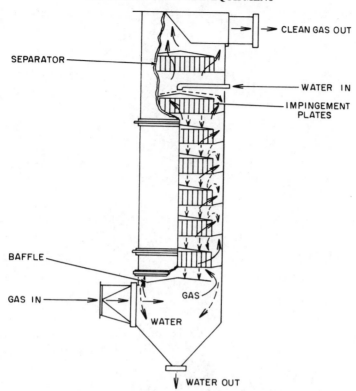

SEPARATOR

CLEAN GAS OUT

WATER IN

IMPINGEMENT PLATES

BAFFLE

GAS IN

GAS

WATER

WATER OUT

Fig. 7.9 Multiwash scrubber. (Courtesy of Clause B. Schneible Co. [20].)

(over 7 fps or about 2 mps) but less than the velocity at which the bed is lifted to the top (13 fps or about 4 mps).

The cross-flow design is best suited for incinerator applications because the collected fly ash is deposited near the front of the bed, where it can be easily washed off by the cleaning sprays. As shown in Figure 7.5, the flue gas moves through the packing in a horizontal direction, while the scrubbing water is introduced at the front and top. The scrubbing-liquid flow requirement is about 10 gallons per 1,000 acfm (1.3 liters per acmm) and the pressure drop on the gas side is about 0.25″ H_2O (62 Pa) per foot of packing.

The flow direction through the floating-bed scrubbers is always countercurrent, and their packing spheres are in continuous, random, and turbulent motion, causing intimate mixing not only on the surface of the spheres but throughout the entire spray zone (Fig. 7.10). The tumbling action of the spheres provides continuous cleaning, which prevents the buildup of solids and the eventual plugging of the bed. The retaining grids of each stage are 18 inches apart and the static bed of spheres

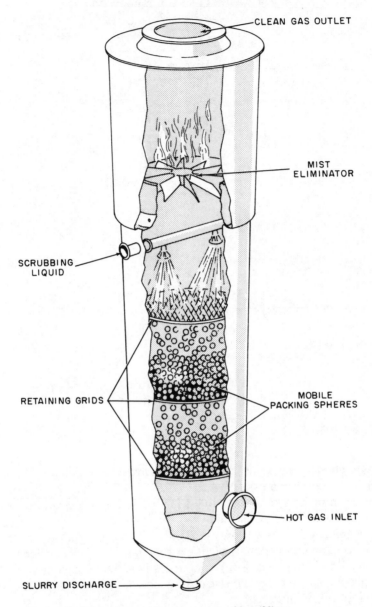

CLEAN GAS OUTLET

MIST
ELIMINATOR

SCRUBBING
LIQUID

RETAINING GRIDS

MOBILE
PACKING SPHERES

HOT GAS INLET

SLURRY DISCHARGE

Fig. 7.10 Turbulent contact absorber wet scrubber (20).

in each stage is about 12 inches deep. The required pressure drop on the gas side is around 3" H_2O (750 Pa) per stage, the gas velocity is 7 to 13 ft/sec (about 2 to 4 m/sec), and the scrubbing-water requirement is 10 to 20 gallons per square foot (around 400 to 800 liters per square meter) of packing area. Collection efficiency increases with the number of stages. On power-plant fly-ash service, 95% efficiency is reported for single-stage designs and 98% for two-stage designs (20). In coal-fired boiler applications, the floating-bed scrubber has been used successfully for the simultaneous removal of both fly ash and sulfur dioxide. One system combines dry additive limestone injection with wet scrubbing and provides 60% SO_2 removal efficiency, in addition to removing 99% of the fly ash from the flue gases (Fig. 7.11). In order to recover most of the heat lost by scrubbing, a stack gas reheater is used to guarantee plume buoyancy (63).

Flooded-bed scrubbers are 99% efficient in removing particle sizes in excess of 2 microns and require about 5" H_2O (1.25 kPa) pressure drop. As shown in Figure 7.12, the dirty flue gases enter below the bed of spherical glass marbles and pass through a spray section before entering the flooded bed of spheres. The bubbles formed in the bed create a 6-inch-deep turbulent layer, with the constant motion of the spheres guaranteeing self-cleaning action. The scrubbing-liquor requirement is about 3 gallons per 1,000 acf (0.4 liters per acm).

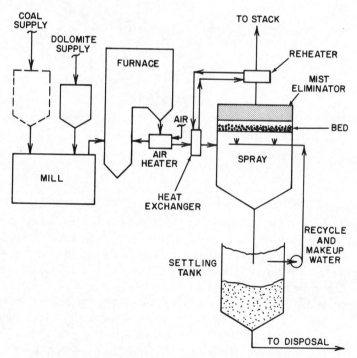

Fig. 7.11 Simultaneous removal of fly ash and SO_2, using the Combustion Engineering process (63).

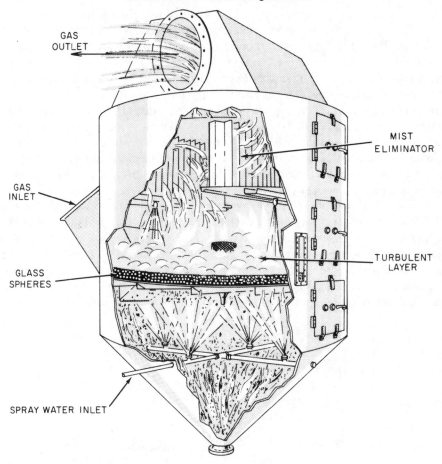

Fig. 7.12 Flooded-bed scrubber (20).

Impingement and Orifice-Type Scrubbers

In impingement scrubber designs, the high-velocity flue gases flow through restricted passages that are partially filled with scrubbing liquid or that direct the gas to impinge on a liquid surface (Figs. 7.13 and 7.14). The efficient wetting and collection of particles is the combined result of centrifugal forces, impingement, and turbulence. The impingement design in Figure 7.13 is used for the simultaneous removal of acid fumes and fly ash. The flue gases enter the bottom section, where they are mixed with the sprayed water, which also wets the "scrubbing tuyere." As the gas passes through the blades of the scrubbing tuyere, the dust particles impinge on the wet blades and are collected. The upper "drying tuyere" separates the liquid from the exiting gas and also performs some additional scrubbing.

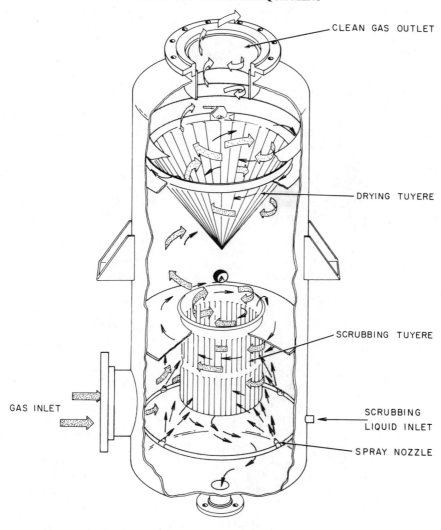

CLEAN GAS OUTLET

DRYING TUYERE

SCRUBBING TUYERE

GAS INLET

SCRUBBING
LIQUID INLET

SPRAY NOZZLE

Fig. 7.13 W-D tuyere scrubber (20).

In the impingement baffle-plate scrubber in Figure 7.14, the flue gas enters the lower spray section, where it is cooled, humidified, and the larger dust particles collected, while the finer dust particulates tend to agglomerate. The spray also keeps the walls and the impingement plate clean. As the flue gas enters the impingement baffle plate, on which a uniform water layer is maintained, the high-velocity flue gas causes a jet-swirl interaction with the scrubbing water. The resulting turbulence produces collection efficiencies in excess of 90% for particles larger than one micron.

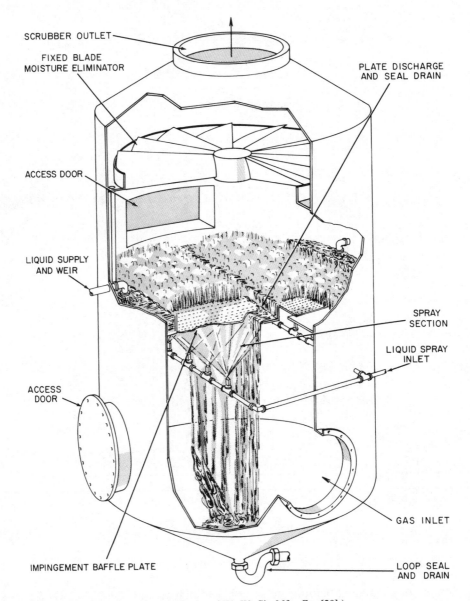

SCRUBBER OUTLET

FIXED BLADE
MOISTURE ELIMINATOR

PLATE DISCHARGE
AND SEAL DRAIN

ACCESS DOOR

LIQUID SUPPLY
AND WEIR

SPRAY
SECTION

LIQUID SPRAY
INLET

ACCESS
DOOR

GAS INLET

IMPINGEMENT BAFFLE PLATE

LOOP SEAL
AND DRAIN

Fig. 7.14 Impinjet scrubber. (Courtesy of W. W. Sly Mfg. Co. [20].)

232

Collection efficiency increases to a maximum of 99% by adding more stages and by increasing the gas-side pressure drop to about 20″ H_2O (5 kPa). The scrubbing-water requirement is between 1.5 and 4 gallons per 1,000 acf (0.2 to 0.5 liters per acm).

In the direct-contact impingement scrubber in Figure 7.6, the hot flue gases enter at the bottom, where the water spray removes the coarser particles while cooling and saturating the gas. The spray also cleans the walls and underside of the agglomerating slots. A nozzle is mounted above each slot; as the gas passes through the slots, it aspirates the water from the blanket covering the plate. The agglomeration of submicron particles, fumes, tars, or resins results from the turbulent contact between the liquid and dust particles. Above the agglomerating plate is the impingement baffle plate, which has about a thousand orifices per square foot and a baffle above each orifice. The impingement baffles are also submerged in a uniform layer of scrubbing liquor; as the jets from the orifices impinge on the wet baffles, the particulates are precipitated and entrapped in the scrubbing fluid. Finally, the gas passes through an entrainment separator to eliminate liquid carry-over. The scrubbing-liquid requirement is about 4 gallons per 1,000 acf (0.5 liters per acm). The gas-side pressure drop ranges from a few inches of water up to 40″ H_2O (10 kPa), depending on the gas velocity used. Collection efficiencies of 97% are reported for particulate sizes in excess of one micron (20). On incinerator applications, the dust loading in the flue gas was reduced from 2 to 5 gr/scf to 0.2 gr/scf.

In orifice-type scrubbers, the high-velocity flue gas disperses the scrubbing water, but the degree of dispersion is less than in the impingement designs. These are an improved version of the early self-induced spray scrubbers, in which the flowing gas induces a spray curtain as it passes through partially submerged baffles and orifices. Figure 7.15 shows a variety of orifice-type designs. These compact and simple scrubbers provide reasonable collection efficiency with low scrubbing-water requirements.

Venturi Scrubbers

Venturi scrubbers (Fig. 7.16) are simple to install and maintain and provide high collection efficiencies when the particle sizes are in the submicron zone. The flue gases are accelerated until they reach maximum velocity in the throat section of the scrubber, where they impact a stream of scrubbing water and atomize it. The high differential velocity between the dust particles and the liquid droplets causes them to collide and agglomerate. The pressure-drop requirement on the flue-gas side rises as the particles become finer; it ranges from 6″ to 80″ of H_2O (1.5 to 20 kPa). The best method of maintaining constant dust-collection efficiency is to keep the pressure drop at a constant value.

The difference between the various venturi scrubber designs is in the methods of introducing the scrubbing liquid and in the shapes of the venturi throats. Their performances tend to be similar if the throat-pressure drop is similar. The throat shapes can be circular, rectangular, oval, or, on large gas-flow rates, annular. If the scrubbing water is introduced by sprays located above a circular throat (the typical

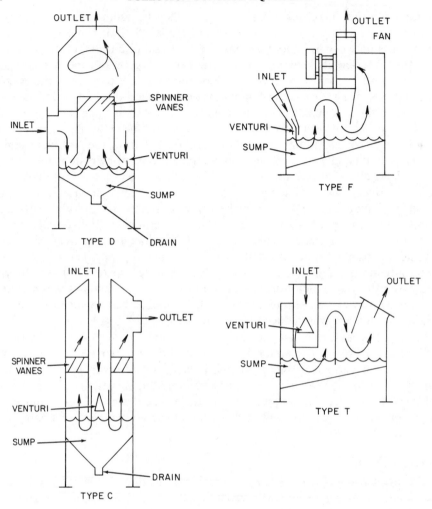

Fig. 7.15 Turbulaire scrubbers. (Courtesy of Joy Mfg. Co. [20].)

case for low-pressure applications), the throat diameter is unlimited. On high-pressure-drop applications, the water is usually introduced in the throat by horizontal sprays and the maximum dimension of the opening should not exceed 20 inches (0.5 m).

Figure 7.16 shows a Pease-Anthony venturi scrubber, where the water is introduced as a curtain in the throat. The water is then atomized by the accelerated flue gas, which reaches velocities of 150 to 400 fps (45 to 122 mps). The scrubbing-water requirement is 3 to 12 gallons per 1,000 acf (0.4 to 1.6 liters per acm). Collection efficiency exceeds 99% for particle sizes in excess of one micron and ranges from

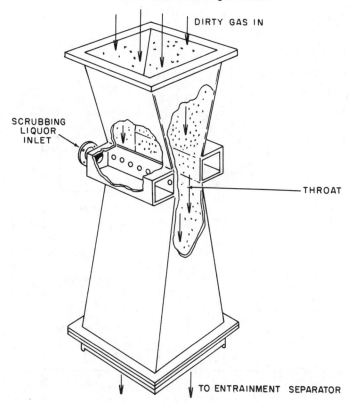

Fig. 7.16 The Pease-Anthony (P-A) venturi scrubber. (Courtesy of
Chemical Construction Corp. [20].)

90% to 99% for submicron particulates. The particulate concentration in the gas
leaving the scrubber has been reported to be in the range of 0.1 to 0.3 gr/acf.

Both the Pease-Anthony and the wet-approach (W-A) venturi scrubbers have
been used in flue-gas applications where the main goal was the scrubbing of sulfur
dioxide (Fig. 7.17). The Pease-Anthony provides a radial liquid feed and the W-A a
tangential liquid feed in the convergent section of the venturi. After leaving the
divergent section, the aerosol enters a cyclonic separator, where centrifugal and
gravitational forces remove the liquid from the gas. The scrubbing (absorbing) liquor
in this case is magnesium sulfite, which absorbs the SO_2 to form $MgSO_3$.

In the wet-approach venturi scrubbers (Fig. 7.18), part of the liquor is introduced by tangential nozzles, and it swirls down the converging walls to form a liquid
curtain at the throat. The rest of the scrubbing liquor is introduced on the central
cone (plumb bob) through open bull nozzles. An annular throat is used for flows
over 100,000 acfm (3,000 acm). The throat area is adjusted by moving the plumb

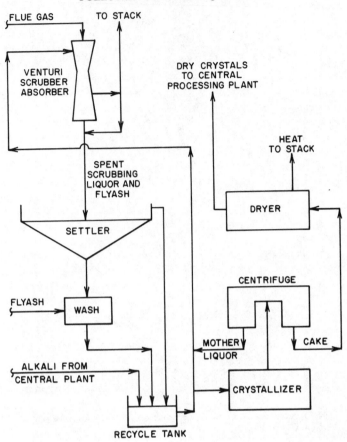

Fig. 7.17 Chemico-Basic recovery system (63).

bob or by dampers. Scrubbing-liquor requirements range from 10 to 20 gallons per 1,000 acf (1.3 to 2.3 liters per acm). The gas velocities are 75 to 250 fps (23 to 76 mps), and the pressure drops range from 5″ to 40″ H_2O (1.25 to 10 kPa).

Variations on the W-A venturi include the flooded-disk design (Research-Cottrel Inc.), in which a wetted disk throttles the opening at the throat, and the Aeromix scrubber (UOP Air Correction Division), where the gas flow is upward and the liquor recirculates by gravity (Fig. 7.19). The pressure-drop requirement for this design is only 4″ to 8″ H_2O (1 to 2 kPa), and the collection efficiency for particles over one micron is 98%. The ventures also can be combined into multiventuri stations or operated as ejector ventures. As shown in Figure 7.20, the ejector venturi uses the energy of the liquid jet not only to atomize the liquor but also to pump the flue gas

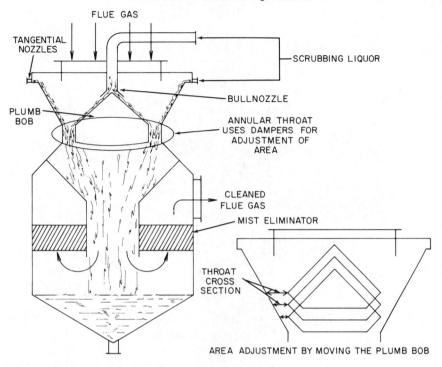

Fig. 7.18 W-A venturi scrubber (20).

through the system. No gas blower is needed, and the total operating energy requirement of the scrubber is in pumping the fluid.

Mechanical Scrubbers

In mechanical scrubbers the sprays are generated by partially submerged rotary elements or by scrubbing liquor sprayed into an impeller. Their main limitations are the operating-power requirement (5 to 20 HP per 1,000 acfm) and erosion and dust buildup on the rotary elements. Their advantages include compactness, low liquor requirements (0.5 to 5 gallons per 1,000 acf), and good collection efficiency (90% to 95% for over one micron).

Figure 7.21 shows a center spray scrubber, where the liquor is atomized both by the rotor and the impact of the high-velocity sprays on the walls. The horizontally formed high-velocity droplets impact the dust particles in the flue gas as it flows vertically upward. Figure 9.23 shows a dynamic scrubber, also called a centrifugal fan scrubber because the scrubbing fluid is sprayed into the fan suction. The dust particles impinge on the wet blades and the spray flushes away the collected dust. The favorable features of this design include a high collection efficiency at moderate

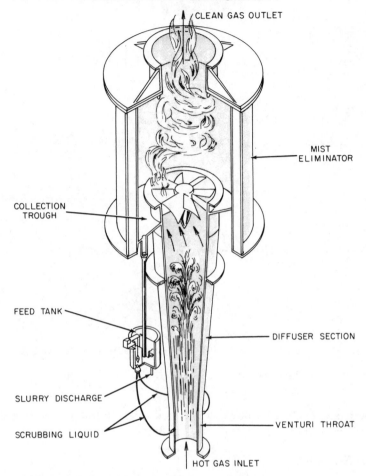

Fig. 7.19 Aeromix wet scrubber. (Courtesy of UOP Air Correction Div. [20].)

power costs and low water and space requirements (20). An in-line version of the centrifugal wet scrubber is made by Western Precipitator Company.

Dry Collectors

Within the group of dry collectors are three major categories of pollution-control equipment: cyclones, or centrifugal dry collectors, filters, and electrostatic precipitators. In incinerator applications, the cyclones are frequently used as precleaners in front of electrostatic precipitators. Electrostatic precipitators operate at efficiencies of 90% to 99.5% and can be found on many existing incinerators. In comparison with baghouse filters, electrostatic precipitators offer lower maintenance and oper-

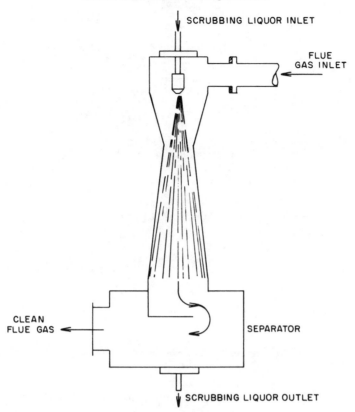

Fig. 7.20 Typical ejector venturi scrubber (20).

ating costs, similar capital investments, and similar temperature limitations, but lower collection efficiencies.

Some of the lowest particulate emissions ever reported (0.0025 to 0.0074 gr/dscf) were from installations using a combination of dry scrubbers followed by baghouse filters (44). For a comparison of flyash contents in the gas discharges from several cleaning devices, refer to Figure 7.1.

Cyclones (Centrifugal Dry Collectors)

When a stream of dust particles is forced to change direction, centrifugal forces act on the dust particles. The separation in a cyclone is illustrated in Figure 7.22.

In industry, cyclones are the most widely used inertial separators used for fly-ash and dust collection. Because of their low efficiencies (Table 7.5), they are used most often as preliminary or first-stage cleaners to remove the larger particles. They are well suited for that task because they are simple, inexpensive, and reliable. Standard designs will remove fly-ash or dust-particle sizes in excess of 5 microns.

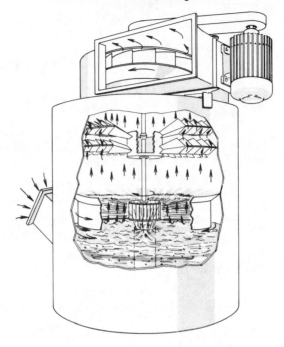

Fig. 7.21 Center spray high-velocity scrubber (20).

The dust content of the cleaned gas is usually in excess of 1 gr/dscf (2.3 g/m³). Table 7.2 lists some of the features of cyclones. The multiple cyclone is the most common for incinerator applications. A cyclone with a diameter under 9 inches (230 mm) is called a "high-efficiency" cyclone.

The operation of a cyclone is illustrated in Figure 7.22. The dust-laden gases tangentially enter the cyclone and move downward in long sweeping spirals. The centrifugal force of this vortex motion moves the dust particles into the outer layers of the gas. The residence time in the cyclone should be long enough for the particles to pass through the gas into the outer layers. The whirling of the vortex is faster near the center and slower at the periphery. As the spiraling gas reaches the bottom

Table 7.5
EFFICIENCY OF CYCLONES (50)

Particle Size Range μ	Conventional	"High-efficiency"
Less than 5	Less than 50	50–80
5–20	50–80	80–95
15–40	80–95	95–99
Greater than 40	95–99	95–99

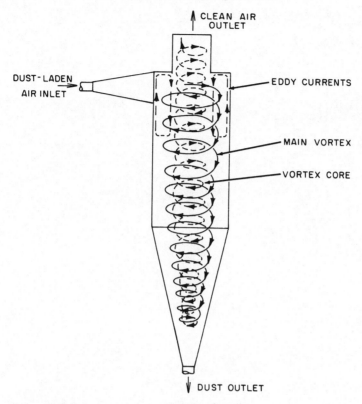

Fig. 7.22 Flow patterns in cyclones (50).

of the cyclone, the vortex reverses its axial direction and, while retaining its direction of rotation, starts rising.

The layer of dust that forms on the inner wall slowly swirls down and collects at the bottom of the hopper. As the bottom of the cyclone is conical, the dust particles contained in the outer layers of the vortex contact the wall and, through friction, lose their velocity and drop out. At the same time, the vortex core continues to spiral upward and the cleaned gas exits at the top of the cyclone. To protect the exiting gas stream from the reentrainment of the dust, a small fraction (10%) of the gas is removed with the collected dust as a purge flow.

In the annular region near the top, between the gas outlet and the cyclone wall, eddy currents form. To maintain cyclone efficiency, this turbulence must be minimized. These currents occur because the tangential gas velocity is higher at the outlet than at the wall. The currents carry some of the dust that was already collected at the walls back to the cyclone outlet. Thus they tend to reduce the dust-removal efficiency of the cyclone. The longer the projection of the gas outlet pipe (Type A

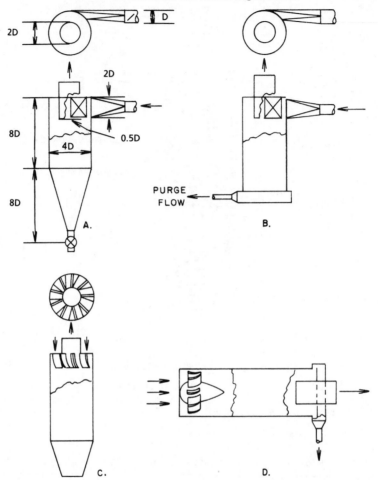

Fig. 7.23 Types of cyclones. (A) Tangential inlet and axial dust discharge; (B) Tangential inlet and peripheral dust discharge; (C) Axial inlet and axial dust discharge; (D) Axial inlet and peripheral dust discharge (50).

design in Fig. 7.23), the greater the loss in efficiency. This problem is eliminated by using axial inlet cyclones with swirl vanes (Type C and D designs in Fig. 7.23).

Figure 7.23 shows various nozzle arrangements; for the "Type A" cyclone, it gives the typical dimensions using the flue-gas inlet diameter (D) as the unit. The tangential inlet nozzles are usually rectangular to keep the inlet gases as close to the perimeter as possible. To facilitate the downward movement of the collected dust, the cyclone has smooth inside surfaces. Fly-ash accumulation or cyclone plugging can occur when moisture condenses on the inner wall or large pieces of caked dust break away from the wall and plug the outlet pipe.

Cyclone Performance Calculations

The smallest particle size that can be collected at 50% efficiency is called the "cut size" for the particular cyclone and is calculated as (50):

$$D^2 = (9 \mu W)/[6.28 N V (dp-d)]$$

Where:

D = cut size in ft
μ = gas viscosity in lb/ft sec
W = inlet width in ft
N = number of effective turns (usually 5–10)
V = gas inlet velocity in ft/sec
dp = particle density in lb/ft^3
d = gas density in lb/ft^3

The efficiency of particle collection rises with increasing gas-inlet velocity (Fig. 7.24). For the same cyclone, the variations in efficiency can be estimated as a function of volumetric flow rate (50):

$$Q_1/Q_2 = (N_1/N_2)^2$$

where Q_1 and Q_2 are volumetric gas flows and N_1 and N_2 are collection efficiencies in weight percentages.

In addition to inlet velocity, separation efficiency also increases with dust loading of the inlet gases, with particle size and particle density, and with cyclone body length, body-to-outlet diameter ratio, and inner-wall smoothness. Separation

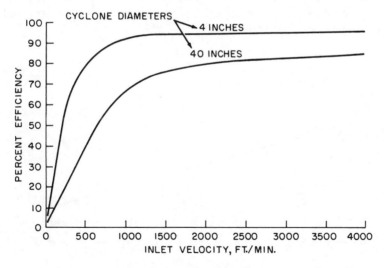

Fig. 7.24 Influence of inlet velocity and cyclone diameter on cyclone efficiency (50).

efficiency decreases with increasing gas viscosity, inlet-duct width and area, outlet and cyclone diameter, and with gas density. Efficiency also drops when inlet or guide vanes are added.

The "separation factor" (S) of a cyclone is related to its efficiency and is calculated as:

$$S = (V^2)/(gR)$$

where:

V = gas inlet velocity in ft/sec
g = gravitational acceleration (32.2 ft/sec^2)
R = radius of gas rotation in ft

The pressure drop across the cyclone usually varies from 1″ to 6″ H$_2$O and can be approximated as (50):

$$dP = (k\ Q\ P\ d)/(T)$$

Where:

dP = pressure drop in ″H$_2$O
k = cyclone constant (Fig. 7.25)
Q = gas flow in ft^3/min
P = absolute pressure in atmospheres
d = gas density in lb/ft^3
T = gas temperature in R

Cyclone Designs

Centrifugal and inertial forces are used in a number of ways to separate dust particles from a gas stream. Described below are inertial collectors, multiple cyclones, air-jet cyclones, and rotoclones.

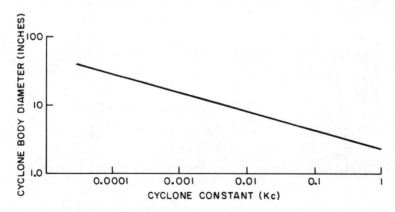

Fig. 7.25 Cyclone constants for pressure drop approximation (50).

Inertial Collectors. When fly-ash-laden air impinges on an object, the air is deflected and the dust, due to its greater inertia, collects on the surface. Inertial separation devices are designed to cause sudden changes in the direction of the gas stream, thereby creating impingement and centrifugal forces to separate the particles by inertia. These units give similar efficiencies as high-efficiency cyclones and require pressure drops of up to 6″ H_2O.

Multiple Cyclones. The smaller the diameter of the cyclone, the more efficient the separation. To handle the large gas volumes in incinerators, several small-diameter cyclones are manifolded in parallel. For a large number of parallel cyclones, a compact arrangement can be obtained by installing a common inlet and outlet plenum. The plenum must be designed to equalize the gas and fly-ash load distribution between the individual cyclones. The design must prevent backflow, plugging, or reentrainment from the common fly-ash bin. The pressure drop through multiple cyclones is usually in the range of 2″ to 3″ H_2O.

A larger cyclone can be installed in front of a small one to remove large or abrasive particles and thereby protect the smaller cyclone against plugging or excessive wear. Multiple cyclones are also used in series to provide backup.

Air-jet-Assisted Cyclones. As shown in Figure 7.26, the rising dirty gas stream forms the inner vortex, while pressurized secondary air (introduced at the top) forms the outer vortex, which is moving downward. The fly ash is collected in an annular ring between the two vortexes as a result of the combined centrifugal forces. A baffle ring at the top increases efficiency by blocking the outer portion of the rising vortex. These separators give better efficiencies at comparable power costs than do conventional cyclones (Table 7.6).

An important advantage of these cyclones is that the fly ash does not come in contact with the wall; thus erosion is not a problem and smooth walls are unnecessary. The main disadvantage is that these cyclones require a large quantity (30% to 40% of the total gas flow) of high-pressure air. This secondary air is frequently provided by a secondary blower, which recycles the clean exhaust gases.

These units can also be arranged for multiple cyclone operation, but it is more difficult to distribute the dust loading evenly between the individual units.

Dynamic Precipitators or Rotoclones. As shown in Figure 7.27, the impellers of this precipitator are hyperbolically contoured concave disks. The impeller blades intercept the fly-ash particles, which slip along the curved blades toward the tip. The concentrated particles are collected from the blade perimeter by a secondary air stream and discharged into a ring-shaped dust chamber. Rotoclone units function in the dual roles of gas transporting and cleaning. Their total power requirement is less than the sum of the power requirements of a fan and cyclone. If the discharge pressure requirement is 10″ H_2O, approximately 3 BHP is required per 1,000 acfm of gas flow. Collection efficiency can be improved by spraying water into the entering air. The minimum dust-particle size removed by dry apparatus is about 8 microns, while wet systems can remove particles as small as one micron.

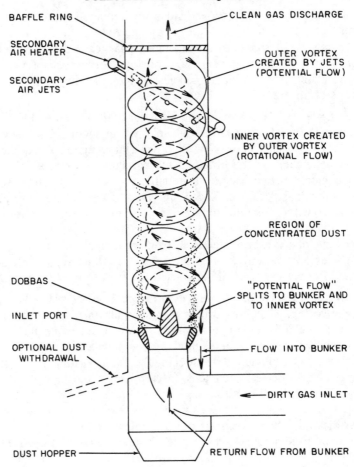

Fig. 7.26 Rotary dust separator (50).

Table 7.6
COLLECTION EFFICIENCY
OF ROTARY STREAM-
DUST SEPARATORS (50)

Particle Size Frequency Maximum, μ	% Collection Efficiency
2.5	92.0
5.0	95.0
10.0	98.5

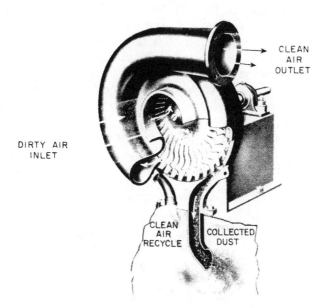

CLEAN
AIR
OUTLET

DIRTY AIR
INLET

CLEAN
AIR
RECYCLE

COLLECTED
DUST

Fig. 7.27 Dynamic precipitator. (Courtesy of American Air Filter Co., Inc. [50].)

Baghouse Filters

Filters remove fly ash and other particulates from the flowing gas not only by trapping particles but also by inertial and centrifugal action. Thus filters can remove dust particles that are smaller than the pores in the filter. In some designs the filter medium travels at a slow speed while the gas impinges on it. In more frequently used designs, the gas passes through stationary filter bags. For temperatures below 250° F, thickly woven cotton or wool felt is used, while woven glass or metal cloth are recommended for higher temperatures (Table 7.4). Cotton is the least expensive filter material. Dimensional stability is an important factor for all filter materials. Cotton and wool require preshrinking, synthetic materials need to be heat treated, and glass must be treated with silicone. (When coated with silicone, glass-fiber filter materials can operate up to 575° F [300° C].) Because condensation would plug the filter pores, saturated gases must be dried or preheated to keep the gases above their dewpoints.

In addition to their high overall dust-removal efficiencies, fabric filters also remove respirable submicron particles. In comparison with electrostatic precipitators, they are less sensitive to variations in flue-gas flows or changes in dust loading (57). Another advantage is their modular construction, which allows for inspection and maintenance without a full shutdown.

Filtration is a relatively new cleaning method (47). Fabric filters have produced flue-gas emissions with fly-ash concentrations of 0.01 gr/scf (22.5 mg/m³) (57), and even better results were reported from the full-scale HRI in Würzburg, Germany (44), where the fly-ash emission concentration, using a combination of dry lime injection and fabric filtering, ranged from 0.0025 to 0.0074 gr/dscf when using the EPA method of testing. The performance range for condensibles was 0.0035 to 0.0071 gr/dscf. At the time this was the lowest particulate emission reported by any method of dust collection (Fig. 7.1). At the baghouse outlet the flue-gas temperature ranged from 360° to 370° F. Because of such good performance reports, flue-gas filtering is being used more frequently.

The main working element in a baghouse filter is a porous, cylindrical fabric bag open at one or both ends. In a common configuration, the bag is suspended vertically, with its open end at the bottom (Fig. 7.28). The flue gas passes up, through the inside of the bag, while the particulates collect on the inside surface. The filter cloth can be formed into 6- to 12-inch-diameter bags up to 40 feet in length and grouped. Long tubes made of porous fabric collect the particulate matter as the flue gas passes through the bag. The gas is cleaned by the fabric as well as by the cake that builds up on the surface. In the design in Figure 7.28, the tubes are suspended in parallel from a steel structure, or baghouse. The dust-laden gas enters the lower end of the tubes and leaves at the top through the baghouse housing. Shaking or rapping at frequent intervals dislodges the accumulated dust from the bags. Design variations are differentiated by the various methods of bag cleaning and the types of filter materials.

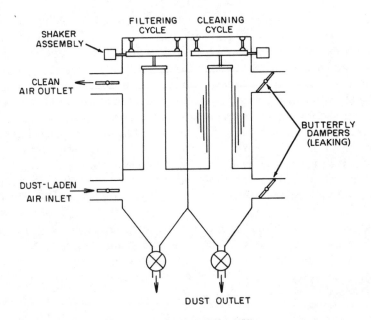

Fig. 7.28 Filter cleaning by mechanical shaker (50).

The main design criterion in sizing a baghouse is the pressure drop across the filter fabric. The resistance of the clean cloth seldom exceeds $0.1''$ H_2O. Pressure drop varies directly with the frequency of cloth cleaning and exponentially with the air/cloth ratio or filtering velocity (50). The maximum baghouse pressure drop in industrial applications is between $6''$ and $12''$ H_2O. On incinerator applications, it is less. In high-temperature applications, the gas must be cooled below the maximum operating temperature of the filter fabric (see Table 7.4) while making sure the temperature stays above the dewpoint so that condensation does not occur. In incinerator applications (see Fig. 6.5), the baghouse is usually located on the suction side of the induced-draft fan. The storage hoppers are usually provided with rappers or vibrators to eliminate bridging as the collected fly ash is continuously removed. The fly ash is transported either by gravity flow or by screw or belt-type conveyors.

In larger installations, baghouses are operated on an automatic, programmed cycle (Fig. 7.29). The main control devices are the pressure-drop detectors (dPIS–1, 2), which actuate when the pressure drops across the bags reach their preset limits. At this point, the section with the high pressure drop is isolated and its shaker drive is started. This mechanical cleaning action also can be assisted by reverse airflow through the shut-off valves UV–5 and UV–6. At the end of the cleaning cycle, the section automatically returns to normal operation and another section is isolated for cleaning. The cleaning can take from three to six minutes; the operating time period between cleanings can range from a fraction of an hour to several hours.

Because baghouse filters represent a potential fire hazard, they should be provided with safety interlocks, which will shut off the airflow at high temperatures and activate a sprinkler system when smoke is detected.

Filter Materials and Forms

The deposition of fly-ash or dust in a bag filter is the result of impingement, diffusion, and electrostatic effects. The form of the filter as well as its material affects collection efficiency. If the dust collects on the outside of a filter bag, an internal screen support must protect it from collapsing (Fig. 7.30). The outside-in designs make bag replacement more difficult, and contact with the frame reduces bag life. The best ratio of filtering area per unit volume is obtained by using envelope-type bags, a common choice when the filter medium is paper. Two disadvantages are the likelihood of dust bridging and the time required to change the bags.

Tubular bags are cyclinders with one or both ends open. They can be sewn together into "multibags" or configured as "unibags" (Fig. 7.31). In the Unibag design, several bags open at both ends are held between two cell plates, one at the ceiling and one at the floor. The gas enters at the top and leaves at the side. Reentrainment is least likely with this design.

The filter material can be paper, woven or felted fabric, metal, Teflon (47) or glass (Table 7.4). Paper filters are not used in larger, industrial units. Cotton is the least expensive filter material, while glass cloth is frequently used for high-temperature applications.

The allowable gas flow rate per unit filter area is called the air/cloth ratio. For woven bags, the ratio ranges from 1.5 to 6 cfm/ft^2, while the more expensive wool-

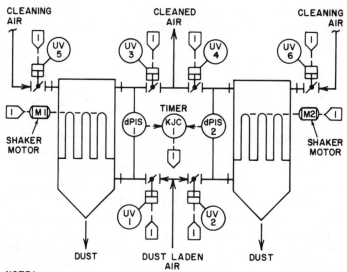

NOTE:
INTERLOCK FLAG ⬡I⬡ REFERS TO THE FOLLOWING OPERATING LOGIC:

OPERATING STATES ╲ CONTROLLED DEVICES	dPIS-1	dPIS-2	KJC-1	M-1	M-2	UV-1	UV-2	UV-3	UV-4	UV-5	UV-6
dPIS-1 AND-2 ARE NOT HIGH, KJC-1 IS OFF	LOW	LOW	OFF	OFF	OFF	O	O	O	O	C	C
dPIS-1 STARTED KJC-1 CYCLE BECAUSE HIGH DIFFERENTIAL	HIGH	LOW	ON	ON	OFF	C	O	C	O	O	C
dPIS-2 STARTED KJC-1 CYCLE BECAUSE HIGH DIFFERENTIAL	LOW	HIGH	ON	OFF	ON	O	C	O	C	C	O

O – OPEN, C – CLOSED

Fig. 7.29 Automatic baghouse cleaning cycle (50).

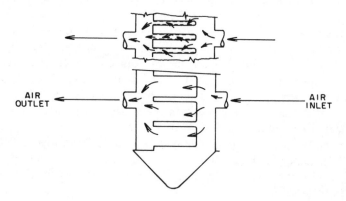

Fig. 7.30 Envelope-type filter bag (50).

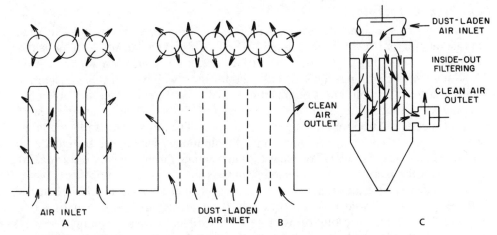

Fig. 7.31 Filter-bag design variations (50). (A) Tubular-type filter bag; (B) Multifilter bags; (C) Uni-bag filter (top entry type).

felted bags give substantially higher ratios, in the range of 4 to 12 cfm/ft². While the air/cloth ratio is much better with felted bags, the bags are not suited for fly ash or other fine dusts because the particulates become embedded in the felt and make cleaning difficult. The air/cloth ratio also depends on the bag-cleaning method applied. Reverse air cleaning (Fig. 7.32) reduces the air/cloth ratio to 1 to 3 cfm/ft². The lifespan of the filter bags is several months. Fabric life is affected by operating temperature, method and frequency of cleaning, the nature of the gas being cleaned, and the type of fly ash or dust being removed.

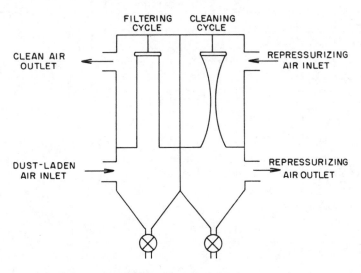

Fig. 7.32 Filter cleaning by repressurizing (50), applied by the "Varia-Pulse" design of Kemix, Hungary.

Bag-Cleaning Techniques

To dislodge accumulated dust or fly ash, the bags are shaken, pulsed by air jets (air shaking), or reverse-air-cleaned ("Puff-back"). Reverse air cleaning has the advantage of providing a large filtering surface area for secondary acid-gas removal (57). For woven fabrics, mechanical shaking is the most widely used cleaning method. The oscillary shaking motion is obtained by converting the rotary motion of a motor by cams or other eccentric elements. As shown in Figure 7.28, the baghouse section being cleaned is isolated from the operating portion. The advantage of this technique over reverse air cleaning is a reduction in the amount of flexing the bags undergo, thus increasing their lifespan. To achieve good cleaning through mechanical shaking, the inside of the bags must be completely depressurized. This can not be achieved solely by shutting the butterfly valves on the gas inlet (UV–01 or UV–02 in Fig. 7.29); even in the closed position they will leak enough to interfere with proper cleaning (Fig. 7.28). The solution is either to introduce a small amount of reverse airflow or shut off the inlet blower during cleaning.

Other methods of bag cleaning (Fig. 7.33) include air-shaking (wind-whipping), by blowing air between rows of bags, and bubble cleaning, where the cleaning air forms a bubble that travels downward in the bag, creating a rippling action. Reverse air cleaning is achieved by using a repressurizing or backwash blower (Fig. 7.32). Because the baghouse is usually located at the suction side of the induced-air-draft fan and therefore is operated under vacuum, a separate backwash fan is not necessary; the air flow through the bags can be reversed by opening a vent valve.

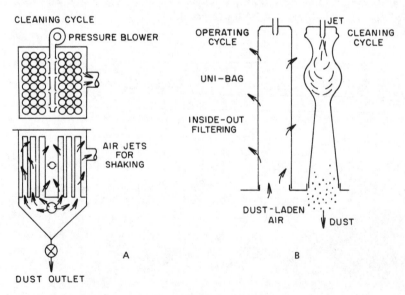

Fig. 7.33 Filter cleaning by (A) air shaking and (B) bubble action (50).

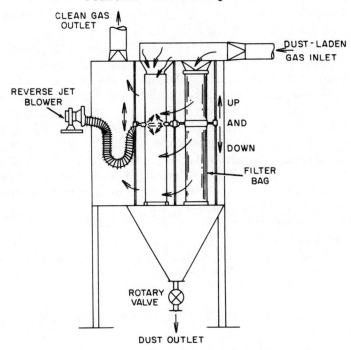

Fig. 7.34 Reverse-jet filter (50).

Reverse air cleaning is satisfactory only if the fly ash is easily removable from the bag.

The reverse-jet method of cleaning is used on felted bags with high air/cloth ratios (Fig. 7.34). A compressed air ring is moved up and down the outside of the bag and the dust removed by pressurized air jets as the air is forced through the felt in the reverse direction. With fine dust particles such as fly ash, reentrainment prevents this method from being completely effective.

Electrostatic Precipitators (ESPs)

In an electrostatic precipitator the suspended fly ash or dust particles are electrically charged by a high-voltage field. As a result, they are attracted by collector electrodes of opposite polarity. In wet ESPs, the collected dust is washed, while in dry precipitators a rapping device shakes the dust off the electrodes into a dust hopper. ESPs have a proven record of success on incinerator applications (Fig. 7.1). They can reduce particulate emissions to a level of 0.03 gr/scf (67 mg/m³). In other, industrial applications they have achieved even lower emission concentrations; for example, 0.01 gr/scf (22 mg/m³) for blast-furnace gas and 0.005 gr/scf (11 mg/m³) for tar removal from coke-oven gas.

ESPs are unique because the collecting force is applied only to the fly ash or dust being collected and *not* to the total gas stream. This and the low system pressure drop (under 1" H_2O) results in relatively low operating costs. The power requirement is 50 to 300 watts per 1,000 acfm, which on a yearly basis comes to a power requirement of 400 to 3,000 kWH per 1,000 acfm of capacity. At 10 cents per kWH, this corresponds to a yearly cost of $40 to $300 per 1,000 acfm of gas-discharge capacity. Added to this operating cost is the yearly maintenance of $20 to $50 per 1,000 acfm. ESPs can handle large gas flows (up to 4 million acfm) and can remove small particles at high efficiencies (90% to 99.5%). In contrast to other collection devices, they also remove tar and acid mists, and some special designs can operate at temperatures up to 1500° F (800° C) and pressures up to 700 PSIG (50 atmospheres) (51). In incinerator applications, chloride corrosion usually limits the operating temperature to 500° to 600° F (260° to 320° C).

ESPs require the highest initial investment of all collection devices—between $1 and $3 per acfm of capacity in larger sizes. Other disadvantages include the safety risks from using high-voltage equipment and the potential for a loss in collection efficiency when operating conditions change.

Operating Principles

The process that takes place inside the ESP consists of three fundamental steps: (1) electrical charging of the fly-ash particles, (2) collection, and (3) removal. Charging is achieved by passing a fly-ash-laden gas stream through a high-voltage electrostatic field. To ionize the gas, the strength of the electrical field must (at least locally) exceed the electrical-breakdown strength of the gas. This field is usually generated by unipolar high-voltage corona discharge, which is at a potential difference of 30,000 to 90,000 volts. (A corona is a local electrical discharge that does not propagate.) An advanced state of a corona discharge is sparking. This occurs when the gas is broken down along a particular path. Table 7.7 lists the sparking potentials of moist air at 100° F (38° C). Most industrial ESPs operate with a negative corona (the ionization electrode is operated at a negative polarity) because higher voltage differentials can be achieved without causing sparking.

The corona is usually established between a fine wire (the ionizing or discharge electrode) and a grounded plate or cylinder (the collector electrode is shown in Figs.

Table 7.7
SPARKING POTENTIALS IN
MOIST AIR (51)

Pipe Diameter (cm)	Sparking Potential, volts	
	Peak	Root Mean Square
10	59,000	45,000
15	76,000	58,000
23	90,000	69,000
30	100,000	77,000

7.35 and 7.36). The negative-discharge electrode attracts the positive ions formed in the corona glow. An equal number of negative ions formed in the corona are attracted to the ground electrode. Therefore most of the space between the two electrodes contains negative ions. Fly ash and dust particles are bombarded by these negative ions and become charged.

In the corona field of the ESP, particle charging occurs both by ion attachment (impact charging) and ion diffusion (diffusion charging). The diffusion mechanism charges particles under 0.2 micron, while the impact mechanism charges particles larger than 0.5 micron. The saturation charge that a particle can acquire increases linearly with the strength of the field and exponentially (by the square) with the diameter of the particle. The fractional approach to the saturation charge increases with the time the particle spends in the corona field. The diffusion charge also increases with residence time. Typically a 10-micron particle might achieve a charge of 30,000 electron charges, while a one-micron particle will obtain 300 electron charges.

Particle Collection

The separating force acting on the particle is the coulomb force, which is proportional to the product of the particle charge and the electric-field intensity. Doubling the charge doubles the coulomb force and cuts the required precipitator size in half. Thus the design goal is to maximize the particle charge. In cleaning industrial gases, the single-stage Cottrell design is used (Figs. 7.35 and 7.36), where the collecting field is a continuation of the corona. The use of two-stage precipitators

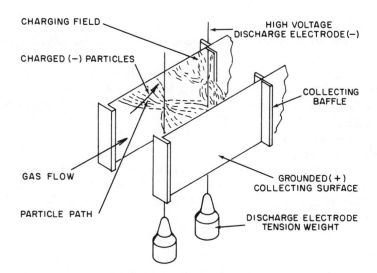

Fig. 7.35 Flat-surface electrostatic precipitator (51).

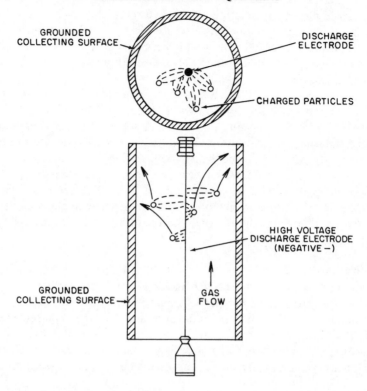

Fig. 7.36 Tubular-surface electrostatic precipitator (51).

is limited to applications where low ozone generation is necessary, such as in air-conditioning services.

The movement of particles toward the collection electrodes is assisted by the coulomb force and resisted by the viscous forces (Stokes' Law) of the gas. The migration velocity of the particle depends on the balance between coulomb and Stokes forces. Under typical precipitator conditions, a 0.5-micron particle has a migration velocity of about 30 mm/sec (1.2 inch/sec) (51). As load (volumetric gas flow) increases, the same particle will spend less time between the collector plates and therefore will have less time to migrate to them. Thus collection efficiency drops as volumetric gas flow increases. The following equation illustrates the relationship between collection efficiency (E), or the fraction of dust left in the gas (1–E), and residence time (66):

$$\log(1-E) = t \log K$$

Where:

t = residence time of the gas in the electrical field
K = apparatus constant, which ranges from 0.05 and 0.8

Because small variations in residence time result in large variations in collection efficiency, ESPs are sensitive to load variations and to overloads.

The collection of the charged particles can be upset by gas-flow changes, by variations in the corona, and by reentrainment of the collected particles. Another factor that can affect performance is the change in the resistivity of the fly-ash particles. A drop in resistivity tends to increase reentrainment. If resistivity rises above 2×10^{10} ohm-cm, the voltage and current at which the ESP operates must be reduced. This in turn reduces performance. Figure 7.37 shows the effect of temperature and moisture on fly-ash resistivity.

Adding sulfur trioxide at a dosage of 10 ppm has been used to reduce flyash resistivity. Ammonia has also been injected into the gas to modify particle resistivity (51). Another way to reduce fly-ash resistivity is to maintain constant collecting electrode temperatures and keep a thin film of water on the electrode surface.

The drift rate of the dust particle (W) is directly related to the corona power (Pc) and is inversely related to the collecting area (A). With a single power supply, the precipitator performance will increase as its size is reduced (51). This is because

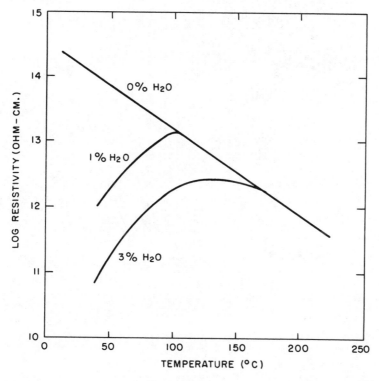

Fig. 7.37 Effects of temperature and humidity on particle resistivity of typical fly ash (51).

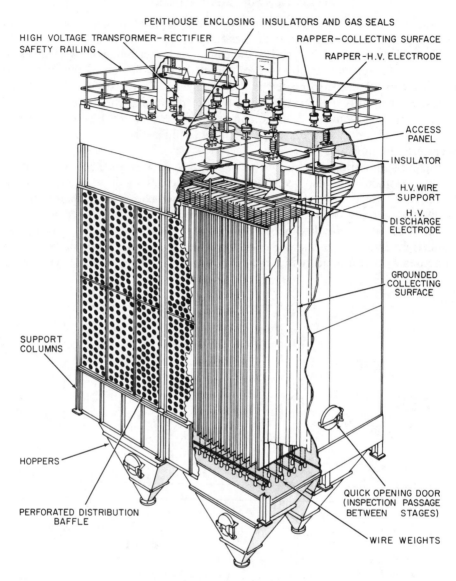

Fig. 7.38 Flat-surface electrostatic precipitator with accessories. (Courtesy of UOP Air Correction Division [51].)

258

at a fixed voltage the sparking rate and intensity increase with size. For this reason, it is desirable to divide the larger precipitators into electrically independent sections. Sectionalization also allows for adjusting to local conditions since different voltages can be maintained in the different sections.

Construction Details

A flat-surface ESP is shown in Figure 7.38. The precipitator shells and hoppers can be made of mild steel, alloy steel, lead-coated steel, plastic-lined steel, acid-resisting brick, cement, concrete, or aluminum. The discharge electrodes (Fig. 7.39) are about $\frac{1}{4}''$ in diameter and can be made of steel alloys, copper, aluminum, nichrome, silver, hastelloy, or monel. The shapes are selected according to application. For example, ribbon and twisted-rod designs are recommended for heavy dust services, where keeping the electrodes clean is a problem.

The desirable characteristics of the collector electrodes include high sparkover voltage, aerodynamic shielding to minimize reentrainment, high strength, and good mechanical rapping. They can be formed as perforated or expanded metal plates, as rod curtains, or as hollow electrode shapes (Fig. 7.40). In incinerators, plate-type collector electrodes are used for collecting dry solids as flyash. Pipe-type electrodes are used for collecting liquid droplets or for smaller gas-flow rates.

Several methods can be used for detecting particle reentrainment. The most effective technique is to examine particle size in the ESP's outlet gas; large or agglomerated particles are a clear indication of reentrainment. Another method is to plot the reciprocal of the volumetric gas-flow rate (Q) against the logarithm of the

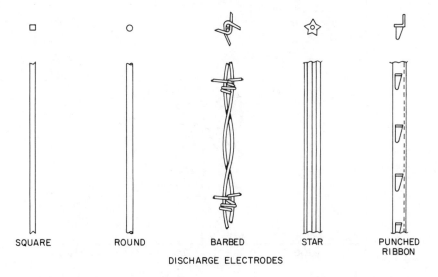

DISCHARGE ELECTRODES

Fig. 7.39 Electrostatic precipitator collecting plates and discharge electrodes (51).

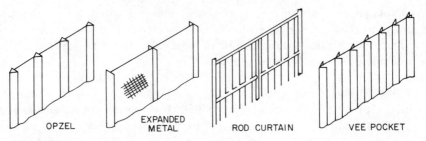

Fig. 7.40 Collecting plates (51).

fraction of the dust that is left in the gas $(1 - E)$. This plot of Q versus $\ln(1 - E)$ should yield a straight line (51). If the experimental data result in a curve instead of a straight line, particle reentrainment is occurring (Fig. 7.41). A third method is to plot migration velocity or precipitation rate (W) against volumetric gas-flow rate (Q) and compare the result to the theoretical relationship described by the equation (51) below, which also considers the collection surface area (A):

$$1/Q = -\ln(1 - E)/AW$$

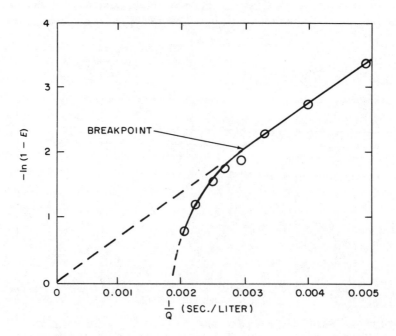

Fig. 7.41 Onset of reentrainment (51).

If the test data show that an increase in Q results in a sharper drop in W than predicted by the theoretical equation, then reentrainment is occurring. In this method, therefore, a loss in collection efficiency is detected and is interpreted as an indication of reentrainment.

The main protection against reentrainment is to protect the collection zones from the direct blast of the gas stream by shielding the collector plates (Fig. 7.42).

Gas Distribution

ESP performance is optimum when the gas flow is smooth and uniform, with little or no turbulence or eddy currents. Nonuniform flow reduces the overall collection efficiency because the increased efficiency of the low-velocity zones is *not enough* to compensate for the drop in efficiency in the high-velocity zones. In addition, reentrainment is likely to occur in the high-velocity zones. When gas-distribution is unbalanced, the collection efficiency can drop to 60%. Unbalanced flows can also cause reentrainment from the dust hoppers, bypassing the precipitation zone, and excessive sparking and high dust collection in regions where the gas velocity is low.

Gas-flow velocities can be measured by hot-wire anemometers, pitot tubes, or area-averaging pitot traverse stations, where higher precision is required (67). Gas-flow distribution and flow patterns also can be analyzed in scaled-down transparent mockups using visual gas-tracer injections or in wind tunnels.

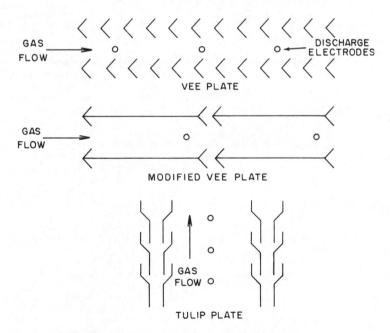

Fig. 7.42 Shielded collecting electrodes (51).

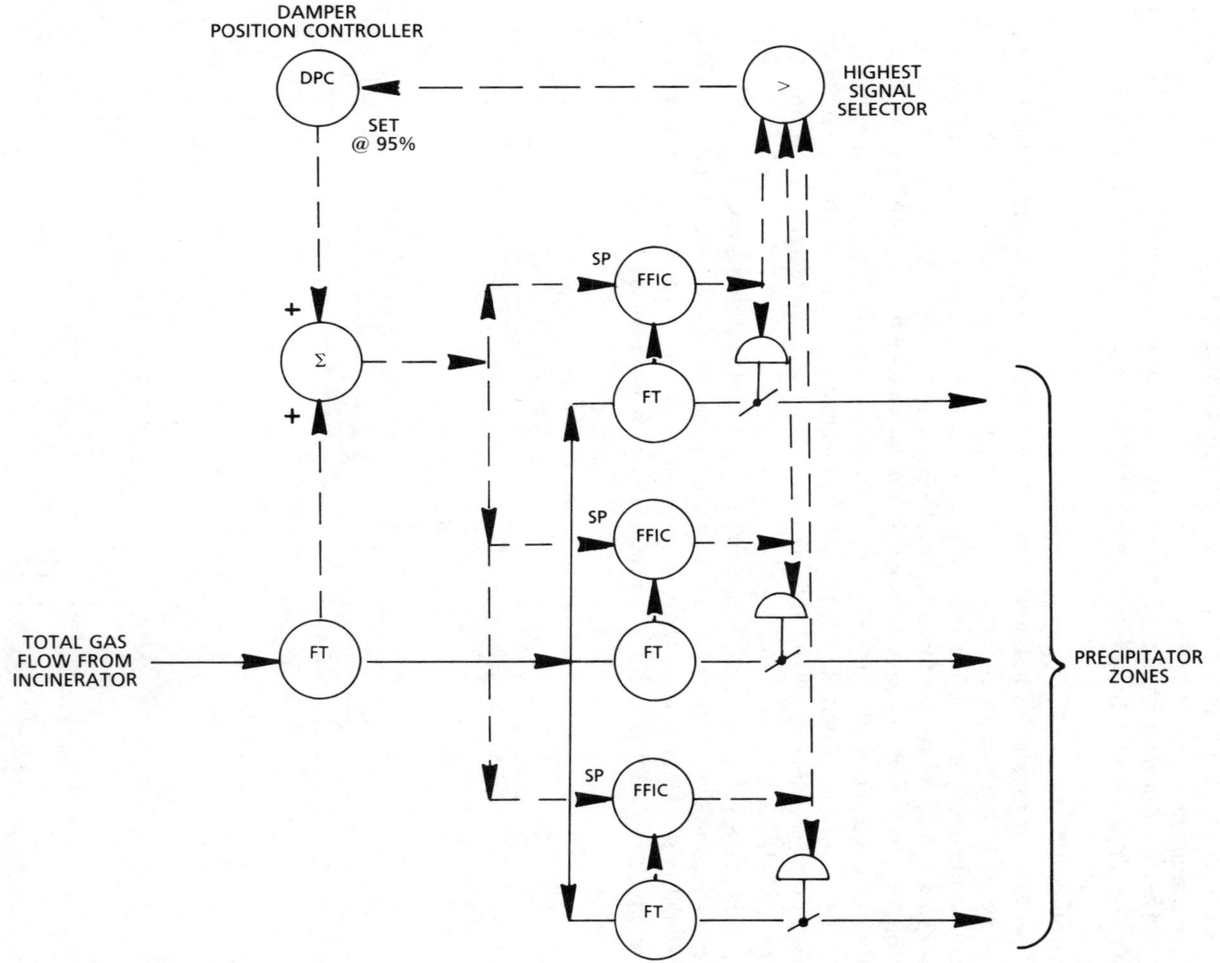

Fig. 7.43 Automatic gas-flow balancing controls.

Figure 7.43 shows a control loop that will automatically balance the flows into each of several zones. Each zone is provided with its own flow-ratio controller (FFIC), which in the illustrated case of three zones receives as its setpoint 33.3% of the total gas flow from the incinerator. The purpose of the damper position controller (DPC) is to keep increasing all the FFIC setpoints until the most open damper (selected by the high-signal selector) is 95% open. This way, accurate balancing is achieved at minimum transportation cost since all dampers are opened up until the most open damper is almost fully open. This minimizes pressure drop and therefore reduces the cost of operating the induced-draft fan.

The Industrial Gas Cleaning Institute defines gas distribution as acceptable when no local velocity reading differs by more than 40% from the average velocity of all zones and 85% of the zones are within 25% of the average. This requirement can be met and even exceeded by the flow-distribution balancing controls in Figure 7.43.

In addition to flow balancing by throttling dampers, gas-flow patterns also can be improved by using guide vanes, diffusion screens, plenum chambers and flue transitions. As shown in Figure 7.44, guide vanes protect against the deterioration of flow patterns in sharp turns. Normally three to five vanes are installed in a 3- to 6-feet-wide (1 to 2 meters) duct. They must be long enough to cover the projection of the opening of the approaching gas stream.

Diffuser elements are installed at the inlets and outlets of ESPs and serve to break up large localized swirls or eddies. These elements can be plates perforated

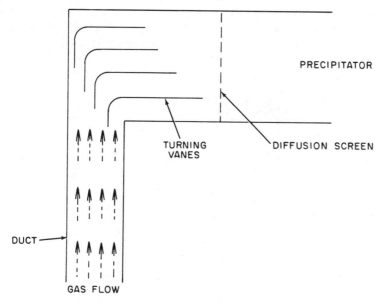

Fig. 7.44 Gas distribution system (51).

with 1- to 2-inch-diameter holes (24 to 50 mm) (Fig. 7.38) or provided with a grid of large rectangular openings that cover 25% to 50% of the area of a flat sheet. Diffusers should be provided with rappers so that they can be cleaned if they become covered or plugged by dust.

Solids Removal

In dry ESPs, rapping devices shake the particles from the collector plates into a hopper. Dry collection is usually assisted by the injection of a conductive gas, such as SO_3. In wet ESPs, water washes the collected particles from the plates. The collected particles must be removed before they reach a thickness of 6 to 25 mm to minimize reentrainment since the collected material will fall off as larger agglomerates.

Improper operation of the jarring, vibrating, or rapping devices can cause reentrainment or buildup on the discharge electrodes, which will affect the corona discharge. Discharge electrodes are usually cleaned by vibration, while impact or knocking rappers are used on the collector electrodes. The rappers can be pneumatic, motor-driven, or magnetic-solenoid devices (Fig. 7.38). The frequency and intensity of the raps can be most accurately adjusted with solenoid rappers. These settings are usually made by visual observation through sight ports provided with high-intensity illumination (Fig. 7.38). In installations where severe reentrainment problems exist, the section being rapped can be isolated during this cleaning period. This isolation is not necessary with well-designed ESPs.

Hoppers are usually located directly below the collector electrodes. Vertical baffles prevent the gas from bypassing the precipitator (Fig. 7.45). The slope of the hopper walls must exceed the angle of repose of the collected particulates so that they can slide down the hopper without bridging; a wall angle of 60 degrees is usually

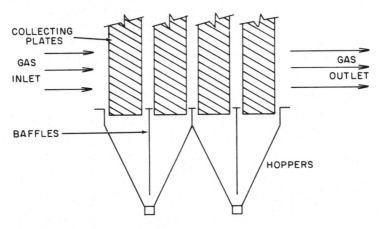

Fig. 7.45 Dust hoppers with baffles (51).

sufficient. The fly ash from the hopper can be removed through knife, rotary, or slide-gate valves or through screw- or belt-type star feeders or conveyors. Fly-ash removal is usually continuous and is under automatic solids-level control. Vibrators installed in the walls of the hoppers will ensure that the dust moves freely.

Electrical Controls

The key sensors provide readings on the high-voltage, current, spark-rate, and rectifier equipment that converts the alternating-current supply into direct current. Precipitator operation is optimized by maximizing the voltage at the discharge electrode, as shown in Figure 7.46. Some ESPs operate without sparking, while the majority perform best when the spark rate is between 50 to 100 per minute. Spark-rate meters provide the operator with readings on this important parameter.

In most existing incinerators, ESPs are operated under automatic-feedback control, maintaining either the current, voltage, or spark rate of the discharge electrodes. A more advanced control is to maintain all three simultaneously within their desirable operating ranges (22). ESP instrumentation should also include measurements of gas flow (67), temperature, pressure, and relative humidity. The balancing of flows between precipitator sections can be controlled by the loop described in Figure 7.43.

Flue-gas composition can be measured by continuous on-line analyzers. Since the intensity of the scattered light is proportional to the mass concentration of dust, photoelectric devices can be used to measure the cleanliness (dust content) of the

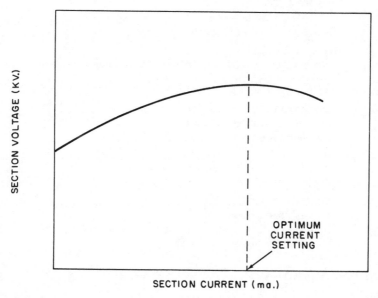

Fig. 7.46 Optimum electrical set adjustment (51).

outlet gases (23). Other instruments detect the intensity of the rappers and the level of the fly ash or dust in the hoppers (25).

ESP Troubleshooting

The specification for a new ESP usually includes the kind of data and requirements listed in Table 7.8. Performance is established on the basis of the difference between the mass of entering and leaving particulates. The filter-thimble technique is used for particle-sampling purposes on dry dusts. The method used to obtain representative samples is shown in Figure 5.50.

Troubleshooting can involve operational or mechanical problems or design errors. Due to the high-voltage sparks, precipitators should never receive combustible gas mixtures. Operating voltage has a direct effect on collection efficiency and electrode alignment affects that voltage. Off-center electrodes can cause as much as

Table 7.8
ELECTROSTATIC PRECIPITATOR SPECIFICATION
CHECKLIST (51)

Particle Properties:
  source; composition
  particle concentration; particle structure
  particle-size distribution; in situ resistivity

Gas Properties:
  total gas flowrate; composition
  pressure; corrosive properties
  temperature

Minimum Collection Efficiency Required

Mechanical Details:
  types of access facilities for maintenance and inspection
  structural details (dimensions and weights)
  number of dust hoppers
  maximum pressure drop through system
  effective collection plate area
  maximum precipitation zone gas velocity
  type of rapping
  type of wire for discharge electrode

Electrical Details:
  type of rectifiers
  type of energization, half-wave or full-wave
  type of stabilizing circuit; resistance, monocyclic,
  saturable reactor
  rectifier power input
  corona power, maximum and average expected
  number of high-voltage bus sections

Control System: Current, Voltage or Spark Rate Control

Table 7.9
COMMON ELECTROSTATIC PRECIPITATOR
PROBLEMS (51)

Operational Problems:
  poor rapper adjustment, intensity or frequency
  poor electrical adjustments
  excessive gas rates during process upsets
  excessive particle concentration in gas
  overflowing dust hoppers

Mechanical Problems:
  poor electrode alignment
  vibration or movement of discharge electrodes
  excessive dust deposits on electrodes
  large dust deposits in gas ducts
  air leakage into system

Basic Design Problems:
  undersized precipitator
  high-voltage set instability
  inadequate rapping equipment
  turbulent gas flow
  re-entrainment
  high-resistivity particles
  inadequate electrical equipment

a 15% drop in operating voltage. Some of the common problems found during troubleshooting are listed in Table 7.9.

ESPs are one of the more promising candidates for pollution control in incinerator applications. One advantage is that many incinerator installations are already in operation. Others include their low operating costs and reasonable collection efficiencies on small fly-ash particles (90% to 99.5%). Their limitation to operating temperatures of 500° F (260° C) because of corrosion are similar to the temperature limits on baghouse filters. Their disadvantages in relation to baghouse filters include their high initial cost, sensitivity to load variations, and lower collection efficiencies (Fig. 7.1).

Removing Acidic Pollutants

The acidic pollutants emitted by incinerators include HCl, HF and SO_x. Sulfur oxides cause acid rain and can damage the leaves of sensitive vegetation at concentrations as low as 0.3 ppm. Coal- and oil-burning power stations are the main sources of SO_x emissions. Municipal incinerators do not discharge large quantities of SO_x because the sulfur concentration of MSW is low. The main cause of acid-gas emissions from incinerators is the burning of plastics, which can result in the discharge of hazardous concentrations of HCl and HF.

Permitting acid-gas emissions from incinerators is based on BACT consistent with regulatory agency definition. For SO_2 and HC1, the EPA considers BAT as 90% to 95% collection efficiency (202). In Germany the limit for sulfur dioxide is 200 mg/m^3 (75 ppm), for hydrogen chloride 50 mg/m^3 (33 ppm), and for fluorides 5 mg/m^3 (see Table 6.7). The unabated incinerator flue gas contains 700 ppm of HCl, 150 ppm of SO_2, and a few ppm of HF (Tables 6.5 and 6.8). In addition to environmental and health hazards, the acidic fumes will also corrode the metallic components and thereby reduce the life and increase the operating cost of the incinerator.

In the past, the ground concentrations of acidic gases were controlled by dispersing them through tall stacks. Today, in most areas, this practice is no longer acceptable. Removal can be achieved by either dry or wet scrubbing methods. The main disadvantage of wet scrubbing is that it cools down the flue gases, which in turn necessitates the use of regenerative heat exchangers to reheat the gases. Because the scrubbing liquor is highly corrosive, special corrosion-resistant materials, such as glass-tube heat exchangers, must be used on these regenerative units (53).

Acid removal through dry scrubbing has been pioneered at the Tennessee Valley Authority, where pulverized coal and limestone (or dolomite) was injected into the combustion zone of boilers. This process removed over 99% of the SO_x in the flue gases and proved to be the most reliable and least costly of the sulfur oxide removal techniques. The success of direct lime injection into the boilers also prompted their use on incinerators. The dry scrubber is located between the furnace and the baghouse filter (Figs. 6.5 and 6.8). The anticipated removal efficiencies of some of these pollution-control devices are given in Table 6.21.

Dry and Wet-Dry Scrubbers

The acid-neutralizing agent can be introduced into the incinerator flue gases either as an injected powder or as an atomized slurry spray. Both systems have been operated successfully (47, 57). In the *"dry"* system the hot gases are first cooled by a water spray and then are introduced into the bottom of a cyclone, where dry, hydrated lime powder is injected into the gas stream countercurrently (the second scrubber in Fig. 6.8). The flue gas, containing both fly ash and lime, is then sent to a fabric-filter dust collector. In addition to efficiently removing particulates and acid gases, this combination also removes dioxins (Chapter 6).

In a "wet-dry" system, a finely atomized lime slurry is sprayed into the hot gas stream, which simultaneously cools and neutralizes the stream while the slurry droplets dry into solid particles (the first scrubber in Fig. 6.8). The reagent can be lime or soda ash (sodium carbonate), which can be atomized either by a high-speed rotary atomizer or by pressurized air nozzles. The chemical reaction between the incinerator flue gas and the fine mist causes fine particulates to form (Fig. 6.5). The evaporation of the slurry cools the exhaust gases to a range of 250° to 400° F (120° to 200° C) at the outlet of the spray drier (57).

If baghouse filters are used for particulate removal, lime scrubbers are a good choice for acid removal. Or lime scrubbers can be followed by ESPs, but they cannot match the combined performance of a dry scrubber followed by fabric filters. The SO_x and HCl removal efficiency of a spray dryer alone is 70% and 90%, respectively.

When followed by a fabric filter, the unreacted lime collects on the bags and provides a secondary surface for acid neutralization. The combined system achieves an SO_2 removal efficiency of 80% to 90% and an HCl removal efficiency of 90% to 95% (57). Another advantage of the combined system is the resulting greater utilization of the reagent and hence a reduction in the lime supply and disposal. The combination also protects the baghouse, as the scrubber cools the gases, which makes the use of less expensive filter fabrics possible.

Dry Absorption Processes

Another acid-removal technique is to inject the neutralizing reagent powder directly into the combustion zone of the furnace. The process involves feeding pulverized raw limestone or dolomite into the hot furnace gases at a location where maximum distribution of the particles can be achieved. The heat of combustion converts the injected limestone or dolomite into the CaO and MgO forms and then reacts to form sulfates with the SO_2 of the flue gases:

$$CaCO_3 + Heat = CaO + CO_2$$

and

$$CaO + SO_2 + 0.5(O_2) = CaSO_4$$

The solid-reaction products are carried with the fly ash into the particle-control equipment. The dry limestone injection process is relatively inexpensive and easy to operate, but it does have some limitations: CaO is capable of reacting with SO_2 *only* in the temperature range of 1200° to 2300° F (650° to 1250° C), and *only* if sufficient time is available for the reaction. Figure 7.47 shows the temperature ranges at which CaO and MgO are effective scavengers of SO_2. Because incinerators operate at around 1800° F and because the residence time of the gas in this high-temperature zone is short, the limestone must be dispersed early and well enough to maximize contact with the combustion gas. The powder dispersion system should be designed for a gas velocity in this zone of around 60 fps (20 mps).

Tests on boiler applications (68) used ball mill pulverized dry limestone with 70% of the particles smaller than 200 mesh. The powder was injected by pressurized air at controlled velocities of 33 to 67 fps (10 to 20 mps) through 2.5-inch-diameter (60 mm) nozzles. At maximum injection velocity and at a limestone-to-sulfur ratio of twice stochiometric, the weight ratio of air to limestone is 1:3. A total of sixteen nozzles were used in this test (Fig. 7.48). When dry limestone injection was used in combination with ESPs, the load on these units increased severalfold because of the large quantity of lime added to the fly ash; in addition, the collection efficiency dropped due to the increased resistivity of the particles as SO_2 was removed.

The main advantage of dry limestone injection for SO_2 removal is that it does not significantly reduce flue-gas temperature or velocity and therefore does not

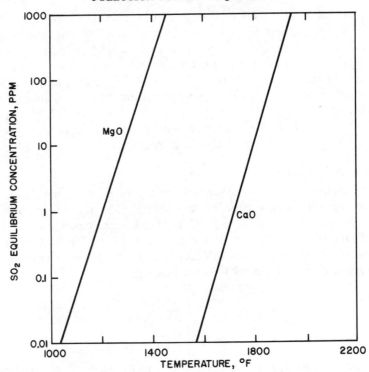

Fig. 7.47 Sulfur dioxide concentrations in equilibrium with CaO and MgO at furnace temperatures (68).

interfere with plume buoyancy. Also, because the process is dry, there is little corrosion in the superheaters, air heaters, and other metallic equipment. The main limitation is in the difficulty to achieve maximum dispersion of particles within the correct temperature zone and for a time period sufficient to complete the reaction of limestone with SO_2. In incinerator applications, acidic gases are present in forms other than SO_2, which adds another limitation of this technique.

Many other dry absorption processes using materials other than lime are too expensive to consider for incinerator applications. One process, using sodium bi-carbonate, which also reacts with HCl or HF, can be regenerated and injected after the last heat recovery unit because it will react with SO_2 at low temperatures.

Nitrogen Oxide Control

The term NO_x refers to the combined total of all forms of nitrogen oxides. The emission of NO_x is undesirable because it contributes to the destruction of the ozone layer in the stratosphere. It also triggers a series of reactions that produce photo-

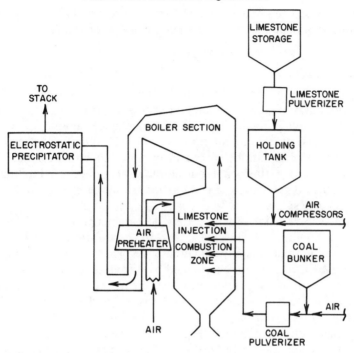

Fig. 7.48 Dry limestone injection (68).

chemical smog. NO_2 is lethal to most animal species at concentrations exceeding 100 ppm.

The incineration of each ton of MSW releases about two pounds of NO_x into the atmosphere. The concentration of uncontrolled NO_x emissions from incinerators ranges from 150 to 300 ppm of dry flue gas, corrected to 3% O_2 (see Table 6.5). These values are not much higher than the limits set by various emission standards and regulations, usually around 225 ppm (57). When NO_x emission controls are incorporated into the incinerator design, the emission concentration is reduced to about 50 ppm (see Table 5.18).

The NO_x concentration in incinerator emissions can be reduced by modifying the combustion and flame-propagation processes, by flue-gas recirculation, improved excess-oxygen control, catalytic decomposition, and ammonia injection. Most of these techniques are relatively new, so operating experience is limited. NO_x generation increases as the combustion temperature rises and as the amount of excess air increases. The combustion-modification technique, which redistributes the combustion air to reduce the formation of NO_x has not been investigated extensively in the United States but was found to be effective in Japanese incinerator installations

(57). The catalytic decomposition techniques have not been adequately demonstrated, while ammonia injection has been applied in only a few cases (58).

Furnace Modifications

The combustion-modification methods that are known to reduce the emissions of NO_x include: (a) operating at minimum excess air, (b) completing combustion in two stages, (c) recirculating the flue gas, (d) reducing the amount of air preheat, and (e) injecting steam or water into the furnace. NO_x emissions can be reduced by about 25% by reducing the amount of excess air provided to support combustion (20). In incinerator applications, this is not a desirable strategy because limiting oxygen can interfere with the completeness of combustion, resulting in an increase in unburned MSW and CO emissions.

Two-stage combustion is achieved by providing a primary and secondary combustion chamber (Figs. 5.4, 5.9, and 5.26). In these furnaces 90% to 100% of the theoretical air requirement is introduced into the primary combustion chamber. The complete burning of the fuel is guaranteed by introducing additional, secondary air in the secondary combustion chamber, where the temperature is lower and therefore NO_x formation is limited by kinetics. Such two-stage designs have achieved up to 90% reductions in NO_x emissions.

In one incinerator installation (Fig. 5.29), some 260 pressurized air nozzles are placed near the region of maximum velocity and temperature. The main purpose of this design is to maximize mixing and therefore combustion efficiency, but a side benefit can be reduced NO_x emissions (34). In some cases the cooling effect of the air jets might suffice to prevent overheating, while in others supplementary water or steam cooling might be needed. The air-jet manifold should be made of high-alloy steel to protect it from corrosion caused by high-temperature chloride attack (31).

NO_x formation can also be reduced by recirculating the cooled flue gas and reintroducing it into the combustion zone. This reduces the maximum combustion temperature, which in turn limits the formation of NO_x. The relationship between flue-gas recirculation and NO_x concentration has been studied on oil- and gas-fired boilers (Fig. 7.49), and it is likely that similar relationships will be found on incinerators. Reducing the preheating of the combustion air also reduces combustion temperature and therefore NO_x formation, but the price of this method is a loss in thermal efficiency. The injection of water or steam has similar effects and limitations.

Two-stage combustion and flue-gas recirculation can be combined to achieve higher NO_x reductions than can be obtained from only one method. In general, NO_x will be reduced when excess oxygen and combustion temperatures are reduced. Tangential firing, where the furnace itself is used as the burner, also tends to reduce flame temperatures and hence NO_x formation. The most efficient heat-transfer rates, and therefore the lowest combustion temperatures, are obtained in fluidized-bed combustion processes.

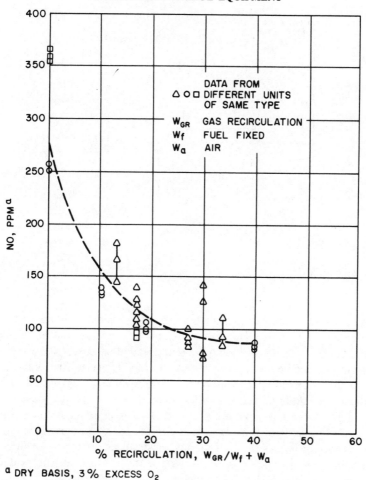

Fig. 7.49 Gas recirculation with natural gas firing: 320-MW corner-fired unit (20).

Catalytic NO$_x$ Control

Furnace modification minimizes the *formation* of NO$_x$, while catalytic decomposition reduces the quantity of *already-formed* nitrogen oxides. The decomposition of NO$_x$ on a catalytic surface is not easily achieved because the concentrations are low, the gas volumes are high, and interfering substances (moisture, CO, SO$_x$, and oxygen) are present. None of the catalytic NO$_x$-removal processes is commercially available, and most require the conversion of the less reactive NO into the NO$_2$ or N$_2$O$_3$ forms.

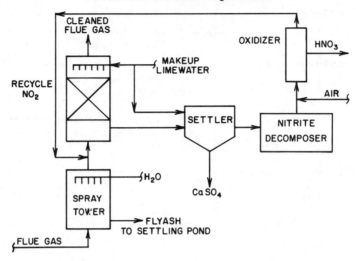

Fig. 7.50 Lime scrubbing for NO_x removal (20).

Catalytic reduction necessitates either the use of air deficiency or the injection of reducing agents (CO, H_2, CH_4). Both approaches reduce thermal efficiency and increase the emissions of the reducing substances. Physical separation of NO_x is not a likely solution either because the properties of NO_x are not sufficiently different from those of the other flue-gas components. Solids such as activated carbon, silica gel, manganese, or alkalized ferric oxide do adsorb NO_2, but physical separation is impractical due to the attrition of the sorbent. Similarly, adsorption by liquids (such as sulfuric acid) has not yet proved to be economical or technologically feasible.

Lime scrubbing is an effective means of simultaneously absorbing both NO_x and SO_x. In the process shown in Figure 7.50, the calcium nitrite is decomposed and NO is oxidized into NO_2; the by-products are gypsum and nitric acid. Other processes use magnesium hydroxide or acidic urea solutions as the scrubbing media, which are more expensive than lime.

Ammonia Injection: Thermal DeNO$_x$

Ammonia injection is a selective, noncatalytic NO_x reduction process that was invented by Exxon Research in the 1970s (58). NH_3 is injected into the combustion gases at a point where their temperature is between 1600° and 2200° F (870° and 1200° C). The ammonia reaction occurs between 1600° and 1650° F (870° and 885° C) and so the injection of ammonia must take place at a point in the furnace where the temperature is above this range. Here the ammonia reacts with the NO_x to reduce it to the N_2 form. About sixty power-plant installations have achieved NO_x reductions between 60% and 80%. Incinerator applications of ammonia injection are few and recent and efficiencies are estimated to range from 20% to 50% (57).

Ammonia is injected into the hot flue gas either by air or steam jets (at 5 to 40 PSIG pressure) at points where both the reaction temperature and residence-time requirements are satisfied. In the temperature range of 1300°–2200° F (700°–1200° C), the reaction is:

$$NO + NH_3 + O_2 + H_2O + (H_2) \rightarrow N_2 + H_2O$$

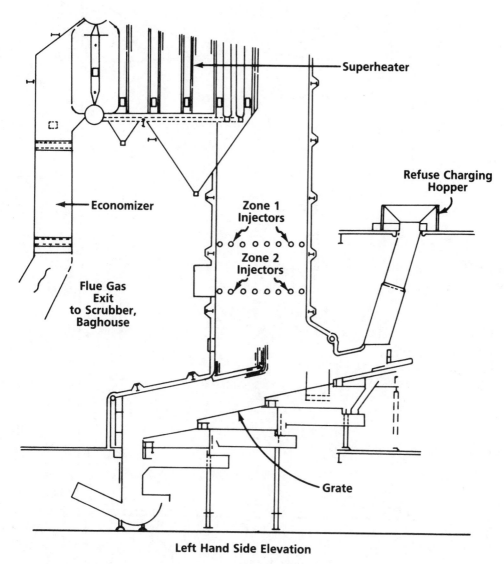

Fig. 7.51 Thermal DeNO$_x$—300 tons/day municipal refuse incinerator (58).

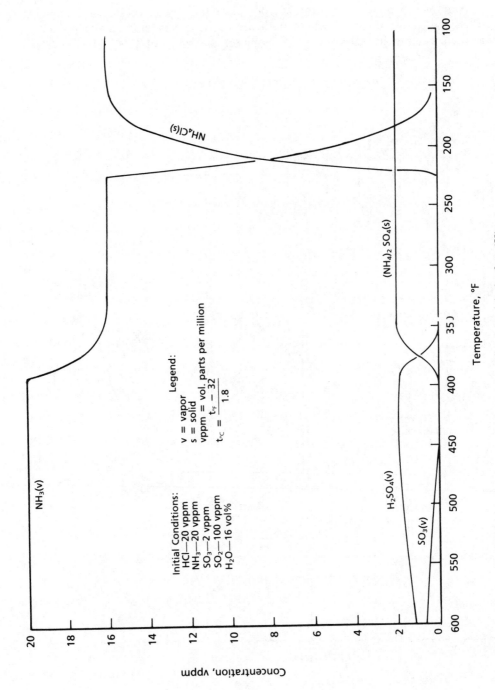

Fig. 7.52 Equilibrium reactions in cooled flue gases from combustion of municipal refuse (58).

The injection of hydrogen is required if the temperature range of the DeNO$_x$ reaction is reduced to 1300° F (700° C). At temperatures above 2200° F (1200° C), the addition of ammonia alone is sufficient, as the following reaction governs the process: $NH_3 + O_2 + H_2O \rightarrow NO + H_2O$.

In larger incinerators, it is difficult to achieve a good mix of ammonia with the flue gases. Two sets of ammonia injectors are provided, which consist of large jets located at the walls of the injection zone (Fig. 7.51). The jets in Zone 1 are for high loads and the jets in Zone 2 are for low loads.

There are some concerns about the consequences of the thermal DeNO$_x$ process because of the residual NH_3 in the flue gas. Ammonia can react with H_2SO_4 and HCl to form ammonium sulfate, ammonium bisulfate, and ammonium chloride. These salts can foul and corrode heat-recovery equipment, scrubbers, and baghouses and also can contribute to plume formation. Fortunately the sulfur concentration in MSW is low and the salts formed are water-soluble and can be removed by washing. For a typical MSW incinerator, the equilibrium reactions in the low-temperature flue gas are shown in Figure 7.52. Figure 7.53 gives the material balance for a thermal DeNO$_x$ system for a 1,000 ton/day MSW incinerator. The operating cost per ton of refuse is estimated at $1.25 (58).

In a test on an industrial boiler package (69), the NO$_x$ reduction was found to be 41%, while the NH_3 emission increased from 11 to 430 ppm and the particulate emission increased substantially. Such results show that ammonia injection systems are not yet capable of tracking load and process variations and the user cannot expect firm guarantees on system performance. Furnace modification and flue-gas recirculation techniques, however, have been used successfully, achieving up to 90% reduction in NO$_x$ emissions.

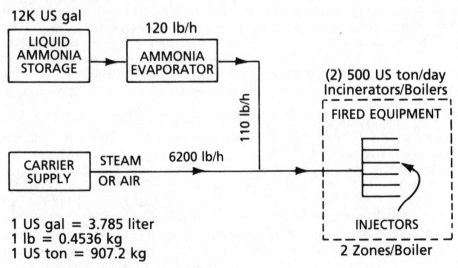

Fig. 7.53 Simplified thermal DeNO$_x$ supply system flow diagram (58).

Chapter 8
Heat-Recovery
Incinerators (HRIs)

□
■ ■

THE TRADITIONAL APPROACHES to solid waste disposal are dumping, burning and burial. In recent decades, it has become clear that longer-range, more permanent, and less polluting solutions are needed. The only solution is to recover and recycle the raw materials in the MSW. This recognition has resulted in public resistance to building incinerators and public support for recycling. Partly in response to this change in public opinion and partly for economic reasons, the incinerator industry has come up with a compromise solution: "resource-recovery plants." These plants are usually mass-burn incinerators with heat recovery and therefore are referred to here as heat-recovery incinerators, or HRIs. The steam generated in the HRI is converted into electricity and sent to the local utility for distribution. Materials-recovery plants, which separate out various materials from the MSW, are described in Chapter 10.

Unfortunately there is a basic conflict between recycling and incineration. In December of 1990, the Bush Administration had overruled the EPA recommendation, which would have required 25% recycling of MSW received at incinerators (200). The state of New York, for example, plans to reduce its MSW production by 50%, through recycling, by 1997. At the same time, the state also plans to incinerate 50% of its MSW by 1997 (186). These goals are in conflict, because it is the combustible fraction of the MSW (paper, plastics, yard wastes) that is most likely to be removed through recycling and composting. The state of Maine faces a similar conflict between its 50% recycling goal by 1994 and its contractual obligations with its incinerator operators. The problems of the city of Hempstead, Long Island, illustrate the resulting conflict: If Hempstead meets its recycling and composting goals, it will not
278

have enough MSW to meet its contract with its incinerator operator. Today, Hempstead spends some $140/ton to send its incinerator ash to Buffalo, New York, because it has no landfill capacity for it. Therefore, what the Hempstead city fathers are looking for is a city that would send its MSW to the undersupplied Hempstead incinerator, and in exchange would accept Hempstead's incinerator ash. Not an easy task.

The public is concerned about both the cost and the health aspects of MSW incineration. Specific concerns are the emission of dioxin, heavy metals, and acidic gases, as well as the greenhouse effect of carbon dioxide emissions. There is yet no federal standard for dioxin emissions, and estimates of cancer cases caused by dioxin emissions from incinerators vary widely (see Chapter 6). Some states are beginning to enforce their own mercury emission limits for incinerators (185). The lead and cadmium content of incinerator ash frequently exceeds federal limits (202). In fact, if EPA recommendations were followed, the ash would have to be placed in toxic-waste dumps. Yet, for reasons of economy, most states are treating incinerator ash as nonhazardous. In some areas hazardous-waste dumps, with double linings and leachate collection and monitoring systems (Chapter 3), simply do not exist. New York City, for example, has no hazardous-waste landfill at all and is left with only one major nonhazardous landfill, at Fresh Kills on Staten Island.

For these reasons, the public in some areas is opposed to incinerators. In Lowell, Massachusetts, and in San Francisco, well-advanced plans for building incinerators have been scrapped. In San Diego, voters passed a zoning amendment that effectively bars building incinerators within a certain distance from schools. While these are telling cases, the building of new incinerators is expected to accelerate.

The Status of the HRI Industry

In 1986, 80% of the MSW generated in the United States was landfilled, 11% was recycled, 3% was incinerated without heat recovery, and 6% was sent to HRIs (156). By 1990 recycling had reached 13%, and the National Solid Wastes Management Association predicts that incineration will reach 20% by 1992 (185). This compares with only 1% in 1970. Today, 14% of MSW is being incinerated (200) and there are some 500 mass-burning HRIs around the world (31). At the end of 1990 there were 130 HRIs and 50 regular incinerator plants in operation in the United States, 31 under construction, and 74 in the planning stages (185). (For a list of HRIs in the United States, see Tables 5.4 and 8.1; a typical HRI is shown in Figure 5.5.) More than one hundred additional HRIs are to be built in the United States before the turn of the century, at an estimated cost of nearly $30 billion.

In 1980 HRI plants handled 2.7 million tons of MSW, and it is projected that by 2000 they will handle 32 million tons (Fig. 8.1). In 1990, 25 million tons of MSW were incinerated in the U.S. (200). According to Kidder, Peabody & Co., the capacity of operating HRI plants in the United States was 50,000 tons/day in 1988, and an added 85,000 tons/day capacity was on order, of which 35,000 tons/day capacity was

Table 8.1

HEAT RECOVERY INCINERATORS IN THE UNITED STATES, 1983 (18)

Plant location	No.	Size (tons/day)	Status	Type
Akron, OH		900	Oper. 1983	RDF w/suspension burning
Albany, NY		600	Oper. 1981	RDF w/suspension burning
Ames, IA		200	Oper. 1975	RDF w/suspension burning
Braintree, MA	2	120 each	Oper. 1970	Mass burning, water-wall furnace
Chicago, IL Northwest	4	400 each	Oper. 1970	Mass burning, water-wall furnace
Columbus, OH		2,000	Start-up 1983	RDF w/suspension burning
Dade Co., FL		3,000	Oper. 1983	RDF w/semisuspension burning
Gallatin, TN	2	100 each	Oper. 1981	Mass burning, water-wall furnace
Hamilton, Ontario	2	500 each	Oper. 1972	RDF w/semisuspension burning
Hampton, VA	2	100 each	Oper. 1980	Mass burning, water-wall furnace
Harrisburg, PA	2	360 each	Oper. 1973	Mass burning, water-wall furnace
Hempstead, NY	2	350 each	Oper. 1974	Mass burning, water-wall furnace
Lakeland, FL Oceanside		300	Oper. 1983	RDF w/suspension burning
Montreal, Quebec	4	300 each	Oper. 1970	Mass burning, water-wall furnace
Nashville, TN	2	360 each	Oper. 1974	Mass burning, water-wall furnace
Niagara Falls, NY		2,200	Start-up 1981	RDF w/suspension burning
Norfolk Navy Station, VA	2	180 each	Oper. 1967	Mass burning, water-wall furnace
Pinellas Co., FL	2	1,000 each	Oper. 1983	Mass burning, water-wall furnace
Quebec City, Quebec	4	250 each	Oper. 1974	Mass burning, water-wall furnace
Saugus, MA	2	600 each	Oper. 1975	Mass burning, water-wall-furnace

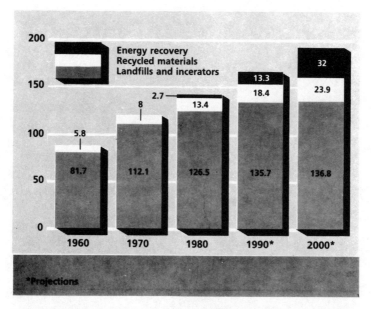

Fig. 8.1 Discards and recovery of household, commercial, and industrial waste in millions of tons. *Projections: Note that the actual total in 1988 was 180 million tons. Of this, 14% or 25 million tons were burned for energy recovery and 13% or 23.5 million tons were recycled. The recycled portion was 78% paper, 9% metals, 6% glass and 7% other material (199). (Franklin Associates, Ltd. [156].)

under construction. In addition to new orders, lately a fair number of cancellations have also been received (155).

Some reports suggest that HRIs can generate enough power to meet 4–9% of total energy requirement (78). Such numbers seem to be exaggerated. If one estimates the total yearly energy consumption in the United States as 8×10^{16} BTUs and the total yearly quantity of MSW as 1.8×10^8 tons, this represents a potential that covers little more than 1% of our total energy needs. (HRIs generate between 1.5 and 3.0 tons of steam per ton of MSW, which corresponds to about 3 to 7 million BTUs per ton of MSW. For the heating value of MSW, see Figures 1.8 and 5.30.)

MSW-based contribution to the total electric power consumption in the U.S. is similarly small. Yearly consumption is around 3.0×10^{12} kWH. Assuming an average electric-power generation of 500 kWH/ton of MSW, the total yearly electric-energy potential of all the MSW in the United States is 9.0×10^{10} kWH. This amount of electricity would be sufficient to cover 2% to 3% of the country's total electric-power consumption. These calculations are based on the assumption that 100% of the produced MSW is incinerated, when in fact in 1986 only 6% of the MSW was being incinerated in HRIs (156).

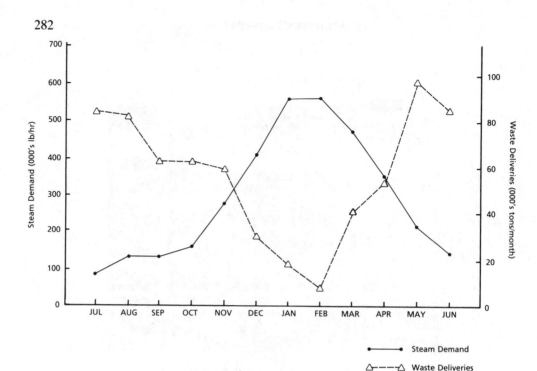

Fig. 8.2 Steam demand and waste-delivery patterns (72).

HRI Costs

The cost of HRIs is a major concern. The construction cost of incinerators has risen from $5,000 to well over $100,000 per ton/day of MSW-burning capacity in the last twenty to twenty-five years (Tables 5.3 and 5.4). In 1977 the total operating expense in fifteen mass-burning HRI units in Europe averaged $24/ton. Of this sum, 50% was spent on interest and depreciation, 33% on operation and maintenance, and the rest on ash disposal, tax, insurance, and overhead. From the total cost of $24/ton, $8/ton was recovered through the sale of energy (31). While the capital investment is still a large cost element, the operating and disposal costs also have increased—and these increases have not been compensated by the income from the sale of recovered energy or salvaged materials. In 1981, for example, at the Windham Energy Recovery Facility in Connecticut, the tipping fee was $7.50/ton. By 1988 the tipping fee had risen to $53/ton, and a rate rise to $82/ton was already being planned. This trend will no doubt continue, because Connecticut has instituted a new rule for ash disposal—requiring lined landfills with leachate collection—that will further increase costs.

Another, somewhat unexpected economic problem is related to the sale of the recovered energy. In 1983 Connecticut passed a law requiring utilities to purchase the power generated by resource-recovery plants at the "municipal rate," which was 8.5 cents per kWH in 1988 and had risen to 10.5 in 1990. The Connecticut Light and Power Company is reported to be suing to overturn this order. If the law is upheld in the courts, the electricity revenues in a few years will outstrip the maintenance costs, and once the original investment is paid off, the tipping fees will start to drop. If the law is upheld, the tipping fee projection for the year 2014 will be $22/ton of MSW, but if the law is overturned and the schedule proposed by the utility is accepted, the tipping fee will reach $147 per ton by 2014. Such legal decisions will have a major impact on the economics of HRIs.

If the recovered energy is distributed in the form of steam or hot water instead of electricity, there are still problems. The first is finding a customer for the steam. The second is that the piping network required to distribute the steam usually exists only in densely populated areas, while incinerators preferably are located in more remote areas. Even if the incinerator is located near the potential steam users—say, on a university campus, which heats its buildings with waste heat—seasonal heat-load fluctuations become a problem. If the primary use for the steam is in central heating, the steam-demand curve might look something like the one in Figure 8.2. Even though the demand for steam is greatly diminished during the warm months, MSW collections are at their maximum. Even with an ideal proximity between MSW source and steam user, it is likely that some of the steam will not be saleable unless it is converted to electricity.

HRI Design and Performance

MSW can be burned in several different ways:

1. Mass-burning the as-received waste.
2. Burning refuse-derived fuel (RDF) after the noncombustibles have been mechanically removed from the MSW.
3. Burning thermochemically processed refuse, where gas is produced by blowing air or oxygen through the fluidized-bed MSW gasifier (pyrolysis).
4. Burning the gases generated by biochemically processed waste (anaerobic digestion).

In each of these processes, the amount of energy recovered is less than the total heating value of the waste. Table 8.2 gives the amounts of energy that can be obtained through combustion in each of these processes, assuming that the heating value of the MSW is 4,400 BTU/lb. The table also indicates the percentages of energy lost due to the energy needs of the processing equipment and the inefficiency of combustion.

Table 8.2
HEAT RECOVERED BY VARIOUS PROCESSES FROM MSW,
HAVING A HEATING VALUE OF 4,400 BTU/LB (18)

Process	Energy Loss, % Processing	Combustion	Total	Total Net Available Energy (BTU/lb)
As received	1	39	40	2,640
Dry shredding	18	30	48	2,288
Wet shredding	35	21	56	1,936
Pyrolysis, oil	62	9	71	1,276
Pyrolysis, gas	32	25	57	1,892
Pyrolysis with oxygen	37	15	52	2,112
Anaerobic digestion	72	6	78	968

The types of HRI designs in use today include:

1. Mass-burning in refractory-walled furnaces, where the waste-heat boiler, located downstream of the furnace, receives the heat from the flue gases (Figs. 8.3 and 8.4).
2. Mass-burning in waterwall furnaces, where most or part of the refractory in the furnace chamber is replaced by waterwalls made of closely spaced steel tubes welded together to form a continuous wall (Figs. 5.5 and 8.5).
3. Combustion of RDF in utility boilers. The RDF is often burned in a partially suspended (fluidized) state, where some of the RDF stays on the grate during combustion (Table 8.1).

Older HRIs tend to be refractory-walled designs, and their steam production is usually limited to 1.5 to 1.8 pounds of steam per pound of MSW burned. In the newer waterwall and RFD-suspension designs, the steam production is around 3 pounds per pound of MSW. The increase in efficiency is mostly due to the reduction in excess air (from about 150% to about 80%). This reduces heat loss through the chimney. In the older, refractory-walled design, large quantities of excess air were needed as a cooling medium to control the combustion temperature. In the two newer designs, temperature control can be achieved without excess air.

Refuse-Derived Fuels (RDFs)

Fuel preparation is common in all fossil-fuel plants. Coal, for example, is first purified by washing and then is pulverized into a powder to make it more combustible. It is also desirable to convert the rather heterogeneous MSW into a more uniform fuel by removing the noncombustibles and reducing particle size through shredding. In such "front-end" processing, the capital cost of incineration is lower because the furnaces and grates can be made smaller and less rugged. The combustion efficiency increases because the shredded fuel requires less excess air to achieve complete

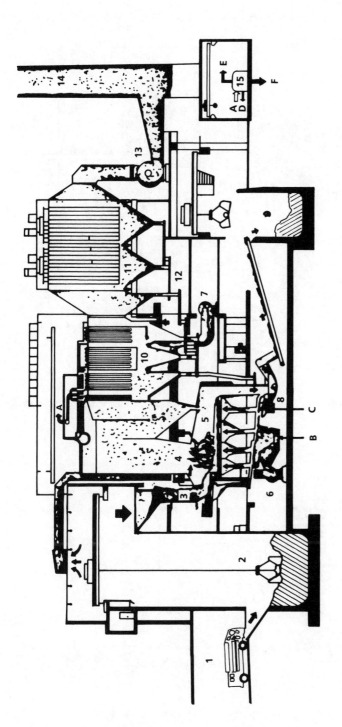

1 Tipping area
2 Waste bunker with charging crane
3 Charging hopper and feeding device
4 Combustion chamber
5 W + E grate
6 Combustion air system with air preheater
7 Secondary air system
8 Residue discharger

9 Residue bunker
10 Steam boiler
11 Electrostatic precipitator
12 Fly ash transport system
13 Flue gas fan
14 Stack
15 Turbine-generator room

A HP- steam to turbine
B LP-steam to air preheater
C Industrial water
D LP-steam
E Electricity
F District heating

Fig. 8.3 Three-pass incinerator boiler section (76).

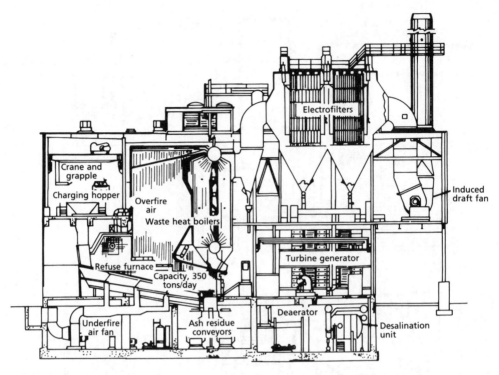

Fig. 8.4 Cross-section of the refuse furnaces, waste-heat boilers, and electrofilters at the Oceanside Incinerator, Town of Hempstead, New York. (Charles R Velzy Associates, Inc. [18].)

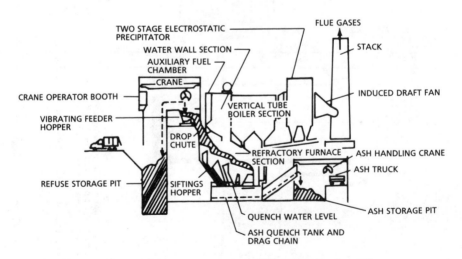

Fig. 8.5 Incinerator with both waterwall and vertical tube boiler sections (47).

286

combustion. Other potential advantages of RDF burning include the possible revenue from the recovery of iron, aluminum, glass, plastics, and organics and the reduction in ash-disposal costs since less ash is produced. In terms of weight percentages, the quantity of ash produced in mass-burning plants is 20% to 30%, while in RDF plants it is 10% to 15%.

When the first RDF plants were designed several decades ago, it was assumed that tree bark and bagasse shredding equipment would work on MSW and that the material handling and separation techniques developed for the chemical, mining, and mineral industries would be applicable to RDF preparation. These assumptions proved invalid because of the unusually difficult plugging, packing, erosive, abrasive, adhesive, corrosive, and sometimes even explosive nature of MSW. The consequence was that a number of first-generation RDF plants were shut down (Table 8.3). Thus the learning curve of RDF technology has turned out to be longer than expected, and more reliable and less expensive methods of cleaning, separation, and shredding are still being developed. For these reasons, startup time (not construction time) for RDF-based HRI plants in the U.S. averages about 3 years.

Table 8.3
HISTORY OF PRE-1982 RDF PLANTS WITH
CONVENTIONAL BOILERS (104)

Facility Location	Capacity (ton/day)	Year Started	1987 Status
Ames, Iowa	200	1975	Operating
Cockeysville, Maryland	1200	1978	Operating
Bridgeport, Connecticut	1800	1979	Shut down
Chicago, Illinois (Southwest)	1000	1978	Shut down
Lakeland, Florida	300	1982	Operating[a]
Lane County, Oregon	500	1979	Shut down
Los Gatos, California	200	1976	Shut down
Madison, Wisconsin	400	1979	Operating[a]
Milwaukee, Wisconsin	1000	1977	Shut down
Rochester, New York (Monroe County)	2000	1980	Shut down
Tacoma, Washington	750	1978	Shut down

Primary source: Schlotthauer, D. J., and Boyhan, G. E. "The Solid Waste Handbook, a Practical Guide." William D. Robinson (ed.). New York: John Wiley & Sons, 1986, Sections 12.1, 12.2, 12.3 and 12.4, 377–423.
[a]Intermittent operation.

As more inerts are removed from RDF, fuel-preparation costs increase and combustibles are lost, but plant performance improves only with improved fuel quality. A plant in Columbus, Ohio, where the RDF quality was found to be low, with a 25% ash residue, had a series of operating problems. In a Madison, Wisconsin, plant the fuel contained only 12% ash, but the RDF preparation was found to be too costly and did not result in a cost-effective boiler fuel. The safety record of RDF plants has not been good either. The plant in Niagara Falls, New York, for example, has reported explosions. Multimillion-dollar RDF plant failures include those in Milwaukee, Rochester, Hempstead (New York), Bridgeport, and Chicago-Southwest. In Ames (Iowa), Albany, Akron, Columbus, and Dade County (Florida), costly modifications were required. (For a discussion of second-generation RDF plants and other types of recycling plants, see Chapter 10.)

Waterwall Furnaces

Furnace volume and capacity increase when the average dimension is cubed, whereas the wall surface area rises only when that dimension is squared. Therefore the larger the furnace, the lower the surface-to-volume ratio; consequently, there is less surface to cool the flames. This tends to cause the walls to overheat, which in turn results in slagging and deterioration. Waterwall furnaces, first introduced in 1965 on MSW service, are designed to overcome this problem (31). In these designs, most of the refractory is replaced by closely spaced water tubes (usually 2.5 inch/63.5 mm-diameter tubes on 3 inch/76.2 mm centers) that are welded together to form a continuous wall. To protect against fireside pitting, the deaerated, softened water in the tubes is held at temperatures in excess of 300° F (130° C). This design reduces refractory maintenance and increases the permissible flame temperature that can be allowed without causing slagging.

The size of waterwall furnaces has been increased over the years and today some units have capacities that exceed 1,000 tons per day of MSW. The design of these large units is still not completely satisfactory, particularly because of insufficient heat absorption in the primary furnaces, which results in excessive temperatures at the superheater. This in turn causes chloride corrosion of the superheater tubes (31). The primary combustion chamber in a waterwall incinerator is tall and the flue gases leave it at the top, where they enter the vertical tube (convection) boiler section (Fig. 8.5). Combustion air is supplied both under and over the grate. The area of underfire air ports is usually restricted to 2% of the total grate area. In a well-designed plant the unburned combustibles amount to about 5% in the HRI bottom residue. Overfire (or secondary) air is introduced to increase combustion efficiency through increased mixing and turbulence. The overfire air is usually pressurized to 20″ H₂O or more and is introduced through closely spaced air-jet nozzles (1.5″ to 3″ diameter) located in a restricted furnace throat area (Fig. 5.29), or through a large number of small, well-distributed air jets at several elevations, supplying high-velocity air to assure good mixing and a uniform air blanket in the combustion chamber. The introduction of overfire air through well-distributed air jets minimizes the emissions of CO and NO_x (see Table 5.18). The required amount of overfire air is about 30% of the total air used by the furnace.

The waterwall tubing, superheater tubing, and supports are subject to corrosion. The alkali metals, heavy metals, and chlorides in the MSW can result in deposits such as zinc chloride, which is highly corrosive (73). In addition, a light coating of fouling also can lower heat transfer and thereby cause the tubes to overheat. Tubes can be kept free from deposits by using soot blowers or tube rappers. Rapping is usually preferred because soot blowers using steam can cause erosion. Corrosion also can be reduced by keeping the waterside temperature below 500° F (260° C). Above this temperature, the corrosion rates increase exponentially (73). To keep the tubes from overheating, the waterside as well as the fireside of the tubes must be kept clean.

In locations where the waterwall-tubing temperature is likely to rise above 500° F in a mass-burning HRI unit, coating the wall with refractory materials is advisable. Because of erosion from the ash-laden gases and chloride corrosion from the products of combustion, the lower-wall tubes are protected with silicon carbide or other refractory materials, which are held in place with ceramic sleeves welded to the tubes with alloy studs. Another problem with the waterwall tubes is fouling and slagging from high-ash fuel. Flue-gas velocities are usually kept under 15 fps in the primary combustion chamber and below 30 fps in the area where the gas enters the boiler convection bank.

Steel or 304 stainless steel does not provide sufficient protection against corrosion in the combustion zone. Promising weld overlay (cladding) materials to protect furnace tubes from chloride attack are high-chromium nickel-base alloys such as Inconel 625 (Alloy 625), or iron-based alloys such as Incoloy 800 and 825 (73). Another recommended tube material is the seamless ASTM A213 Grade T11, also known as 1.25 Cr–0.5 Mo low-alloy steel (107).

Superheater Operation and Corrosion

To generate electricity from steam efficiently, the steam must be heated to at least 700° F (371° C) in superheater tubes. This results in more fireside corrosion in MSW-fired boilers than in regular boilers. (Table 8.4 lists some of the standard superheater tube materials for regular boilers). Because of the chlorinated-plastic content of MSW, chloride corrosion is a problem. High-nickel-alloy superheater tubes (Inconel 825) minimize superheater corrosion in addition to protecting the furnace from overloading and providing rugged furnace walls. HCl corrosion begins by penetrating a slag layer on the superheater tubes. The tubes must be kept clean by soot blowers or mechanical rapping. Chlorides in hot gases become corrosive and can destroy a superheater made of steel. In larger HRIs (more than 1,000 tons of MSW per day), the problem of chloride corrosion has not yet been eliminated.

HRI units can generate steam successfully at 900° PSIG (6205° kPa) and 830° F (443° C) (70). In well-designed furnaces (Düsseldorf), even a superheat temperature of 930° F (500° C) will not cause excessive corrosion (31) because the flue gases can be cooled sufficiently as they approach the superheaters. Other installations, how-

Table 8.4
SUPERHEATER AND REHEATER TUBES
MAXIMUM ALLOWABLE DESIGN STRESS, lb/in² ($\times$ 0.070307 = kgf/cm²) (71)

Material	ASME Spec. No. and Type	900 (482)	950 (510)	1000 (538)	1050 (566)	1100 (593)	1150 (621)	1200 (649)	1300 (704)
Carbon steel	SA210, Grade C	5,000	3,000						
Carbon moly	SA209, T1a	13,600	8,200						
Croloy ½	SA213, T2	12,800	9,200	5,900					
Croloy 1¼	SA213, T11	13,100	11,000	6,600	4,100				
Croloy 2¼	SA213, T22		11,000	7,800	5,800	4,200			
Croloy 5	SA213, T5				4,200	2,900	2,000		
Croloy 9	SA213, T9				5,500	3,300	2,200	1,500	
Croloy 304H	SA213, TP304H				9,500	8,900	7,700	6,100	3,700
Croloy 32H	SA213, TP321H				10,100	8,800	6,900	5,400	3,200

Source: ASME Code, 1983.

ever, have problems even at substantially lower temperatures. It seems that chloride corrosion in superheaters is not caused solely by the superheat temperature of steam. Corrosion can also occur at lower superheat temperatures when the gas temperature entering the superheater is excessive. Installing air jets to provide intense mixing in the furnace throat helps to reduce superheater corrosion.

Design Changes to Minimize Superheater Corrosion

By improving superheater designs, the operating superheater temperatures can be increased from the traditional 750° F to 825°F to 900° F (107). This can be achieved by keeping gas velocities between 15 and 18 ft/sec to minimize the erosion caused by the impact of particulates. In addition, tubes should be liberally spaced to mitigate the increases in velocity as ash buildup occurs. Figures 8.6 and 8.7 show the recommended superheater design criteria for velocities and temperatures.

Another design improvement would be to eliminate the harmful effects of sootblowing by steam or air, such as damage to the protective oxide film, the creation of hot spots from nonuniform cleaning, and the reentrainment of ash into the flue gas. Rapping, rather than blowing, can help eliminate these effects (107). Figure 8.8 illustrates pneumatically actuated mechanical rappers that allow the deposits to slide down the tube surfaces into the ash hoppers below.

The boiler design should also protect against stratification (which can result in a reducing atmosphere) by forcing the flue-gas stream to make multiple 180° turns prior to entering the superheaters (Fig. 8.6). High levels of CO concentration caused by incomplete combustion can be prevented by maintaining excess air levels at 80% to 85%. Another desirable design feature would be a ceramic lining for the postcombustion zone, thereby providing a one-second (minimum) residence time for the flue gases at temperatures in excess of 1800° F (980° C) before they enter the superheater section.

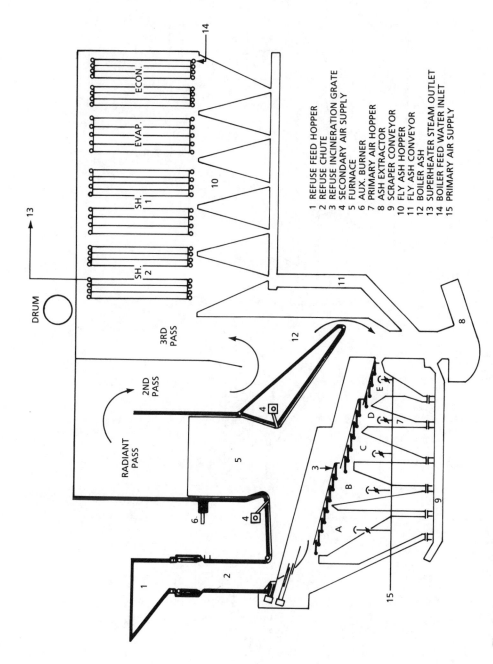

1 REFUSE FEED HOPPER
2 REFUSE CHUTE
3 REFUSE INCINERATION GRATE
4 SECONDARY AIR SUPPLY
5 FURNACE
6 AUX. BURNER
7 PRIMARY AIR HOPPER
8 ASH EXTRACTOR
9 SCRAPER CONVEYOR
10 FLY ASH HOPPER
11 FLY ASH CONVEYOR
12 BOILER ASH
13 SUPERHEATER STEAM OUTLET
14 BOILER FEED WATER INLET
15 PRIMARY AIR SUPPLY

Fig. 8.6 Mounting tubes vertically in a horizontal superheater section prevents particle velocity increases (107).

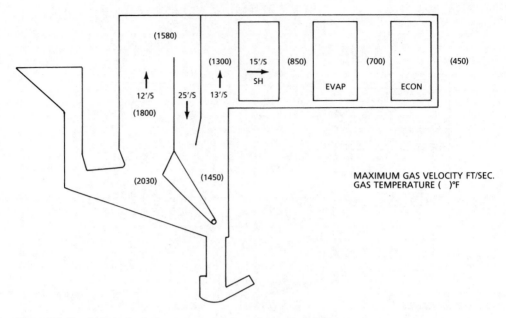

MAXIMUM GAS VELOCITY FT/SEC.
GAS TEMPERATURE ( )°F

Fig. 8.7 Boiler design criteria for corrosion and erosion control (107).

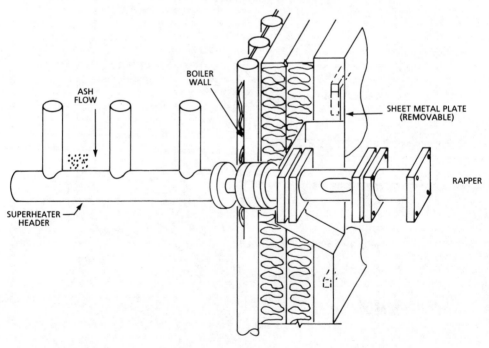

Fig. 8.8 Rapper boiler superheater headers (107).

292

A number of steps can be taken to minimize the "below deposit" corrosion caused by chlorides. These steps include maintaining low gas velocities (not exceeding 13–20 ft/sec), providing multiple radiant gas passes with 180° deflections, and providing radiant cooling of the flue gases so that they enter the superheater section at temperatures below 1300° F (700° C). Other steps include designing coflow and counterflow water/steam circuits through the convection section. Thus metal temperatures can be limited to 860° F (460° C) while producing 830° F (443° C) steam (107). Superheaters should be sized so that at 70% design steam flow a minimum of 100° F superheat is maintained. Finally, the proper selection of tube materials is also important. For superheater tube applications in incinerators, the ASTM A213 Grade T11 seamless tube, also known as 1 1/4 Cr–1/2 Mo low alloy steel has been found to be effective (107).

HRI Start-up and Operation

In smaller HRI plants, with steam production rates under 20,000 lb/hr (9,072 kg/h), the most frequently reported operating problem (71% of fifty-two plants studied) is refractory damage caused by forcing bulky items through the system (55). Frequent start-ups and shutdowns also damage the refractory. Other operating problems reported include the plugging of underfire air ports (35%), the undersizing of tipping floor areas (29%), the warping of charging doors and dampers (29%), the failure of charging ram hydraulics (25%), and the soot buildup on infrequently cleaned boiler tubes (25%) (55). The majority of the HRI plants reviewed operated in the "starved air" mode, at capacities under 50 TPD, for 16 hrs/day.

After construction is completed in larger RDF-burning plants, the startup and debugging period usually exceeds three years (74). A typical case is Hooker's RDF plant in Buffalo, New York, which required substantial redesign. In Columbus, Ohio, an 80% RDF/20% coal, 3,000 TPD plant with a nameplate power generation capacity of 90 MW had to be completely redesigned, at a cost of over $10 million. In this plant, 32% of the major problems were associated with the bottom-ash quench basins, 26% with the grates, and 23% with fuel handling from the surge bins to the boiler chutes (74).

In Columbus, the most frequent problems had to do with oversize materials, shredder plugging, ineffective magnetic separators, bridging in bins and hoppers, inconsistent boiler feed, unstable furnace draft, jamming, plugging, and warping of the grates, and jamming of bottom-ash quench basins. Corrective steps included using stronger magnets, increasing crane grapple capacities, and adding vibrating conveyors, Heil 92B shredders, and enlarged belt and submerged drag-chain conveyors for the bottom ash. General surveillance was improved by adding TV monitors on the boilers.

A review of the Columbus plant concluded that the simultaneous firing of coal and RDF is not practical and that plant availability and reliability is more important than the achieved efficiency in heat recovery. The review also concluded that both the underfire and overfire airflows must be adjustable and accurately controllable over wide ranges. Also, it was found that the reinjection of fly ash has caused more problems in waterwall erosion and combustion interference than it contributed to improved efficiency. The operators were satisfied with the design concept of multiple boilers (six smaller boilers instead of fewer, larger ones). The review also pointed to the need for longer operator training periods (74).

Performance Testing

The performance of an HRI is evaluated on the basis of its MSW throughput and its energy output. Performance is directly dependent on the heating value of the feedstock, which is a difficult parameter to determine. There are two basic methods for determining heat input. The American Society of Mechanical Engineers (ASME) Performance Test Code for Large Incinerators (PTC 33) is known as the *direct*, or *input-output*, method. In March 1986 ASME formed a new code committee (PTC 33.1) for large incinerators, applying the *indirect*, or *boiler/calorimeter* method.

The efficiency of an HRI is the ratio between the input and output of heat:

$$\text{Efficiency} = E = \text{Steam produced/Heat contained in the MSW} = Qo/Qi$$

The *direct* or *input-output* method assumes that the heat input (Qi) is the product of the mass flow rate of the waste charged (W) and the higher heating value (H) of the MSW burned. With this approach, efficiency is calculated as:

$$E = Qo/(W)(H)$$

This method has drawbacks because it is difficult to determine the heating value of the waste in a calorimeter since it is almost impossible to obtain a representative MSW sample.

The *indirect* or the *boiler/calorimeter* method, assumes that the heat input (Qi) is the sum of the steam generated (Qo) and the losses in the HRI system. With this approach the efficiency is calculated as:

$$E = Qo/(Qo + \text{Losses})$$

This method's main advantage is that it does not need a representative MSW sample since the incinerator itself is used as the calorimeter. The disadvantage is that a substantial amount of data-taking and calculation is required to calculate accurately the many forms of heat losses. This method, however, is about twice as accurate as the direct method (78).

Once the efficiency is determined by the indirect method, the heating value (H) of the MSW can be calculated, if the mass flow of the waste (W) is known:

$$H = (Qo + \text{Losses})/W$$

Both methods of determining heat input will be discussed in detail below.

Direct Testing of HRI Performance

The starting point for the direct method is to determine the heating value of the MSW. One approach is to sort a large sample of MSW into ten or eleven types of components and individually weigh and examine each component (79). This is a time-consuming and unpleasant process. An alternative method, used in the 1960s, was to use a large calorimeter that is capable of receiving the contents of a "Munich type" dustbin, which is about 70 liters (19 gallons) in volume (Fig. 8.9). The calorimeter consists of a cylindrical combustion chamber, a ring-type gas burner, underfire and overfire air inlets, and a heat exchanger. The heating value (H) is calculated as the sum of the heat transferred to the water (Qw) and the losses, divided by the weight of the sample (W):

$$H = (Qw + \text{Losses})/W$$

In acceptance test measurements, differences between heating value test results were more than 50%. On the average, the calorimeter readings were 6% lower than the results from the analysis of sorted samples (78). These poor results are due to the heat capacity of the calorimeter itself (which needs to be calibrated), as well as to the heat input from the burning gas and the incompleteness of the combustion.

ASME's procedure for MSW sampling (Test Code for Large Incinerators, PTC 33) is, first, to accumulate a 1,600–3,200-pound sample, taking 100–200 pounds of material every half hour over an eight-hour period. Next, this sample is quartered and one of the quarters is hand-picked to remove the noncombustibles. Then the sample is shredded to produce laboratory samples for ultimate analysis. This method was applied in a test where two 15–20-pound boxes prepared from the same sample were each sent to different laboratories. The heating values reported by the two

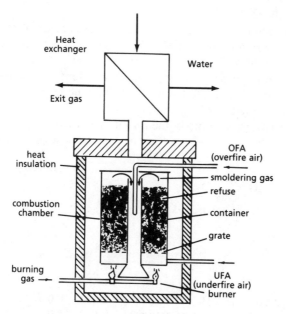

Fig. 8.9 Schematic of large calorimeter for a 25 kg sample of refuse (78).

laboratories were in complete disagreement and therefore of little value (Table 8.5). Because of such poor results, the direct method is no longer in favor.

Indirect Testing of HRI Performance

The indirect boiler/calorimeter method of efficiency calculation considers the incinerator itself as a calorimeter. This method measures the waste and steam flow rates and calculates the heat losses, but it eliminates the need to determine the heating value of the MSW. By separately measuring the fuel mass flow (W), and the useful heat output (Qo) and by calculating the various losses, the heating value of the waste can be determined as:

$$H = (Qo + \text{Losses})/W$$

Where:

> Qo is detectable within an error of 1–2%
> Losses are detectable within an error of 2%
> W can be measured within an error of 3–5%

Therefore the overall error in the heating value (H) measurement is about 6%, which is less than the error of the direct method (78). In calculating the heat losses, the total amount of ash and moisture leaving the HRI can also be determined, so the heating value of the waste can be recalculated on a moisture- and ash-free basis. This is a useful check because the heating value of ash and moisture-free MSW is known to be around 9500 BTU/lb (77).

MEASUREMENTS REQUIRED

The indirect method of testing requires a large number of measurements and calculations. The *mass flow rate of the waste* can be based on the truck weight-bridge readings or on the weighing sensors on the cranes. Better results are obtained when the test refuse is preweighed and stored in a separate pit or when the filled grabs are weighed on a calibrated scale. In more modern installations, accurate weight-belt feeders are used (25).

Airflow rates can be accurately measured (see Table 5.9) with area-averaging pitot stations (25). *Flue-gas flow rates* can be measured with anemometers and jet-

Table 8.5
HIGHER HEATING VALUE,
btu/lb (80)

Date	Laboratory A Sample		Laboratory B Sample	
	1	2	1	2
10/3/84	4900	5400	5450	4200
10/4/84	4250	5100	4850	4150
10/5/84	5500	5900	5400	3700

deflection probes (25). The reference method for *stack-gas flows* uses two pitot tubes, one in the center of the stack and another to detect the velocity profile across the cross section of the stack (Fig. 8.10). *Steam flow rates* can be detected by vortex shedding flowmeters at temperatures up to 750° F (399° C). At higher temperatures the shunt or orifice-type flowmeters with multiple d/p cells are recommended (25).

For measuring *stack-gas temperature*, the detector most often used is the thermocoupler, but RTDs are also used. Flue-gas composition is analyzed at the economizer outlet. Its moisture concentration is detected by infrared analyzers and its excess oxygen content is measured by zirconium oxide probes (Fig. 5.51). The concentration of *particulates* or *unburned hydrocarbons* is detected by opacity or flame ionization analyzers. The concentration of *carbon dioxide* and *carbon mon-oxide* is measured by infrared analyzers. In the past, the detection of moisture and

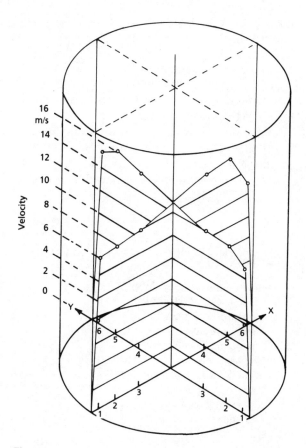

Fig. 8.10 Velocity distribution (Ingolstadt Incinerator) mea-surement with reference method (78).

hydrocarbons could not be done reliably (77). Water added to the flue gases by the pollution-control equipment must be corrected for in order to determine the true moisture content of the flue gas. This moisture measurement should only reflect the moisture content of the fuel and the formation of water vapor due to the combustion of hydrogen.

Modern stack-gas-analyzer packages, sometimes combining on-stream fiber-optic detection techniques with computer analysis, have made it possible to use a single instrument (Fig. 5.54) to measure several flue-gas constituents (25). Recently, continuous-emission-monitoring (CEM) packages have been developed for incinerator applications that are capable of simultaneously measuring CO, CO_2, SO_2, NO_x, total hydrocarbons, opacity, and HCl.

In addition to the above measurements, it is also necessary to detect: (a) steam pressures and temperatures, (b) feedwater flows and temperatures, (c) boiler-drum pressures, (d) boiler blowdown rates, (e) turbine throttle flows, temperatures, and pressures, (f) bottom ash and fly-ash quantities, (g) barometric pressure and ambient wet/dry bulb temperatures, (h) net and gross power outputs, and (i) turbine exhaust pressure.

The unburned-combustibles content of the various ash streams is determined by bomb calorimeters. The sensitive heat in the clinker and fly ash can be measured by pyrometers or by measuring the temperature rise in the quenching water (if the quenching water flow is high enough to avoid excessive evaporation).

Table 8.6 lists nine different forms of heat losses that need to be calculated in order for the indirect boiler/calorimeter method to be accurate. The largest loss—about two-thirds of all the losses—is the heat that is carried out by the flue gases into the atmosphere. This can represent as much as 20% of the total heat input into the HRI.

Table 8.6
HEAT OUTPUT AND LOSSES (77)

Symbol	Quantity	Percent
Q_{out}	Heat Output (Boiler duty)	60–80
L_1	Sensible heat in dry flue gas	10–15
L_2	Moisture in flue gas (hydrogen and moisture in fuel; moisture in air and moisture from residual quench)	12–18
L_3	Unburned combustible in ash streams	0–5
L_4	Sensible heat in ash streams	0–1
L_5	Unburned combustible in gas. Carbon monoxide and hydrocarbons	0–1
L_6	Radiation	0.2–1
L_7	Unaccounted for	0.5
L_8	Margin	1.0

In 1986, 6% of the total MSW production of the United States was sent to HRIs. Today 25 million tons a year—14% of all MSW produced—is incinerated (200) and it is projected that in the year 2000 HRIs will handle 32 million tons. There is little question that the HRI industry is here to stay. There is also little question that major problems remain unresolved, from dioxin and heavy metal emissions, to the increased generation of "greenhouse effect" gases, to corrosion and ash disposal. Major controversies continue regarding mass-burning versus RDF burning and recycling versus incineration.

Chapter 9
Recycling

□
■ ■

AS THE WASTE-ABSORBING CAPACITY of the oceans, atmosphere, and land is exhausted, we will need to find newer, permanent solutions. The only permanent solution to the solid-waste problem is recycling and reuse. This chapter will discuss some of the broader issues of recycling, both its national and international as well as its political and commercial aspects, and the change in social attitudes toward it. Specific recycling strategies will be outlined, including new concepts of labeling and packaging, the design of a recycling-oriented kitchen, and the concept of "separation at the source." The specific aspects of paper, metal, glass, plastics, rubber, organic, and incinerator-ash recycling also will be discussed.

The Status of Recycling

One of the most effective ways to achieve the goals of recycling is through source reduction or "precycling," which means that we reduce the quantity we make in the first place. The yearly quantity of packaging in the United States increased from 24 to 45 million tons between 1960 and 1990 (191) and now represents 33% of the total MSW generated. The yearly amount of plastics used in packaging increased between 1974 and 1990 from 3.3 to 8 million tons (191). Packaging is an ideal candidate for precycling and government is in an ideal position to do something about wasteful packaging of goods. In addition to packaging, precycling can reduce the quantity of poorly made goods that will last only a season, telephone books, and junk mail. In some ways precycling has already started. For example, plastic soda bottles that weighed 60 grams in 1970 weigh only 48 grams today (179).

300

Table 9.1

GENERATION AND RECOVERY OF SELECTED MATERIALS IN MUNICIPAL SOLID WASTE, 1960 TO 1986 (178)

Item and Material	1960	1965	1970	1975	1980	1981	1982	1983	1984	1985	1986	1988 (199)
Gross waste generated:												
Paper and paperboard	29.8	37.9	43.9	42.6	53.9	55.0	52.4	57.7	62.5	61.7	64.7	72.0
Ferrous metals	9.9	10.1	12.6	12.3	11.6	11.4	11.3	11.4	11.3	10.7	11.0 ⎫	15.3
Aluminum	.4	.5	.9	1.1	1.8	1.9	1.9	2.1	2.2	2.3	2.4 ⎬	
Glass	6.5	8.6	12.7	13.9	14.9	15.0	14.5	14.2	13.8	13.2	12.9	12.6
Plastics	.4	1.4	3.0	4.4	7.6	7.8	8.4	9.1	9.7	9.8	10.3	14.4
Materials recovered:												
Paper and paperboard	5.4	5.7	7.4	8.2	11.8	11.4	11.1	12.0	13.2	13.0	14.6	18.4
Ferrous metals	.1	.1	.2	.2	.4	.3	.3	.3	.3	.3	.4 ⎫	2.2
Aluminum	(NA)	(NA)	(NA)	.1	.3	.5	.6	.6	.6	.6	.6 ⎬	
Glass	.1	.1	.2	.4	.8	.8	.8	.9	1.0	1.1	1.1	1.5
Plastics	(NA)	(NA)	(NA)	(NA)	(NA)	(NA)	(NA)	(NA)	.1	.1	.1	.1
Percent of gross discards recovered:												
Paper and paperboard	18.1	15.0	16.9	19.2	21.9	20.7	21.2	20.8	21.1	21.1	22.6	25.5
Ferrous metals	.5	1.0	1.6	1.6	3.4	2.6	2.6	2.6	2.7	2.8	3.6 ⎫	14.3
Aluminum	(NA)	(NA)	(NA)	9.1	16.7	26.3	31.6	28.6	27.3	26.1	25.0 ⎬	
Glass	1.5	1.2	2.3	2.9	5.4	5.3	5.5	6.3	7.2	8.3	8.5	11.8
Plastics	(NA)	(NA)	(NA)	(NA)	(NA)	(NA)	(NA)	(NA)	1.0	1.0	1.0	1.1

Source: Franklin Associates, Ltd.. Prairie Village. KS. *Characterization of Municipal Solid Waste in the United States, 1960 to 2000*, 1988. Prepared for the U.S. Environmental Protection Agency. Note: In millions of tons, except as indicated. Covers post-consumer residential and commercial solid wastes which comprise the major portion of typical municipal collections. Excludes mining, agricultural and industrial processing, demolition and construction wastes, sewage sludge, and junked autos and obsolete equipment wastes. Based on material-flows estimating procedure and wet weight as generated.

While precycling is a relatively new concept, recycling is not. In 1986, 80% of the MSW in the United States was landfilled, 11% (Table 9.1) was recycled, and 9% was incinerated (156). By 1990 recycling had increased to 13% (Table 1.13) and incineration reached 14% and is projected to reach 20% by 1992 (185). The EPA target for 1992 is 55% landfilling, 25% recycling, and 20% incineration. As of the end of 1990, recycling laws are in effect in some 1500 communities in 35 states in the U.S. (200). The recycling law of New York City requires that 25% of all MSW be recycled by April 1994. The pressure to recycle is the result of a shrinking number of landfill sites and the rising costs of disposal. In New Jersey, for example, half the household waste is trucked to out-of-state landfills up to 500 miles away. In Union County, the annual dumping costs per household exceed $400, and "billing by the bag" is being considered to achieve a fairer distribution of the expenses. The maxim "If you hit people in the pocketbook, it will open their eyes" also applies to recycling.

In Japan 50% of the MSW is recycled; in Europe the range is 20% to 50%. In the United States, while MSW recycling has doubled in the last twenty years (Fig. 9.1), it was still only 13% in 1990 (200). In Japan 95% of beer bottles are reused (on average, twenty times), as is 66% of other glass bottles (at least three times). Overall, 42% of all new glass manufactured in Japan is made from recycled glass and 40% of all aluminum cans are recycled (108). Japan also recycles 95% of its newspaper and 51% of all other paper products; it produces 90% of its toilet paper from recycled paper (159). In Denmark 99% of beer and soft drink containers are refilled (191). In contrast, the U.S. recycles 63.6% of its aluminum, 26% of its paper, 33% of its

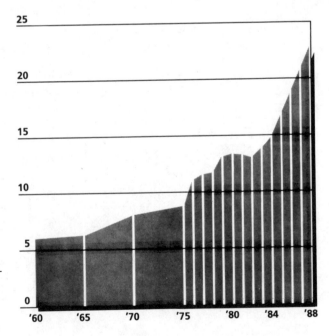

Fig. 9.1 Yearly quantities of materials recycled in the United States in millions of tons. (*Statistical Abstract of the United States,* 1988 [118 & 200].)

newsprint (but only 10% of it into newspapers; the rest is used in cereal boxes and automobile interiors, some 60 pounds per car), 12% of its glass, 14% of its metals, and little more than 1% of its plastics (160, 197, 199, 200). In 1988, California recycled about 10% of its waste, Seattle about 28%, and San Francisco 24%, while New York State recycled only 4% and New York City less than 2% (129). Today, the recycling percentage of New York State is 10–15% and New York City recycles about 6% (190). Curbside recycling of glass, metal, and paper in Davis, California, has reduced the total MSW of the city by 19%. Davis also composts separately collected yard debris, which further reduces the total municipal waste stream. In Connecticut the recycling leader is the town of East Lyme, which already recycles about 30% of its waste. In Connecticut 147 of 169 towns already have a recycling program in place and all must have them by 1991, when a statewide program mandates the recycling of glass bottles, cans, and newspapers. As part of this program, five regional recycling centers are being built.

While Japan is a leader in recycling, the country's strong economy has contributed to the evolution of a more lavish and wasteful lifestyle, which will reverse the progress made in the previous decades. Japan's strong currency enables its industry to import raw materials more cheaply than buying recycled steel or paper— a practice that has restricted the market and lowered the prices of recyclables (159).

Recycling Versus Incineration

While neither incineration nor recycling completely eliminates the need for landfilling, both reduce it substantially. The incinerator residue that requires landfilling is about 25% of the total MSW by weight in mass-burning incinerators and 10% when RDF is burned. The recyclable proportion of the MSW can approach 75% by weight (3) if it is separated into paper, glass, metal, and compost. According to an Environmental Defense Fund study, recycling creates thirty-six jobs per 10,000 tons of MSW against six for landfilling and one for incineration.

Therefore, incineration and recycling can both achieve 75% waste reduction by weight. The unit cost of building a new incinerator is over $100,000 per TPD of capacity. A fair comparison therefore is to evaluate if a recycling program can also be established at that investment. The economics of recycling should not only be based on the market values of the recycled goods. The total cost-benefit analysis of these programs should also consider the savings from not having to pay the tipping fees at landfills and the savings in not having to build incinerators.

Unfortunately, incineration and recycling do not complement each other. For example, if paper is removed for recycling, the remaining content of the MSW may need additional fuel for combustion. Thus recycling diverts energy-rich components from the incinerator fuel and reduces its heat output and its income from the sale of steam or electricity. The conflict between recycling and incineration has resulted in contractual arrangements where the incinerator operator is given exclusive rights to all MSW in the area. This "flow control" contract clause has been used to prevent recycling in some locations, such as Akron, Ohio (3). Maine and New York find that their goals of 50% recycling by 1994 and 1997, respectively, are in conflict with their

Fig. 9.2 Curbside container used in the Dutch glass-recycling program. (Courtesy of B. V. Handelsonderneming "Maltha," Rotterdam, the Netherlands.)

contractual arrangements with incinerator operators (186). In other locations long-range landfilling contracts are in conflict with recycling. Los Angeles, for example, must pay a penalty if it does not provide a minimum of 6,000 daily tons of MSW, destined to fill a number of canyons in the area (186). In contrast, separation at the source and MSW recycling through mechanical separation do complement each other. Any material that is removed from the trash by household separation only simplifies the subsequent MSW separation and recycling steps.

Markets for Recycled Materials

As incineration depends on the availability of landfills for ash disposal, so recycling depends on a market for recycled materials. This is a critical element in recycling. As more states and local governments gear up for recycling, the market will get, at least periodically, overloaded. While the metal, glass and plastics industries have not yet experienced such problems, the paper industry (particularly in the Northeast USA) has (160). Today the market for glass and aluminum is good because reusing these materials is still cheaper than starting afresh (180). The plastics industry, which is afraid of being legislated out of existence, is financing its own search for markets for recycled plastics; both Coke and Pepsi are developing programs to use recycled bottles. As of this writing government has failed to stimulate a steady demand for recycled paper and for steel cans.

In 1989 there were regions where recycling resulted in such oversupply that the demand for recycled newspapers completely collapsed (180). In New Jersey the sales value of a ton of used newspaper in 1989 has dropped from $40 to − $25 (179). In New York the price of newsprint fell in 1990 from $7 to $1 per ton (200). In 1990, due to an oversupply of used paper from the U.S., the sale price in Holland has dropped from 8 cents to 1 cent per kilogram (197). Recent legislation in several states mandating the use of a fixed percentage of waste paper in newsprint has started to reverse this trend and hopefully it will result in a dependable and steady market for recycled paper. Tables 9.2 and 9.3 give 1989 prices for recycled materials in the United States.

Maintaining a demand for recycled materials requires cooperation between government, industry and the public. Municipalities are good at collecting garbage and at legislating monetary incentives or at mandating the use of recycled materials. The public and industry are good at responding to monetary incentives, while scrap dealers are good at selling used materials, if there is a market for them. Therefore, while uncoordinated, poorly planned programs, which disregard monetary incentives and market forces can self-destruct by flooding the market, well-coordinated legislation cannot only motivate recycling, but can also guarantee a market for the recycled goods. In the future, government must help find a way both to stimulate the demand and stabilize the market for recycled products. Providing tax incentives may be necessary to allow the market for recycled materials to expand sufficiently to handle the growing flood of collections. If a stable and steady supply of high-quality recyclable materials is achieved, the market will probably respond and industry's investment in recycling will probably grow.

Recycling Laws and Regulations

The effectiveness of recycling programs depends on monetary incentives and on education. The lowest participation in recycling programs is usually reported from the lowest- and the highest-income neighborhoods. Not only the public needs to be educated. American industry has also been seduced into excessive reliance on virgin materials rather than scrap. Government needs to provide incentives and to find a

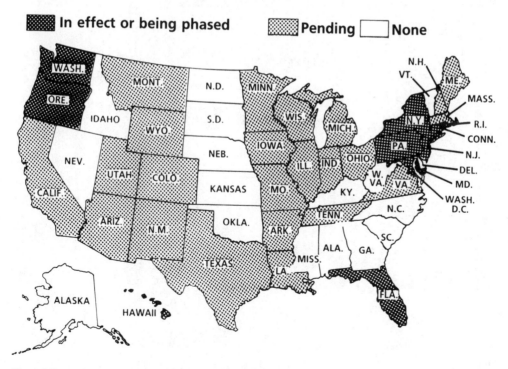

Fig. 9.3 Statewide recycling legislation. Does not reflect voluntary programs or municipal programs. (National Solid Waste Management Association [156].)

Table 9.2
PRICES OF RECYCLED PAPER, GLASS, PLASTICS, RUBBER, AND WOOD, 15 MARCH–15 APRIL 1989, IN TRUCKLOAD LOTS, PAID AT SELLER'S DOCK, FOR ONE TON OF MATERIAL (Unless Otherwise Stated) (161)

PAPER PRICES—FROM PROCESSORS/DEALERS

	Northeast	Mid Atlantic	South	South Central	East Central	West Central	West
Newspaper*	($ −25b)–17b	$5b	$5b	$20–25s	$15s	$12–30b	$5–15b
Corrugated	0–20s	186s	10b	35s	NA	10–40b	10–60b
Mixed waste paper	3.75–8	5b	10s	—	7s	7.6b	5b–10b
Computer Print-Out*	110–200	300b	—	180s	—	70b	130b–220b
White ledger	135–210s	60b–195s	—	80s	—	20b	135b–210b
Colored ledger	45–115s	40b–125s	—	10s	—	20b	135b
Magazine paper	5s	—	—	—	—	—	—

*Newspaper prices for various grades. Notes −40 price reported from one source in New Jersey. Computer Print-Out high prices generally for laser-free CPO.
"b" = price paid at the buyer's dock.
"s" = price paid at the seller's dock.
1. All prices for baled material in large quantities, relatively free of contamination, unless otherwise noted.
2. Where more than one figure is given, prices are low-high reported for that commodity in region. When one figure is given, it may or may not reflect price from one or more sources.
3. In many cases wide range of prices may reflect reports from municipalities with small quantities and irregular shipments (lower prices) and reports from dealers with high volumes to sell (higher prices).

PAPER PRICES—FROM CONSUMING MILLS

	Northeast	Mid Atlantic	South	South Central	East Central	West Central	West
Newspaper*	0	—	20b	20–25b	25b	30	35
Corrugated	10–15b	—	30–50	35–40	—	35	65
Mixed waste paper	—	—	15–20	15–20	—	12.50–30	25
Computer Print-Out*	100	—	225–235	—	225	70	300–340
White ledger	80	—	150–205	—	205	20	235–245
Colored ledger	80	—	65	—	160	—	135–140
Magazine paper	—	—	0	—	—	—	30–35

*Newspaper prices for various grades. Computer Print-Out high prices generally for laser-free CPO.
All prices paid at the dock of the buyer—a consuming paper mill.
1. All prices for baled material in large quantities unless otherwise noted.
2. Where more than one figure is given, prices are low-high reported for that commodity in region. Where one figure is given, it may or may not reflect price from one or more sources.

GLASS PRICES—FROM PROCESSORS/DEALERS

	Northeast	Mid Atlantic	South	South Central	East Central	West Central	West
Clear (flint)	$25–45s	$40s–80s	$40s–80s	$70s	$40s–80s	$40s	$80s
Brown (amber)	10–15s	40s–80s	40s–80s	70s	40s–80s	40s	80s

	Northeast	Mid Atlantic	South	South Central	East Central	West Central	West
Green (emerald)	10–15s	40s–80s 10b	70s	40s–80s	40s–80s	40s	80s
Mixed							

*West Region high price reflects affect of California's redemption law.

"b" = price reported paid at buyer's dock.

"s" = price reported paid at seller's dock.

Some glass buyers (consumers) have a two-tiered price structure for glass, in which they pay significantly more for waste glass bottles from buy-back centers than they pay for curbside-collected material. This two-tiered price structure may explain the range of prices in some of the regions above.

Note distortion in West Region prices caused by California container redemption law.

GLASS PRICES—FROM CONSUMERS

	Northeast	Mid Atlantic	South	South Central	East Central	West Central	West
Clear (flint)	$10–15	$55–65	$55–60	$50–60	$70–80	—	$95–100
Brown (amber)	10–15	50–55	55–60	50–60	55–60	—	95–100
Green (emerald)	30–50	45–50	30	50–60	50–55	—	95–100

All prices paid at buyer's dock. Prices obtained from consumers (end-using glass manufacturing plants).

PLASTICS PRICES

	Northeast	Mid Atlantic	South	South Central	East Central	West Central	West
Clear PET baled	8–11¢ ($160b–220s)	9¢ ($180b)	8.5–9¢ ($170s–180s)	—	6¢ ($120s)	—	8¢–9¢
granulated	—	10.5¢ (210b)	—	—	6¢ (120s)	—	—
Green PET baled	5–9¢ (100s–180s)	6.5¢ (130b)	8.5¢ (170s)	—	5¢ (100s)	—	7–8¢ (140–160)
Mixed PET & HDPE	2¢–4¢ (40s–80b)	5¢ (100b)	2.5¢–7¢ (50s–140s)	—	4¢ (80s)	—	—
HDPE—"natural" color only	10¢ (200s)	7¢ (140b)	8¢ (160s–160b)	—	—	—	3¢–10¢ (60s–200s)
Rigid Containers, Mixed	12¢–18¢ (240b–360b)	—	—	—	—	—	—

"b" = price paid at buyer's dock

"s" = price paid at seller's dock

All prices for baled material in large quantities (such as 40,000 pounds) unless otherwise noted. Prices in cents per pound and dollars per ton. Dollar prices have buyer/seller dock designation.

Note distortion in West Region prices caused by California container redemption law.

307

Table 9.3
PRICES OF VARIOUS METALS AND OTHER RECYCLABLES, 15 MARCH–15 APRIL 1989 IN TRUCKLOAD LOTS, PAID AT SELLER'S DOCK PER ONE TON OF MATERIAL (unless otherwise stated) (161)

STEEL CAN PRICES

	Northeast	Mid Atlantic	South	South Central	East Central	West Central	West
Clean steel/tin cans price per gross ton*	$70–80	$52–80	—	$75–80	$40	$70–80	$55–65

*Gross ton = 2,240 pounds. All prices at buyer's dock.

OTHER ALUMINUM SCRAP PRICES

	Northeast	Mid Atlantic	South	South Central	East Central	West Central	West
Borings	—	15¢–20¢	—	—	—	—	25–40¢
Cast	30¢	18¢–56¢	30¢–42¢	33¢–43¢	37¢–52¢(s)	17¢	43¢
Clips	35¢	38¢–44¢	65¢	65¢	39¢	—	40–41¢
Extruded	33–50¢	36¢–42¢	22¢–60¢(s)	—	39¢	44¢	43–47¢
Foil	10¢	12¢	5¢	10¢	15¢	10¢	10¢
Mixed	—	—	25¢–46¢	—	60¢–61¢	—	55–63¢
Old Sheet	26–41¢	36¢–38¢	22¢–60¢(s)	56¢	39¢(s)	40¢	40–54¢
Painted	35¢	42¢	—	—	39–54¢(s)	19.5¢	40–41¢

All prices in cents per pound only. All prices at buyer's dock unless otherwise indicated by (s). Prices for used beverage cans listed elsewhere on this page.

COPPER & BRASS PRICES

	Northeast	Mid Atlantic	South	South Central	East Central	West Central
Copper						
No. 1	60–78¢	55¢–75¢	75¢–85¢	97¢	89¢	65.7¢
No. 2	56–54¢	55¢–65¢	—	85¢	65¢–85¢	56.5¢
Brass						
Red	33¢	30¢–45¢	30¢–40¢	—	61¢	28¢
Yellow	43¢	30¢–40¢	—	—	39¢–40¢	27¢

Prices from five regions only. Prices in cents per pound only. All prices reported paid at the buyer's dock.

ALUMINUM BEVERAGE CAN PRICES

	Northeast	Mid Atlantic	South	South Central	East Central	West Central	West
Used beverage cans							
processors/dealers	30¢ ($600b)	35¢–40¢ ($600b–880b)	43¢–57¢ ($860b–1,140b)	58¢ (1,160b)	46¢ ($1,110s–1,200s)	40–43¢	42–68¢ ($1,180s)
consumers	68¢ ($780–900)	39¢–45¢	43¢–52¢ ($860–1,040)	48¢–54¢ ($960–1,080)	48¢–53¢ ($960–1,060)	40¢–45¢ ($800–900)	50¢ ($720–1,600)

"b" = price paid at buyer's dock.
"s" = price paid at seller's dock.
Prices in cents per pound and dollars per ton. Note prices divided in this table by reporting entity: processors/dealers and consumers (end users).
Processors/dealers dollar prices provide buyer/seller's dock location in which prices are paid. All consumers prices paid at consumer's (buyer's) dock.

Market Data: Waste Generation

Generator Segment	Paper	Corrugated	Plastic	Metals	Other
Office Buildings	65%	15%	6%	2%	12%
Industrial Cos.	35%	20%	25%	6%	14%
Retailers	35%	40%	8%	1%	16%
Tran/Comm/Utes*	20%	15%	15%	5%	45%
Whol/Ware/Dist**	25%	32%	25%	7%	11%
Public & Inst.***	45%	10%	5%	6%	34%

Weekly Waste Generation by Occupied Square Foot

Office Buildings	0.05 pounds
Industrial Companies	0.06
Retailers	0.22
Trans/Comm/Utes*	0.10
Whol/Ware/Dist**	0.06
Public & Institutional	0.04

*Transportation, communication, and utilities businesses.
**Wholesale, warehouse, and distribution businesses.
***Public and institutional buildings.
A paper by David Cerrato, project manager for Malcolm Pirnie Inc., presented data on materials in the commercial waste stream, based on work the firm has performed in Westchester County, N.Y. The above is taken from two tables included in the paper.

Commingled Recyclables

	Glass & Aluminum	Glass, Tin & Aluminum	Glass, Tin, Alum & Plastics
Northeast			
New Jersey	$13	$0–12	$0
Mid-Atlantic			
Philadelphia			$2*

Prices in New Jersey are from privately owned MRFs. Philadelphia price is from non-profit facility on a city contract.
Prices paid by MRFs for commingled recyclables delivered to facility.

Used Oil Prices

	East Central	North East
Any quantity		(−20c)/gal.
5 drums or less	(−$10)/drum	
6 drums or more	$0	
Truckload*	+ 5c/gal.	

*Truckload specified as 5,800 gallons.

309

working substitute for the "recycling class" of the past—the ragman, the junk collector, and the milkman. In 1988 the federal government set a goal of eliminating 25% of the country's MSW by 1992 through a combination of recycling and waste reduction (158). Waste-reduction methods include making more durable products and reducing the amount of packaging. More than 500 cities and 1500 communities in 35 states provide curbside collection or require some type of separation at the source; 32 states (Fig. 9.3) have comprehensive recycling laws (156, 200). These numbers have roughly doubled in the last five years (160).

Each state and city has its own recycling strategy. New Jersey has had a recycling program since 1987 and Connecticut started in 1991 (156). In 1988 New York State issued recycling quotas for municipalities, with the aim of 50% reduction in MSW generation by 1997. In 1988, the New York State legislature also passed a law requiring each community to develop a recycling plan by 1992 (156) and mandating a statewide recycling program by 1992 (130).

In December 1990 the legislature authorized manufacturers to use the official state emblems on labels designating products as "recycled," "reusable," or "recyclable" if the packaging meets state standards (193).

In 1989 New York City recycled about 1,000 tons of MSW a day while generating 28,000 tons daily (129,190). New York State in 1989 recycled more than three million tons of trash, yard debris, sewage sludge, and commercial waste, about 15% of its MSW, but the statewide net production of solid waste is still growing (193). The mandatory recycling law, passed in April 1989 by the New York City Council, calls for two-thirds of the city to recycle some materials by July 1992 and 25% of all MSW by April 1994 (130). It will take years for the mandatory recycling program to reach all parts of the city, and at the end of five years fines will be levied if recyclable materials are found in regular garbage. At the end of 1990 mandatory recycling was in effect in only a small part of the city, and cutbacks are delaying the expansion of even this limited program. Since November 1989 the city has issued 44,000 warnings and levied fines ranging from $25 to $10,000 (181). The sanitation police force of 178 officers, however, is facing cutbacks. In 1990 New York City also started a modest program of curbside collection, using blue recycling cans but this program was cut back in 1991.

Eventually all New York City households will be required to separate bottles and cans and bundle newspapers for collection on specific days. The city is also experimenting with new compartmented garbage trucks and with different collection routes (156). By 1996 it will be mandatory for all city residents to separate and set aside all newspapers, magazines, paper and corrugated cardboard, glass, metal, and plastic in color-coded containers (red for plastic, blue for newsprint, white for white glass, green for green glass, brown for amber glass).

In 1987 New Jersey passed a law requiring its twenty-one counties to develop their own recycling plans. Each county is supposed to recycle at least three materials in order to reach the goal of 25% recycling in two years. Some communities use three-way separation (paper, other recyclables, and the rest), while others require four or five containers. The methods of enforcement vary from community to community. In some, peer pressure and pride are the only incentives; in others, the delinquent families forfeit garbage collection, while in Woodbury noncompliance results in fines up to $500.

A Connecticut law requires all municipalities to recycle 25% of their MSW. Residents must separate trash into recyclable and nonrecyclable segments. Landfills and incinerators are not allowed to accept recyclable items (162). Anyone who does not recycle may be fined $100 (162); haulers of unseparated garbage are also fined and unseparated garbage will not be picked up. Special pickups are provided for leaves and branches, scrap metal, waste oil, and car batteries.

The modest goals of 25% recycling set in the Northeast have been exceeded in some areas. Philadelphia and Los Angeles County have set a recycling goal of 50% and Washington, D.C., has set a goal of 45%. A new law in Florida mandates municipalities to recycle 30% of their MSW by 1994 (129), while Seattle has adopted a 60% recycling goal.

The Politics of Recycling

State and local governments could encourage recycling by providing tax incentives to support the use of recycled materials, reduce unnecessary packaging, and stabilize scrap markets. These markets are notoriously volatile for scrap iron, copper, and newsprint. Governments also could ban the use of harmful substances or apply tax incentives to eliminate toxic substances, nonrecyclable plastics, and chlorofluorocarbons from industrial products.

Government can increase the market for some recycled goods, such as paper, by mandating its use in all government offices by levying virgin-materials taxes, or by requiring a preset percentage of recycled materials in goods sold. Government can also increase the market for reusable or recycled goods by labeling them as such. It could also modify the freight rates to make the transportation of recycled materials less expensive than the cost of transporting virgin raw materials.

In densely populated urban areas the options available to politicians for solving the MSW-disposal problem are fewer. The New York state legislature, for example, had no choice but to close all landfills on Long Island by 1990 because rain was seeping through toxic landfills into the subterranean drinking-water reservoirs. At the end of 1990 four Long Island landfills remain open. In other areas incineration is causing health hazards as the burning of plastics, batteries, pesticides, and cleaning and drain-clearing agents combine to discharge toxic substances into the air and poison incinerator ash. As a result, the political leadership is beginning to take a closer look at recycling and to provide funds for curbside bins, collection centers, and specialized trucks to collect recyclable materials. Neighborhood recycling programs have demonstrated that residents will deliver recyclable materials to central sites or to street-corner bins. Locating recycling containers at all train and subway stations, supermarkets, and other convenient central locations has been found to be an effective first step in developing a recycling program.

In 1988 New York City spent only $4 million on recycling pilot projects, while in 1989 this budget increased to $25 million. In 1989 the city recycled about 1,000 tons of its MSW per day (190). Until the 1991 cutbacks, plans called for increasing that amount to 7,000 tons in the next five years. In 1990 New York City passed out

$25 fines to homeowners who did not separate their waste. At the beginning of 1991 additional mandatory recycling programs began in New Jersey and Connecticut. Today, most of New York's waste paper is shipped to Taiwan and South Korea, where it is turned into paper products for export back to the United States. Similarly, most waste paper in Connecticut ends up in Canada.

One of the most successful recycling programs in the country is in operation in Seattle, where the garbage-disposal fees are based on volume; thus residents who do not recycle pay more. About 60% of the households are participating in the program. Each household receives three 90-gallon containers from the city: for paper, glass, and metal, and a fourth is being planned for plastics. Private contractors pick up and further sort the trash about once a month. As a result of this "pay as you throw" approach, the volume of garbage produced per household was cut by more than half (203). Seattle charges $165 per year for the weekly pickup of a 32-gallon can and $273 for a 60-gallon can.

The overall experience with recycling is mixed. In a handful of communities, like Rockford, Illinois, recycling had to be abandoned because of a lack of compliance, while in many other communities, such as Austin, Texas, 85% compliance has been reported as a result of peer pressure and public education (156).

The interdependence of national recycling policies and the need for international coordination was illustrated in 1990, when Denmark wanted to make returnable all beer and soft-drink containers sold in the country. The high court of the European Community ruled against Denmark and in favor of the retail associations, which argued that fair competition and the right to the free flow of goods also guarantees them the right to market nonrefillable bottles in Denmark (191). Such examples, as well as the increased involvement of organized crime in various aspects of the recycling industry (192), will necessitate a higher degree of international cooperation than what exists today.

While uncoordinated, poorly planned programs that disregard monetary incentives and market forces can backfire by flooding the market, well-coordinated legislation can not only motivate recycling but can guarantee a market for the recycled goods.

Separation at the Source

The most effective recycling method is to separate waste in each household. Since this approach can be inconvenient and space-consuming for busy people in crowded living quarters, various elements of the recycling strategy will have to be integrated into a harmonious whole. Municipalities should supply free color-coded containers (Figure 9.4) and curbside bins and conveniently scheduled collection, including special pickups for bulky or toxic items. Municipalities can also assist in stabilizing the local market for recyclables (Tables 9.2 and 9.3) by supporting the recycled paper, glass, metal, plastic, rubber and composting industries through both tax advantages and by using the recycled products.

In addition, it will be necessary to financially motivate the residents in favor of separation. Social behavior is most effectively changed through the pocketbook. Garbage collection should be billed according to the form and quantity of MSW

Fig. 9.4 Household containers manufactured in the Netherlands. (Courtesy of B. V. Handelsonderneming "Maltha," Rotterdam, the Netherlands.)

generated. Just as the installation of individual water meters tends to serve the conservation of water, recycling can be made more effective by separate monthly billings to each household. If separation reduced the monthly garbage collection bill of a household by 50%, cooperation would increase. This was clearly demonstrated in San Francisco, where a rebate was issued to customers who participated in a recycling program (92). Seattle charges its citizens according to the amount of MSW produced.

The Role of Manufacturers

Manufacturers can contribute to successful household recycling by color-coding their products, perhaps with a colored dot under the IBM bar code to correspond to color-coded waste containers. Manufacturers also could eliminate mixed-material packaging and reduce the use of plastics. The type of plastic used could be assigned a code number that would appear inside the color dot. (Color-coding would eliminate the need for using refrigerator magnets to determine if a can is made of aluminum or ferrous metals.)

It would be desirable if manufacturers concentrated on reducing the quantity and toxicity of packaging. Some of the worst offenders in terms of excessive packaging include microwaveable goods, items with tamper-resistant wrappings, and partially filled containers and boxes. In terms of toxic heavy metal content, red and yellow colored packages are the worst offenders. Government could use the monetary incentives of taxation and the indirect incentives of labeling to encourage more responsible packaging practices. Labels could also encourage the use of refillable and reuseable containers, such as plastic detergent bottles that can be refilled by small pouches of concentrates ("enviro-packs") that are mixed with water.

Manufacturers could eliminate toxic materials, such as heavy metals in colored ink and chlorine, which is present both in plastics and in bleached paper. (In 1990 New York State passed a law reducing the amount of heavy metals that are allowed in the ink used on packaging sold in the state.) They could also help recycling in small ways, such as not using plastic strips on cardboard price tags, not putting staples into cardboard boxes (which need to be "mined" before recycling the cardboard), making easily removable bottle labels, designing recyclable containers for ease of cleaning, and making discarded toys or electrical devices easier to pull apart.

Manufacturers could provide a market for recycled goods by purchasing recycled materials. Some industries are planning to do this, such as Coke and Pepsi, and others are already practicing it. Car manufacturers, for example, are using some sixty pounds of waste paper in each automobile produced. While car manufacturers represent a steady and growing market for recycled materials, their own product is not easily recycled. Car and appliance manufacturers will have to redesign their products both in terms of eliminating toxic substances and also to make them easier to take apart and recycle. In Germany, in fact, auto manufacturers are close to having designed a recyclable car!

The "New Kitchen"

Kitchen design is an important ally of the recycler. The greatest hurdle in recycling in the home is usually psychological rather than spatial. A typical household might first try three-way separation—for wet, nonrecyclable trash, cardboard and reusable paper, and recyclable glass and metallic items. An established routine should take about ten minutes a day. Then eventually three-way separation might be replaced with four-, five-, or even six-way separation.

The "new kitchen" would contain sorting closets, cubbyholes, built-in lazy Susan trays for garbage bins or built-in sorting bins. Another convenient item (Fig. 9.5) is a small chute covered with a plastic swivel lid. Vegetable scraps and organic

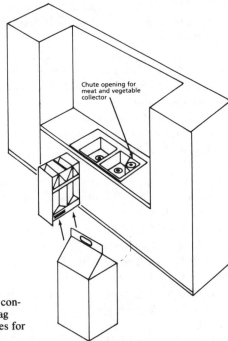

Chute opening for meat and vegetable collector

Fig. 9.5 The "new kitchen" is designed for practical containers that will fit in small spaces, such as a three-bag sorting drawer for foldable shopping bags with handles for lifting from the recycling bin (157).

kitchen wastes can be dropped into this chute, which might empty into a milk carton in the sink and from there be emptied into the nonrecyclable container.

Kitchen designers and suppliers of kitchen equipment will need to become more sensitive to the needs of recycling. Major manufacturers of kitchen equipment should make sorting drawers, lazy Susan sorting bins, and tilt-out bins as standard kitchen equipment. Kitchen designers should keep in mind small convenience items, such as automatic label scrapers, trash chutes, and can flatteners to make recycling more convenient.

The more finely household waste is separated, the greater its contribution to recycling. Figure 9.6 shows an approach where household waste is separated into four containers.

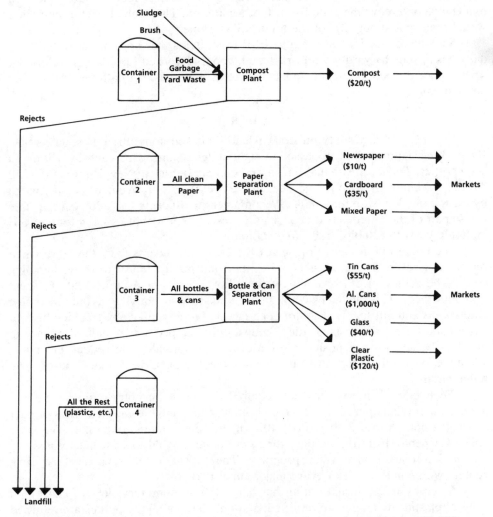

Fig. 9.6 Basic separation scheme (3). Prices shown represent the American market in 1988 (199).

Container 1 would receive all organic or putrescible materials, including food-soiled paper and disposable diapers and excluding toxic substances and glass or plastic items. The contents of this container can be taken to a composting plant that also receives yard wastes and possibly sewage sludge and produces soil additives.

Container 2 would receive all clean paper, newspapers, cardboard, and cartons for paper processing, where contents are separated mechanically and sold to commercial markets.

Container 3 would receive clean glass bottles and jars and aluminum and tin cans free of scrap metals and plastics.

Container 4 would receive all other waste, including plastic, metal, ceramic, textile, and rubber items. (Later, a fifth container could be added for recyclable plastics.) The contents of this container can be considered nonrecyclable and sent to a landfill or a recycling plant for further separation. The contents of this container would represent about 12% of the total MSW (Table 9.4).

Separate collections are required for trash items that are not generated on a daily basis, such as yard waste, brush and wood, discarded furniture and clothing, "white goods" such as kitchen appliances, toxic materials, car batteries, tires, used oil, and paint.

"Bottle Bills"

In 1981 Suffolk County on Long Island outlawed nonreturnable soda bottles. By 1983 legislation had been passed in eight states requiring a 5-cent deposit on all soda bottles. Today, the annual return rate on beer bottles in New York is 87% and 80% of six billion soft drink and beer bottles. Further improvement is anticipated by raising the deposit on nonrefillable containers to 10 cents and allowing the state to use part of the unredeemed deposits (at present kept by bottlers and totaling $64 million a year) to establish recycling stations.

The bottle bill in New York gives the customer 5 cents from the store when returning a bottle. The store gets 6.5 cents from the distributor to cover handling costs. The selling prices charged by the wholesalers already reflect the 6.5-cent reimbursement. This law gives wholesalers a financial incentive to avoid redeeming empty cans and bottles. Distributors can gain unfair advantage by trucking beer to supermarket chains in distant cities, knowing that empty bottles will be redeemed locally. By saving on deposits, the out-of-town wholesaler can reduce his price, while the local competitor must raise his to cover the expense of added out-of-town redemptions.

Since 1986 California has not required a deposit on bottles. Each year, 25 million Californians purchase 12 billion bottles, cans, and plastic containers of beer and soft drinks, representing about 10% of the total American market. The state collects a penny from the bottlers for each container sold and finances voluntary redemption centers and recycling programs. The state has established 2,400 recycling centers within half a mile of every major supermarket.

A survey in California found that only 57% of aluminum, 46% of glass, and 0.4% of plastic containers are returned, and only 30% of all the bottles are returned

Table 9.4
COMPOSITION OF TRASH AND DISTRIBUTION INTO SEPARATE HOUSEHOLD CONTAINERS (3)

Component	Trash Composition (% of Total Trash)			Content of Household Containers (% of Total Trash)			
	Total Trash	Recyclable	Nonrecyclable	(I) Food Garbage & Yard Waste	(II) Paper	(III) Bottles & Cans	(IV) "All the Rest"
Paper							
Newsprint	15.4	15.4		0	15.4	0	0
Magazines	2.6	2.6		0	2.6	0	0
Corrugated	1.5	1.5		0	1.5	0	0
Brown paper	8.3	8.3		0	8.3	0	0
Mail	4.1	4.1		0	4.1	0	0
Food cartons	3.1	3.1		0	3.1	0	0
Tissue	3.0	3.0		3.0	0	0	0
Wax cartons	0.4		0.4	0	0	0	0.4
Plastic-coated	1.1		1.1	0	0	0	1.1
	39.5						
Metals							
Ferrous beverage containers	1.2	1.2		0	0	1.2	0
Other ferrous	6.4	6.4		0	0	6.4	0
Nonferrous beverage containers	0.4	0.4		0	0	0.4	0
Other non-ferrous	0.7	0.7		0	0	0.7	0
	8.7						
Glass							
Beverage containers	5.0	5.0		0	0	5.0	0
Other glass	4.7	4.7		0	0	4.7	0
	9.7						
Plastic	4.7		4.7	0	0	0	4.7
Rubber & Leather	2.0		2.0	0	0	0	2.0
Textiles	2.0		2.0	0	0	0	2.0
Food Garbage	11.4	11.4		11.4	0	0	0
Wood	4.0	4.0		4.0	0	0	0
Yard Waste	16.0	16.0		16.0	0	0	0
Miscellaneous	2.0		2.0	0	0	0	2.0
Totals	100.0	87.8	12.2	34.4	35.0	18.4	12.2

Sources: For percent total trash: Holzmacher, McLendon & Murrell, P.C., 1986; for percent waste of subcomponents except brown paper, mail, cartons, tissue, wax cartons, and plastic-coated paper: US EPA, 1979b; for percent waste of brown paper, mail, cartons, tissue, wax cartons, plastic-coated paper: EG&G, 1982.

through the redemption centers. Under the law, if the return rate for any type of container does not reach 65% by the beginning of 1990, the redemption value will increase to 2 cents, then to 3 cents in 1993. Some centers have already closed because of insufficient volume. Part of the problem is that the value of aluminum (25–60 cents per pound) is greater than the redemption value, so too many bottles are being returned to places other than the voluntary redemption centers.

Bottle bills, while having achieved partial success, need to be improved. As of 1988 only nine states had bottle bills. Bottle bills should be integrated into overall recycling programs, which include office paper and newspaper recycling, cardboard collection from commercial establishments, curbside recycling, establishment of buy-back recycling centers, wood waste and metal recycling, glass and bottle collection from bars and restaurants, and composting programs. Advertising and public education are important elements in the overall recycling strategy. Street signs, door hangers, utility-bill inserts, and phone book, bus, and newspaper advertisements are all useful. The most effective long-range form of public education is to teach school-children the habits of recycling.

Plastics Recycling

Plastics are strong, waterproof, lightweight, durable, microwavable, and more resilient than glass. For these reasons they have replaced wood, paper, and metallic materials in packaging and other applications. Plastics generate toxic by-products when burned and are nonbiodegrable when landfilled; they also take up 30% of landfill space even though their weight percentage is only 7% to 9% (158). Recent research has found that paper does not degrade in landfills either and because of compaction in the garbage truck and in the landfill, the original volume percentage of 30% in the kitchen waste basket is reduced 12% to 21% in the landfill (179 and 188). In addition, plastics foul the ocean and harm or kill marine mammals. Other problems include the toxic chemicals used in plastics manufacturing, the reliance on nonrenewable petroleum products as their raw material, and the blowing agents used in making polystyrene foam plastics, such as chlorofluorocarbons (CFCs), which cause ozone depletion. CFCs are now being replaced by HCFC-22 or pentane, which does not deplete the ozone layer but does contribute to smog. For these reasons, recycling appears to be the natural solution to the plastic disposal problem.

Unfortunately, recycling and reuse are not easily accomplished because each type of plastic must go through a different process before being reused. There are hundreds of different types of plastics, but 80% used in consumer products is either high-density polyethylene (milk bottles) or polyethylene terephthalate (large soda bottles) (1991). It is not yet possible to separate plastics by types because manufacturers do not indicate the type of plastic used. Plastic parts of automobiles are still uncoded, so salvagers cannot separate them by type (158). Even if recycled polystyrene were separated and could be used as a raw material for a plastics recycling plant, such plants are just beginning to be built and we do not know if they will be successful. For these reasons, environmentalists would prefer to stop using plastics altogether in certain applications.

Table 9.5

QUANTITY OF PLASTIC WASTES GENERATED IN THE
U.S. (84)

Type of Plastic Waste	Amount of Plastic Waste (10^6 tons/yr)		
	1971	1975	1980
A. Waste containing a single type of plastic	0.2	0.3	0.6
B. Waste containing several types of plastic	0.3	0.7	1.4
C. Waste containing plastic and nonplastic refuse	1.5	3.0	6.0
Total	2.0	4.0	8.0

By 1984, U.S. plastics production increased to 29.4 billion pounds (13.4 million tons) (158) and by 1988, 60 billion pounds (30 million tons) were produced (188). In 1971, the plastic fraction of MSW was only 2 million tons (Table 7.5); in 1988, Americans discarded some 14.4 million tons of plastics (199). The amount of plastics used in U.S. packaging has increased from 3.3 to 8 million tons per year between 1974 and 1990 (191). The amount of plastic recycled is only 1.2% of the plastic produced (199), mostly soda bottles, valued at $120/ton or about $100 million/year. The most widely recycled products are plastic soft-drink containers (28% in 1990) made of PET (polyethylene terephthalate). Plastics manufacturers predict that PET recycling will rise to 50% by 1992. In 1987 the beverage industry alone used 890 million pounds of plastics and recycled 100 million pounds of that into fiberfill and carpet facing. Coke and Pepsi plan to obtain approval to recycle PET into new bottles.

In 1984 plastics represented 6.8% by weight or 25.4% by volume of the MSW collected in the United States (166). Today, it is 8% by weight (see Tables 1.3 to 1.10). The EPA estimates that the percentage of plastics in MSW will reach 15% by the turn of the century (163). In 1984, 42% of the plastics in the MSW came from packaging, 12% from consumer goods, 8% from electronics, 8% from furniture, 6% from transportation, 2% from the construction industry, and 22% from other sources (158). Today, 17% of all American packaging is plastic and in 1990 8 million tons of plastics were used for packaging purposes only (191).

Bans and Regulations

As the use of plastics increases, so are the laws aimed at curbing it. Suffolk County, Long Island, proposed banning the use of plastic grocery bags and food containers by July 1989; but in May 1989 the State Supreme Court ruled that the law could not take effect until Suffolk officials conducted a thorough environmental-impact study. The Suffolk law prohibits the use of polystyrene and PVC plastics for utensils, plastic-foam coffee cups, and foam hamburger "clam-shells." The law exempts some plastics for which there are no convenient alternatives, like clear plastic wrap (163). The plastics industry has filed suit, and as a result the Suffolk bill's effective date has been delayed.

In March 1989 the city council in Minneapolis adopted an ordinance, effective July 1990, to ban most throwaway plastic food packaging from grocery store shelves and fast-food outlets. The ordinance prohibits any packaging that is not biodegradable (paper), reusable (glass jars), or conveniently recyclable (metal cans). A nearly identical piece of legislation is pending in St. Paul (163). Massachusetts and Berkeley, California, have recently banned cups made with CFCs. Other states are considering restrictions on the use of plastics for egg cartons, disposable diapers, and grocery bags. Denmark is considering a ban on all plastic packaging. In Italy, certain types of plastic bags will be outlawed by 1991. Similar restrictions are being sought in Britain, Switzerland, and Austria.

Biodegradable Plastics

Because plastics are not biodegradable, the plastics industry is beginning to respond to mounting public concern by introducing photodegradable and biodegradable plastics that are claimed to fully decompose in fifteen years. All plastics are photodegradable to some extent, but by adding carbon monoxide, cobalt, and iron, which absorb the sun's ultraviolet rays, they degrade faster. In biodegradable plastics cornstarch is embedded into the molecular matrix. When bacteria and fungi consume the starch, they break the chains, leaving behind a plastic "Swiss cheese."

The production of degradable plastics has not been welcomed by everyone (158). Probably the strongest argument against degradable plastics is that they cannot be recycled because they *do* decay to some degree. Even a small amount of degradable plastics can contaminate a whole batch intended for recycling. Also, because biodegradable plastic is weaker, the products made from it have to be thicker; therefore, the same item uses more plastic. Another problem is that the tightly packed and sealed landfills do not contain enough moisture and oxygen to decompose even degradable plastics (164). Unfortunately, nature works slowly in a landfill. One can dig up newspapers after ten years in a landfill and read them. The same is likely to happen to biodegradable diapers or plastic bags (164).

Another concern involves the chemical fillers, stabilizers, and additives used to strengthen biodegradable plastic products. These include cadmium, lead, chromium, copper, beryllium, and zinc. In addition, the ink or paint used on colored plastics may contain toxic heavy metals. As the plastics biodegrade in the landfill, they may leach toxic heavy metals. In addition, the presence of bacteria can infect the contents held by a plastic container.

For these reasons the "degradable" label has recently been removed from the Hefty trash bag line of Mobil, the Handi-Wrap products of Dow, the Glad trash bags of First Brands, and others (188). However, degradable plastics do have some useful applications, such as bags used to collect yard waste for composting and six-pack rings. Degradable bags made by ADM are supposed to degrade in two months in direct sunlight. Bags used to collect leaves and yard debris could be shredded and composted together with the contents, and six-pack rings would spare marine life.

Reusing Plastics

Part of the difficulty in recycling plastics is in sorting different plastic types. The plastics industry recently initiated a voluntary coding system that would identify the type of resin used in each product. This coding would also help make car parts recyclable. But even if plastic products are coded, the public will need some inducement to sort them. McDonald's has started pilot programs where customers are asked to separate their waste food from the plastic packaging. The problem here is that the shipping costs to the recycler could outweigh the value of the used plastic (158).

Polyethylene terephthalate (PET) and high-density polyethylene (HDPE) are both being recycled. So far, recycling involves mostly PET bottles. In 1990, 28% of PET bottles were being recycled and the Council for Solid Waste Solutions is aiming to recycle 25% of all bottles (18 billion) by 1995 (199). Plastic recycling has increased from 0.1% in 1986 to 1.1% by 1988 and the plastics industry is investing some $20 million per year in building reprocessing plants (200). Plastics recyclers pay about 6 cents a pound for flattened PET bottles. After cleaning and pelletizing, the PET pellets can be sold for up to 30 cents a pound ($600/ton), while virgin PET costs even more (167). In 1987 reconstituted PET pellets could be sold for $200–$300/ton (104). (For the costs of recycled plastic materials in various areas in 1989, see Table 9.2.) Reclaimed PET is used as fill for clothing and pillows, carpeting, sporting goods, paintbrushes, automotive parts, furniture stuffing, flower pots, corrugated pipe, fence posts, bathtubs, shower stalls, sinks, spas, mattresses, decks, piers, picnic tables, park benches, and landfill liners. Unseparated mixed plastics can also be reused to make plastic lumber, drainage troughs, wood panel replacements, industrial pellets, and molded extruded products. Commingled plastics can also be converted into car stops in parking lots, fence posts, park benches, and boat docks.

Plastic Reprocessing

The problems with plastics in MSW include: (a) their high chlorine content, which contributes to dioxin formation when burned; (b) their resilience and non-biodegradability, causing unsightly litter; (c) the pentane content of polystyrene cups and other foam products, which contributes to smog; and (d) the toxic metal content (lead, cadmium) that is released when burned and (e) the low density (800 pounds per bale) and large volume, which increases collection and transport costs. Yet plastics recycling has lagged far behind that of glass, paper, and aluminum, even though recycling is feasible.

The best way to separate plastics is at the source. When MSW is separated according to density, light plastics tend to come out with the paper and heavy plastics with the inerts. Flotation as a separation technique fails because the density differences between many polymers are small and are made variable by the use of pigments and other additives.

Reprocessed plastic is divided into three types. *Type A* plastic is obtained from industrial or in-plant waste streams containing only one type of polymer. *Type B* plastic is obtained from diversified plastics-processing plants (automobile, appliance, and electrical industries) and contains several types of polymers. *Type C* plastic is obtained from randomly collected MSW and must be segregated from nonplastic materials and decontaminated prior to use. Some plastic products contain thermoset materials, highly filled materials, and enclosures of nonmelting substances, which makes separation from MSW more difficult.

Plastics Recycling Plants

Thermoplastics are the easiest type of plastic to recycle because they can be melted and reformed. Type A plastic waste, such as reject cable, bottle waste, or reject film, can be used to make drainage pipe, agricultural or construction fill, or cable splice joint covers. The main problem in reusing Type B plastic waste is compatibility. Chemically different polymers do not form homogeneous blends with adequate physical properties; this problem can be reduced by adding compatibilizers, chlorinated polyethylene, chlorwax, and co- or terpolymers (84).

Cross-linking can be achieved by chemical and irradiative methods. Cross-linking agents include organic peroxides; irradiative techniques include radio-nuclide or electron beam radiation. Unfortunately, the cross-linking process frequently alters the polymer in such a way that it can no longer be remelted or reprocessed, so further recycling becomes difficult or impossible. Table 9.6 lists the properties of some recycled plastics. Figure 9.7 illustrates low-pressure molding equipment used on plastic-scrap applications. Some scrap-molding equipment can be fed by unground and irregularly shaped plastic bottles, fibers, films, or pelletized waste products. The products that can be made by scrap-molding equipment are listed in Table 9.7 and illustrated in Figure 9.8.

Table 9.6
PROPERTIES OF RECYCLED MIXED PLASTICS (84)

Plastic Sample Evaluated	Tensile Strength in PSI (ASTM D638-617)	Izod Impact, ft-lb/in., (ASTM D256-56)
Polyethylene, injection grade mixture of low and high density	3,600	2.8
Polystyrene, impact modified	5,470	1.2
Polyvinyl chloride	7,300	1.5
Ternary blend of sample[a]	1,450	0
Sample[a] with 15% compatibilizer[b]	1,715	0.45
Sample[a] with 25% compatibilizer	1,600	1.6
Sample[a] treated by chemical cross-linking	1,800	1.8
Sample[a] treated by irradiation to achieve cross-linking	1,800	1.7

[a] Sample consists of 40% PE, 40% PS and 20% PVC (incorporating 30% plasticizer).
[b] Compatibilizer is chlorinated polyethylene.

Table 9.7
TYPICAL SCRAP-MOLDING PRODUCTS (84)

Type of Product	Number of Cavities	Shot Weight, oz	Cycle Time, Sec	Type of Plastic in Feed Stock
Shoe soles	2	19	50	PVC
Luggage handles	4	7	45	PE, PS, ABS
Suction pump	1	6	60	PVC
Seals	4	5	50	PVC
Toy animals	6	4	45	PS, PVC
Bicycle saddles	1	12	90	PVC
Bicycle pedals	4	8	60	PE, PVC
Lawn mower wheels	1	9	45	PE, PVC
Door stops	4	7	50	PE, PVC, PS
Dog toys	4	10	60	PVC, PE, PS
Cable housing	2	24	90	PP, PE
Auto parts	various	2–8	30–60	PP, PVC

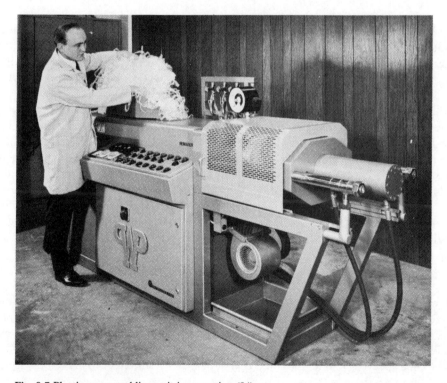

Fig. 9.7 Plastic-scrap-molding unit in operation (84).

Fig. 9.8 Molded products from plastic wastes (84).

The Center for Plastics Research at Rutgers University has developed an automated system to convert soft-drink bottles and milk jugs back into their raw materials (usually PET) and to produce park benches, fence posts, carpet backings, and paintbrush bristles from them (160). The center is also working on polystyrene recycling processes.

Efforts to recycle polystyrene are just beginning. In 1989 seven polystyrene manufacturers established the National Polystyrene Recycling Company at a total investment of $14 million. They hope to establish five recycling centers and achieve 25% recycling by 1995. They have jointly built one recycling center in Leominster, Massachusetts. The polystyrene is collected from schools, hospitals, and restaurants. At the recycling center it is ground, washed, dried, and melted to a waxy consistency (165). After filtration to remove dirt, it is cooled and chopped into pellets. In 1990 another large polystyrene recycling facility was opened in the suburbs of Los Angeles to recycle foam coffee cups and hamburger clamshells from fast-food restaurants, hospitals, and schools (191).

Similar work is being done at the Amoco demonstration plant at Greenpoint, Brooklyn. Here the plastic trash from several cafeterias and restaurants is being processed to make insulating board, cafeteria trays, and in-baskets from foamed coffee cups and clamshell boxes. The Greenpoint plant is designed to separate polystyrene foam from intermixed restaurant trash. This intermixed trash, however, must be rich enough in polystyrene to support recycling. The value of virgin polystyrene is 55 to 60 cents per pound (164), and this might not be sufficient to make polystyrene recycling cost-effective.

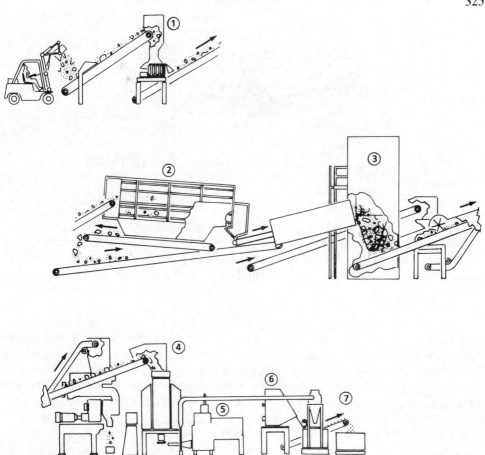

Fig. 9.9 The Amoco demonstration plant for polystyrene recycling (164).

The Amoco plant uses separation equipment designed for the mining industry. The waste is shredded (point 1 in Fig. 9.9), then small items like utensils and straws are removed (point 2) by a rotating screen, called a trommel. Next, the light items such as plastic foam and paper are separated (point 3) by air pressure. This "light fraction" is ground up and pulped with water (point 4), then screened to wash away the paper fibers from the polystyrene (point 5). The resulting ground polystyrene is dried (point 6) and collected (point 7) for shipment to the user's plant, where it is remelted to make plastic products.

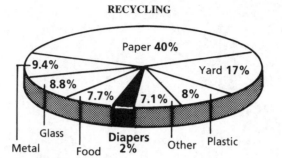

Fig. 9.10 Disposable diapers contribute approximately 2 percent to our total MSW ("Diapers in the Waste Stream," Carl Lehrburger, Sheffield, Mass. [119].)

Besides plastic manufacturers, others are also entering the plastic recycling business. Polymerix Inc. in Lincoln Park, New Jersey, for example, operates a pilot plant that produces "TriMax" lumber from municipal and industrial plastic scrap. The company is also planning full-scale units in Lincoln Park and in Islip, New York.

Disposable Diapers

Every baby requires some 5,000 to 10,000 diapers, which results in 16 billion disposable diapers (3.6 million tons) being dumped yearly in landfills. The yearly cost of diaper disposal exceeds $100 million, and the diaper volume approaches 2% of the total MSW production (Fig. 9.10). Some argue, however, that disposable diapers represent less than 1% by weight of the total MSW in our landfills (179) and that their total environmental impact is similar to that of cloth diapers, once the costs of laundering and transportation are factored in (191). The use of disposable diapers is a symptom of a throw-away culture. In a single generation the cotton diaper (which can be reused up to 200 times) has been replaced by disposable plastic-lined paper diapers. The disposable diaper costs about 22 cents; the reusable one about 15 cents. Disposable diapers also can become a health hazard, breeding viruses and bacteria in the landfills.

Oregon and Washington are considering legislation to ban, label, or tax disposable diapers. Some states are recommending giving the same tax incentives to the delivery services of cotton diapers (a labor-intensive industry that creates jobs for the unskilled) as they give to recycling operations. Some manufacturers like Procter & Gamble, the maker of Pampers, are experimenting with recycling disposable diapers (158), while others are considering using biodegradable plastic liners in their diapers.

Recycling of Toxic Substances

The careless disposal of products containing toxic or hazardous substances can create health hazards if allowed to decompose and leach into the groundwater from landfills or if vaporized in incinerators. Since hazardous-waste landfills are limited, the available options are either to have manufacturers substitute toxic materials with nontoxic substances or recycle the products that contain toxic materials. Municipalities are just beginning to consider the requirements of toxic-waste recycling. Products that are toxic or contain toxic substances include paint, batteries, tires, some plastics, pesticides, cleaning and drain-cleaning agents, and PCBs found in white goods (appliances). Separate collections are also required for medical wastes (Chapter 3).

Batteries play an important role in the recycling of toxic substances. Batteries represent a $2.5 billion-a-year market. At present, practically no batteries are being recycled in the United States. Battery manufacturers feel that recycling is neither practical nor necessary; instead, they feel that all that needs to be done is to lower the quantities of toxic materials in batteries (168). It is estimated that 28 million car batteries are landfilled or incinerated every year. This number contains 260,000 tons of lead, which can damage human neurological and immunological systems. The billions of household batteries disposed of yearly contain 170 tons of mercury and 200 tons of cadmium (168). The first can cause neurological and genetic disorders, the second, cancer. Some batteries also contain manganese dioxide, which causes pneumonia. When incinerated, some of these metals evaporate. The excessive emissions of mercury were the reason why Michigan temporarily suspended the operation of the incinerator in Detroit, the nation's largest.

Some states have recently initiated efforts to force manufacturers to collect and recycle or safely dispose of their batteries. The Battery Council International has prompted several states to pass laws requiring recycling of all used car batteries. In 1988 California, Florida, Minnesota, Pennsylvania, and Wyoming passed some form of model legislation concerning batteries (174). Four states (California, Florida, Minnesota, and Rhode Island) forbid the incineration of car batteries, while a not-yet-enforced law in New York also bans the incineration of household batteries. A battery bill pending in New York would promote recycling by requiring a 25-cent deposit on each household battery and a $5 deposit on each car battery (168). Suffolk County on Long Island also has passed a bill that places a $5 refundable deposit on car batteries. The EPA estimates that only 8% of car batteries produced in the United States are being recycled (174). In many European countries used batteries are returned to the place of purchase for disposal.

The disposal of white goods (appliances such as refrigerators, air conditioners, microwave ovens) is also a problem. Until 1979 appliance capacitors were allowed to contain PCBs (polychlorinated biphenyls). Even after the ban, some manufacturers were granted an extra year or two to deplete their inventories. When white goods are shredded, the "fluff" remaining after the separation of metals (consisting of rubber, glass, plastics, and dirt) is landfilled. When it was found that the "fluff" contains more than 50 ppm of PCBs, the Institute of Scrap Recycling Industries advised its 1,800 members not to handle white goods. The safe disposal of PCB-containing white goods would require scrap dealers to remove the capacitors before shredding. Similar toxic-waste disposal problems are likely to arise in connection with electronic and computing devices, the printed circuit boards of which contain heavy metals.

A long-range solution to toxic-waste disposal might be to require manufacturers of new products containing toxic substances to arrange for recycling *before* the product is allowed on the market, or at least to provide instruction labels describing the recommended steps in recycling.

Paper Recycling

Paper used to be made of reclaimed materials such as linen rags. Rags were the raw materials used by the first paper mill built in the United States in 1690 in Philadelphia. Only in the nineteenth century did paper mills convert to wood-pulping technology. It takes seventeen trees to make a ton of paper. All Sunday newspapers in the United States, for example, require the equivalent of half a million trees every week (191). When paper is made from waste paper, it not only saves trees but also saves 4,100 kWh of energy per ton (the equivalent of a few months of electricity used by the average home), 7,000 gallons of water, 60 pounds of air-polluting emissions, and three cubic yards of landfill space and the associated tipping fees (182). The production of recycled paper also requires fewer chemicals and far less bleaching.

The paper output of the world has increased by 30% in the last decade (197). In 1990 the United States used more than 72 million tons of paper products, but only 25.5% of that (18.4 million tons) is made from recycled paper (191 and 199). This compares with 35% in Western Europe, almost 50% in Japan, and 70% in the Netherlands (197). There are some 2,000 waste-paper dealers in the United States who collect nearly 20 million tons of waste paper each year. In 1988, 20% of the collected waste paper was exported, mostly to Japan. As paper recycling increased in 1989 and 1990, so did the percentage that is exported (110). According to the American Paper Institute, in 1988 the Port of New York exported 1.1 million tons of scrap paper, making waste paper the city's largest single export material (129).

The waste-paper market is very volatile. In 1990 the average value of one ton of waste paper was only about $10 (see Table 9.2). In some locations the mixed office waste or mixed-paper waste (MPW) has no value at all and tipping fees must be paid to have them picked up. Therefore, what pays for collection and processing is not the prices paid for waste paper, but the savings represented by *not* landfilling them at $70/ton on the East Coast (199). A ton of old newspapers in California brings $25 to $35 because of the Japanese market demand (183). In 1988 New York City had twenty-four active waste-paper contracts. They ranged from paying $16/ton to the city to receiving $5/ton from the city for taking the waste paper. In the Northeast an oversupply in 1989 caused the waste-paper price to plummet from $15/ton to about −$10/ton (179 and 183). This oversupply also resulted in increased waste-paper exports to Europe, which in turn caused the collapse of the waste-paper market

Table 9.8
WASTEPAPER STOCK PRICES[a] (83)

Type of Wastepaper	Price Range per Ton of Wastepaper ($)
No. 1 mixed office waste	1–3
No. 1 super mixed	5–6
No. 1 news	10–12
No. 1 old kraft	25–26.50
Double-lined kraft corrugated cuttings	18–20
Old corrugated boxes	16–18
Boxboard clips	2–6
No. 1 books and magazines—graded wood free	3–7
Books and magazines mixed	2–5
White news cuts no. 1	30–35
Ledger stock no. 1	25–27.50
Ledger stock—colored repacked	18–20
Manilla tab cards no. 1	40–42
Colored tab cards	28–30
No. 1 white envelope cuts	50–55
No. 1 hard white shavings	37.50–40
No. 1 soft white	32.50–37.50
New manilla envelope cuts, grade white free	45–50
New manilla envelope cuts, white grade wood	35–37.50
Mill wrappers	7–9

[a] Prices are given FOB, Chicago, based on shipments in the carload size range. Some of the no. 1 mixed paper is nearly worthless and, in some cases, it is necessary to charge the person shipping the paper for removal. No. 1 otherwise refers to the best grade in the particular category.

in Holland, where the value of a kilogram of waste paper dropped from eight cents to one cent (197).

Waste paper can be classified into "bulk" or "high" grade. The highest-grade papers are manila folders, hard manila cards, and similar computer-related paper products. High-grade waste paper is used as a pulp substitute, whereas bulk grades are used to make paper boards, construction paper, and other recycled paper products. The bulk grade consists of newspapers, corrugated paper, and MPW. MPW consists of unsorted waste from offices, commercial sources, or printing establishments. Heavy black ink used on newspapers reduces its value, however. The value of the paper is also reduced by the presence of other substances that interfere with a single-process conversion into pulp, such as the gum in the binding of telephone directories or the chemical coating of magazines.

The most effective way to create a waste-paper market is to attract a pulp and paper mill to the area. To keep such a plant in operation, however, requires a high-grade waste-paper supply of about 300 tons per day. In addition, facilities are also needed for wastewater treatment.

Paper Recycling Methods

Paper recycling might start with a trailer on a parking lot or with recycling centers provided with dumpster containers where paper can be left at any time. Ten states already have mandatory regulations for the recycling of waste paper, while others are in the process of preparing such legislation (Fig. 9.3).

Some 85% of the waste from commercial buildings is high-grade paper (129). In San Francisco, white office paper is recycled by giving a desk-top container to each office employee. The full containers are emptied into a predesignated barrel, which is emptied once a week into bins outside the building. This sorting and collecting is troublesome and requires a change of habits. As of 1989, not even the office paper of the U.S. House of Representatives is separated. Instead, it is sold mixed and baled at $1 per ton.

If waste paper is not separated at the source according to quality, it must be hand-picked and mechanically separated in a paper-processing plant. One such plant operating in Holland is shown in Figure 9.11, where 200 to 400 daily tons of paper are separated into more than 50 categories. The separated paper products are shredded, baled (Fig. 9.12), and stored (Fig. 9.13) prior to shipment to users.

Uses for Recycled Paper

MPW is usually not acceptable as pulp substitute. On the other hand, paper boxes, pressed paper trays, towels, disposable tissues, stationery, and various types of construction materials have been made from scrap paper (Fig. 9.14). In addition to newsprint and fire-starter logs, one of the major uses for salvaged newspapers is wallboard, either with or without a plaster core.

Where no better market exists, MPW can be converted easily into a densified fuel that is suitable for direct combustion and has a heating value similar to that of wood. This fuel has good storage characteristics and a low sulfur content. The end-product can be in the form of pellets, cubes, or briquettes.

The fuel production steps shown in Figure 9.15 include shredding, air classification, magnetic separation, and densification. The baled MPW first passes through a swing hammer mill with a 2″ (51 mm) grate spacing. The shredded feedstock then passes through an air knife classifier, which removes any heavy, typically noncombustible contaminants. The final size distribution is provided by a secondary fixed hammer-mill shredder. A magnetic separator positioned above the shredder removes most of the remaining ferrous materials. The densification process is different for briquetting, cubing, and pelletizing (Table 9.9). In briquetting, for example, a single piston forces material through a single die, while in cubing or pelletizing multiple dies are used.

For a 100 TPD densification plant, having a capital cost of $3 million, no tipping fee, and operated by a staff of eight five days a week in a single shift, the break-even sale price for the product is $45.50/dry ton, representing $2.92 per million BTU

Fig. 9.11 The J. De Paauw & ZN B.V. paper-separation plant in Rotterdam, the Netherlands.

Fig. 9.12 Baling separated paper products. (Courtesy of J. De Paauw & ZN B.V., Rotterdam, the Netherlands.)

of heating value (88). The break-even costs for operating the plant at different tipping fees are shown in Figure 9.16. The higher the market value of the fuel produced, the lower the tipping fees the plant needs to charge to break even. Positive values represent a fee charged for accepting the MPW, while negative values suggest a price paid for the waste. If the produced fuel can be sold for $45.50/dry ton, the tipping fee for accepting the MPW is zero. Other densified refuse-derived fuel (DRDF) plants reported lower break-even sale prices, in the range of $23–$37/ton for single shift and $16–$29/ton for two-shift operations (88). The produced DRDF has a heating value of 7,800 BTU/lb, which compares only to the least expensive fuel, such as low-quality coal.

Newsprint Recycling

A large part of the waste-paper problem has to do with newsprint, which makes up 8% of the total MSW by weight. Some 13 million tons of newsprint are consumed yearly in the United States, 60% imported from Canada. The total North American demand for waste paper in newsprint production was only 1.55 million tons in 1989, or about 12% (183). It is estimated that this demand will double by 1992, to about 3.07 million yearly tons (183), and to 5 million tons by 1993 (200). Recently, over

Fig. 9.13 Storage of the various grades of paper. (Courtesy of J. De Paauw & ZN B.V., Rotterdam, the Netherlands.)

the objection of the American Newspaper Publishers Association, seven states (Connecticut, California, Florida, Minnesota, Maryland, Wisconsin, and North Carolina) have adopted laws requiring the use of recycled paper in newspapers. The leader in this area is Connecticut, which requires the use of 20% recycled paper in the newspapers sold in the state by 1993 and 90% by 1998. Suffolk County on Long Island requires 40% by 1996. New York State reached a voluntary agreement with its publishers to achieve the 40% goal by the year 2000. Florida applies a ten-cent waste-recovery fee for every ton of virgin newsprint used and grants a ten-cent credit for every ton of recycled newsprint used.

The net effect of such legislation will be an increased and steady demand for waste paper, which is essential for the success of recycling. As the demand for waste paper products rises, paper manufacturers will also increase their capacity to produce recycled paper. At the end of 1988 only 9 out of the 43 newsprint mills in North America used recycled fibers. Now there are plans to increase that number to 26 at an investment of nearly $2 billion (183). Atlantic Packaging Ltd. is opening a 135,000 ton/year capacity plant in Ontario, Canada, and the Jefferson Smurfit Corp. is opening a plant in upstate New York that will use 330,000 tons of waste paper yearly. Now that the sustained demand for used newspapers is guaranteed, Canadian paper mills

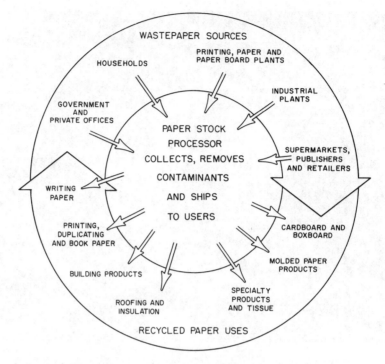

Fig. 9.14 The sources and uses of recycled paper. The circle with two arrows is one of the symbols used to note recycled products (83).

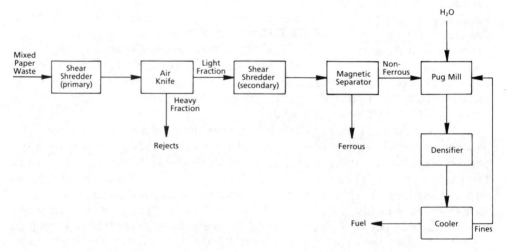

Fig. 9.15 Mixed paper-waste processing scheme for densified fuel production (88).

334

Table 9.9
DENSIFIER CHARACTERISTICS (88)

Equipment Type	Manu-facturer	Throughput Capacity (tons/hr)	Throughput Capacity (Mg/hr)	Installed Horse-power	Power Consumption (kwh/ton)	Power Consumption (kwh/mg)	Feedstock Specifications Optimum Moisture Content (%)	Feedstock Specifications Maximum Moisture Content (%)	Feedstock Specifications Particle Size	Product Size	Capital Cost ($/TPH)
Pellet Mill	Simon-Barrow	6	5.5	300	40	44	18–20	25	95% < 1.5 in (38 mm)	0.75 in (19 mm) Dia	21,100
	Sprout-Waldron	4–5	3.6–4.5	300	—	—	15–20	30	100% < 0.75 in (19 mm)	0.625 in (16 mm) Dia	22,100
	California Pellet Mill	5	4.5	300	30	33	10–15	30	90% < 1.25 in (32 mm)	0.75 in (19 mm) (or 1.0 in (25 mm) Dia	20,000
Cuber	Warren & Baerg	8–10	7.3–9.1	200	14	15	12–18	20	90% < 2.0 in (51 mm)	1.25 in (32 mm) Sq	12,700
	Papakube	6–10	5.5–9.1	150	11	12	13	20	90% < 2.0 in (51 mm)	1.25 in (32 mm) Sq	12,600
	Lundell	3–5	2.7–4.5	150	15–26	17–29	12–15	—	—	0.75 in (19 mm) or 1.75 in (44 mm) Sq	—
Baler	Balemaster	4.5	4.1	25	—	—	—	—	—	72 in × 36 in × 30 in (1.8 m × 0.9 m × 0.8 m)	7,800
	International Baler Corp.	3–5	2.7–4.5	30	—	—	—	—	—	72 in × 48 in × 36 in (1.8 m × 1.2 m × 0.9 m)	7,800

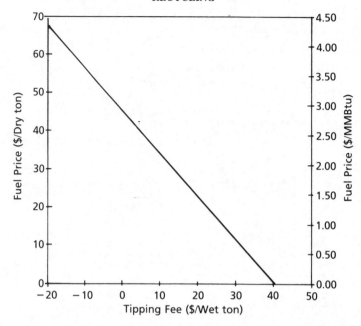

Fig. 9.16 Break-even cost of mixed paper-waste processing as a function of fuel prices and tipping fees (88). With tipping fees exceeding $50 in some areas, mixed paper processing can be a viable business. In 1988, the average price of a ton of waste cardboard in the United States was $35, white ledger $45, and newspaper $10 (199).

are proposing long-term contracts to purchase all the used newspapers that will be collected in Connecticut.

The cost of a new paper mill is about $400–$500 million. The addition of deinking equipment costs only $60–$80 million (183). In a deinking mill the waste paper is dumped into a huge blender. Here hot water, centrifugal forces, chemicals (caustics and detergents), and fine screening separate the ink, clay, and fillers from the paper. The main advantage of this process is that it does not require chlorine, as the lignin from the wood fibers had already been removed during bleaching, when the original paper was made. Because chlorine interacts with lignin in the bleaching process to form dioxins and furans, this is an important advantage. As waste-paper fibers have already been bleached, deinking plants require 75% less bleach and therefore provide less opportunity for dioxin formation.

The sludge from the deinking plant contains ink residues, fillers, clays, and fiber fragments. If the ink contains lead, chromium, cadmium, or other heavy metals, the sludge must be disposed of in hazardous-waste landfills. If the sludge is nontoxic, it can be used as clay-heavy soil conditioner.

Glass Recycling

About 13 millon tons of glass are disposed of in the United States every year, representing more than 7% of the total MSW that is generated (156). But only about 12% of the total glass production is recycled. In comparison, Japan recycled 42% of its "nonreturnable" glass containers in 1983, while in 1987 Germany recycled 47% and Holland 50% (169). In 1986 Europeans recycled almost 3.5 million tons of glass (Fig. 9.17).

Salvaged glass has been used in bricks and paving mixtures. "Glasphalt" can be made from a mixture of glass and asphalt or a mix of 20% ground glass, 10% blow sand, 30% gravel, and 40% limestone (83). In spite of all these other uses, the main purchasers of crushed glass are the glass companies themselves. The use of recycled crushed glass reduces both the energy cost and the pollutant emissions associated with glass making. Crushed glass is easily saleable (Table 9.2), with a market almost as good as that for aluminum. Manufacturers use from 20% to as much as 80% of salvaged glass in their glass-making processes. The first glass re-

Fig. 9.17 Glass recycling in Europe in units of thousands of tons per year in 1986. In 1988 Americans recycled 1.5 million tons of glass, out of the total of 12.6 million tons in the nation's garbage, or about 12%. The market value of a ton of recycled glass, separated by color, was $40 to $60 (199). (Courtesy of B. V. Handelsonderneming "Maltha," Rotterdam, the Netherlands.)

cycling plants in Europe were built in the 1950s and consisted of simple sorting and washing steps.

The first step in the "nonreturnable" glass recycling process is the curbside recycling container (Fig. 9.2). These containers are 2 to 3 cubic meters (70 to 100 cubic feet) in volume and for reasons of convenience are located near shopping centers. Holland alone has more than 10,000 such units. Each truck can carry about 22 tons of glass and collects it from as many as 250 curbside containers. The glass is delivered to the processing plant (Fig. 9.18) and placed into storage prior to processing. In Holland 50% of the glass is green, 40% is white, and 10% is amber and other colors. The processing in the plant in Rotterdam is highly automated and has the capability to separate the glass according to color. When the glass is taken from storage, it is crushed and grinded. In this process its density increases from 275 kilograms per cubic meter (17 pounds per cubic feet) to 1,200 kg/m³ (75 pounds per cubic feet), or about fourfold. The glass is also cleaned and the foreign matter separated out. The specifications in the Rotterdam plant limit the maximum amount of ceramics to 100 grams per ton of crushed glass, the same limit for aluminum is only 6 grams per ton. The cleaned, separated, ground glass is sold to glass companies. Similar glass recycling plants exist in many parts of the world, including ones operated by Owens Illinois and others in the United States.

Fig. 9.18 Glass processing plant. (Courtesy of B. V. Handelsonderneming "Maltha," Rotterdam, the Netherlands.)

Metal Recycling

In the United States over 15 million tons of metals are discarded every year (199). This represents almost 9% of MSW by weight. We recycle 14% of our metallic wastes (nearly 64% of aluminum) (168). Table 9.10 shows the quantities of nonferrous metals that were recycled in the United States in the 1960s. During the last fifty years, more than half of the raw materials used in steel mills was recycled (Fig. 9.19) and at least one-third of the aluminum produced is from recycled sources (83).

Aluminum recycling is profitable and well established because it requires only 5% of the electric power to remelt aluminum as it does to extract it from bauxite ore. In 1990 the average price paid for crushed, baled aluminum cans was $1,050 per ton (199) and some 55 billion aluminum cans (0.96 million tons) have been recycled. The recycling rate of aluminum increased from 61% in 1989 to 63.5% in 1990 (199). Steel has also been recycled for generations, but the recycling of steel cans is relatively new. It was necessary to reduce the rust-preventing tin layer on the steel cans first, so that they might be added directly to steel furnaces. The recycling of steel cans has increased from 5 billion cans in 1988 to 9 billion in 1990 and its market value varied from $40 to $70 per ton in 1990 depending on location (199).

The main sources of scrap metals are cans, automobiles, kitchen appliances (white goods), structural steel, and farm equipment. The value of the noncombustibles in incinerator ash varies from area to area. Table 9.11 provides data on both the unit values and the compositions of MSW in the early 1970s. Scrap-metal prices for 1989 are given in Table 9.3.

Scrap-metal recycling has not always been profitable. In New Jersey, for example, the market value of ferrous scrap in 1987 was about $30 per ton. A waste

Table 9.10

NONFERROUS SCRAP RECOVERED IN THE UNITED STATES
IN 1964 (83)

Type of Scrap	Weight Recovered, Thousands of Tons
Aluminum	552.0
Antimony:	
Antimonial lead	22.3
Secondary smelters	20.8
Primary smelters	0.3
Copper, alloyed and unalloyed (73 to 90% as much as domestic mine output, 1955–64)	1093.0
Secondary lead	541.6
Magnesium alloys	10.0
Mercury (76 lb flasks)	0.9
Nickel	23.0
Tin	23.5

Fig. 9.19 Crane with magnet moving scrap in scrap yard. The scrap in foreground was just dumped from a truck and is being piled in area where burners are operating. In background to the right is a small alligator shear (Courtesy of Bureau of Mines, U.S. Department of the Interior [83].)

recovery plant in Albany, New York, reported that in 1987 the cost of removing the ferrous scrap from a stockpile, reshredding it for densification and air cleaning it, then transporting it to New Jersey cost about $40/ton. Because the operation proved uneconomical, the plant instead landfilled its scrap metals (93). In contrast, a processing plant in Tennessee reported its processing costs as $3/ton, while the ferrous scrap could be sold for $27/ton. This $24/ton profit corresponded to a ferrous income

340

Table 9.11
PRODUCT VALUES IN THE SOLID RESIDUES
OF MUNICIPAL REFUSE (83)

Type of Material	Unit Value ($/ton)	Quantity per Ton of Solid Residue (lb)	Value per Ton of Solid Residue ($)
Ferrous metal	10	610	3.05
Aluminum	240	32	3.84
Copper-zinc	380	24	4.56
Colorless glass	12	552	3.31
Colored glass	5	398	1.00
			Total $15.76

of $1/ton of MSW because the MSW ferrous concentration was 5% and the efficiency of recovery about 80%.

In the early 1970s the value of a pound of recycled aluminum was about 12 cents. In California in 1988 the value ranged from 20 to 60 cents. In the first half of 1989 the price of used aluminum beverage cans began to drop, from about 70 cents per pound early in the year to 56 cents per pound by August (175). This is considered a natural cycle and not a sign of market saturation. According to W. O. Bourke, chairman of Reynolds Aluminum, the demand for recycled aluminum is still rising. In 1985 in Tennessee, aluminum could be sold for $700/ton; the aluminum concentration in MSW was about 0.4%. When recovered at an efficiency of 75%, the aluminum revenue was $2.10/ton of MSW (90). In 1990, the average value in the U.S. was $1050 per ton.

Metal Recycling Stations

Metal recycling stations frequently process a mixture of cans and glass bottles collected from curbside containers. This mixture is separated at the processing plant into steel, aluminum, and various colors of glass. In Camden County, New Jersey, residents rinse the containers and place them in curbside pails once a week.

At the Camden plant, trucks dump the containers onto the tipping floor. An inclined conveyor passes the containers through an air-blower section, where the lighter-weight containers (such as aluminum) are blown off while a magnet holds the steel containers on the conveyor belt. An inspector removes the plastic and paper items that remain on the conveyor. Afterward the cans are cleaned, flattened, and sent to shredding. The heavier containers that stay on the first conveyor are cleaned by tumbling and rubbing actions to remove all food deposits and labels. Next, inspectors remove all bimetallic cans (steel body, aluminum lid). The tin cans are shredded and shipped to steel mills. The glass on the conveyor is separated by color and cleaned and crushed prior to shipping to glass companies.

When older "white goods" are shredded, the resulting residue contains more PCBs than allowed. Even when the scrap market is "good," disposing of older appliances can be a problem. It has been suggested that scrap dealers be instructed on proper removal of capacitors that contain PCBs, so that they may certify that the white goods can be safely shredded.

Rubber Recycling

In the United States some two billion old tires have been discarded, and their number is growing by about 240 million a year. In the past tires were either piled, landfilled, burned, or ground up and mixed with asphalt for road surfacing. These "solutions" were expensive and often caused environmental problems because of the air pollution resulting from massive tire fires. Some newer rubber recycling processes have tried to overcome these limitations. The new processes do not pollute air or water because nothing is burned and no water is used (170). The tires are shredded and the polyester fibers removed by air classification. The steel from radial tires is removed magnetically. The remaining rubber powder is mixed with chemical agents that restore the ability of the "dead" rubber to bond with other rubber and plastic molecules. The vulcanized or "cured" tire rubber loses its ability to bond during the vulcanizing process.

Combining old rubber with "virgin" rubber or plastics results in an economically competitive product. The cost of virgin rubber is about 65 cents a pound and polypropylene costs about 68 cents, while the "reactivated" product is about 30 cents a pound ($600/ton) (170).

Organic Waste Recycling

Organic wastes are present in MSW, sewage sludge, and in food wastes. The yearly MSW of the United States contains 12.5 million tons of food wastes and 28 million tons of yard wastes; the total organic percentage in the MSW is about 25% (156). Food wastes can be recycled as livestock feed and other organic wastes can be converted into soil-conditioning compost products.

The oldest method of food-waste disposal is to feed it to livestock, but this is no longer widely practiced in the developed parts of the world. The value of food wastes is low, the collection and transportation costs are high, and the availability of labor is limited. When used for livestock feed, food wastes must be recooked and pulverized to minimize the danger of glass contamination.

Compost as a Soil Conditioner

Composting is the biochemical stabilization of organic materials into a humus-like substance. It is also the natural way to recycle organic waste and sewage sludge. The organics in MSW can be separated at the source and converted into compost and used as soil conditioners (Fig. 9.6), which is nature's way of returning organic wastes to the soil. Farmers reproduce the natural process in leaf or manure piles.

The introduction of chemical fertilizers has changed farming practices in many parts of the world. After World War II the fertilizer industry mounted an intense promotional effort suggesting that the answer to increased farm production was the use of nitrogen-phosphorus-potassium (NPK) fertilizers. Concurrently, the soil bank program paid farmers *not* to raise crops on certain lands. So farmers used the subsidies (received for not planting on their worst lands) to buy chemical fertilizers to produce even larger crops on the rest of their lands. The overuse of NPK began to pollute surface and groundwater and thereby accelerated the eutrophication of lakes and rivers. Chemical fertilizers have depleted the organic content of the soil to the point where in some areas the soil is no longer productive.

Organic matter can be replaced in the soil by planting green crops and plowing them under or by adding compost. Compost as a low-grade fertilizer or soil conditioner is beneficial to heavy clay or loose sandy soils when 10 tons of compost are added per acre every five to ten years. In the 1970s the value of compost was estimated at $25/ton (98).

Composted MSW can be considered a true fertilizer only if it is enriched by chemicals. The nitrogen content can be increased to the required 6% level by adding urea or anhydrous ammonia; rock phosphate can be added to increase the phosphorus content. Prior to bagging, the compost must be screened to remove glass fragments, bottle caps, and pieces of plastic. The compost can be sold to farmers or used on land reclamation projects, including covering landfills and surface mining sites. In Japan the ash from incinerated sewage sludge is sold as a special fertilizer (Fig. 4.8). As the phosphate supply of the United States diminishes, the prospect of utilizing the phosphate content of sewage will merit further investigation. The zinc and chromium in sludge ash will not damage grass or cereals but can harm other crops (100).

The Characteristics of Compost

The addition of compost improves the friability and porosity of the soil. This increases aeration and supports earthworms, whose burrows help to increase rain percolation and reduce soil erosion and surface runoffs. Another advantage of compost is that it holds 350% of its weight in moisture (98), slowly releasing it during droughts. Compost also prevents the washing out of chemical fertilizers. This reduction in leaching can lower nitrogen and phosphorus fertilizer needs by about 40%. Organic matter also stimulates root growth and breaks down insoluble minerals, such as rock phosphates, to make them available to plant growth.

Compost made from MSW must be free of pathogens, weed seeds, glass, metal, plastics, and pesticides and have a low ash content. It should be dry (but not dusty), odor-free, and uniform in texture. As noted, anhydrous ammonia and rock phosphate can be added to increase nitrogen and phosphorus content. (Nitrogen can be added by mixing sewage sludge in the MSW, but this adds pathogens, which requires further treatment through composting. Proper composting [24 hours at 150–160° F] pasteurizes and stabilizes the organic material and kills all weed seeds and pathogens.) Composting also destroys certain pesticides, such as the commonly used carbaryl, while it does not break down some of the polyaromatic hydrocarbons, such as phenathrene (116).

The Composting Process

Nature converts organic solid wastes into compost in about one year. The main goal of the various composting processes is to produce free and stabilized compost material in a shorter time. The promoters of the Agripost process claim to produce stabilized compost in 21 days, while DANO, another respected composting process, requires 90 days for full compost maturation and curing (173).

The carbon-to-nitrogen ratio (C:N) in the feed material should be less than 50:1, a requirement that is met in most MSW. One should not attempt to compost sawdust or paper, because the C:N ratio is over 100:1 and bacteria will not have enough nitrogen for digestion. Actually, these substances are already stable and do not need composting. The material should be shredded to a size of one to three inches and kept within a pH range of 5.5 to 8.0 to prevent nitrogen loss. Some

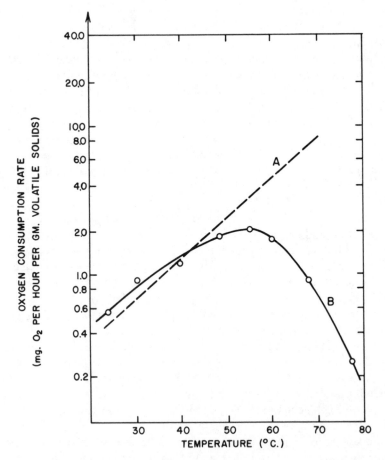

Fig. 9.20 Effect of temperature on oxygen consumption rate. A = theoretical value ($Y = a10^{KT} = 0.1(16^{0.028T})$ for the range of 20° to 70°C); B = experimental value (99).

processes rely more heavily on shredding than do others. The Agripost process, for example, operates two hammermills in series before and one hammermill after composting (173). The ideal moisture content is between 52% and 58% (99). At the end of the process, excess air is added to dry the material to about 30% moisture. The temperature should be controlled between 120° and 160° F (48° and 71° C). For rapid composting, the oxygen consumption rate is maximized and the biological stabilization process optimized at a temperature of 131° F (55° C) (Fig. 9.20). During the first few days of composting, the temperature must be maintained at 120° to 130° F (48° to 54° C), then raised to 140° to 160° F (60° to 71° C) for the remainder of the active digestion period.

Air should reach all parts of the composting material, and slow stirring should be applied to prevent caking or air channeling. An air-supply pressure of 4″ to 6″ H_2O is sufficient to force the air through a 6- to 8-foot depth of composting material (Fig. 9.21). Ground-up MSW with 55% moisture requires between 10 and 30 cubic feet of air per day per pound of volatile solids (99). The air pressure and volume required for composting various depths of material are given in Table 9.12. Usually the airflow rate is adjusted to control the composting temperature. Mixing the contents of the compost pile once a day equalizes the temperature and moisture and

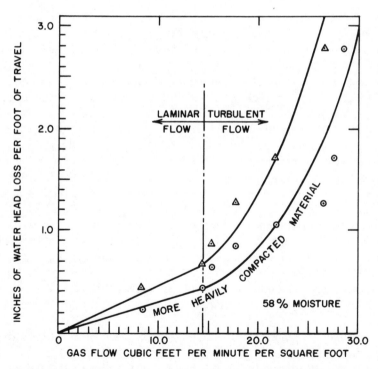

Fig. 9.21 Resistance to mass movement of air through porous spaces of ground garbage (99).

RECYCLING

Table 9.12
AIR PRESSURE AND VOLUME
REQUIRED FOR VARIOUS DEPTHS OF
COMPOST AT PEAK RATE OF
DIGESTION (99)

Total Depth of Material (ft.)	Cubic Foot Air Required per Minute Leaving 50% Residential O_2 per ft.² of Area	Head Loss (inches of water)
1	0.77	0.023–0.036
2	1.54	0.092–0.144
3	2.31	0.207–0.324
4	3.08	0.368–0.577
8	6.16	1.47–2.31
16	12.32	5.9–9.2

prevents channeling and caking. Toward the end of the process, the pile should be mixed once every few days. To speed the process, it is also desirable to recycle 1% to 5% of the partially composted material ("seed-compost").

Composting Plants

Composting plants have been in operation for many decades (Table 9.13). The capital cost of building these plants is shown in Figure 9.22. Recent construction announcements include an 800-TPD Agripost system in Dade County, Florida (173). According to the EPA, 25% of the nation's 7.6 million annual dry tons of sewage sludge is being recycled through some form of composting. Seattle, which used to ocean-dump its sewage sludge, now recycles 100% as commercial fertilizer. Los Angeles mixes its sludge with sawdust and sells it as compost. Milwaukee markets its sludge-based compost under the trade name Milorganite and uses it as a nutrient on grass in parks and on playing fields. In Connecticut, Greenwich and Bristol operate sludge-composting plants that use leaves and wood chips in the mixture. Fairfield County, Connecticut, opened its new $3 million composting facility designed by International Process Systems and known as the Agitated MultiBay Composting System. In the two-acre building, six concrete troughs, each the length of a football field, are loaded with sewage sludge mixed with chipped leaves and other yard wastes. The mixture is churned and aerated, the noxious odors are drawn off through underground filters, and in about 21 days the compost is ready to be used as capping on the 18-acre county landfill. A relatively new sludge-treatment process, called N-Viro, mixes the sludge with cement kiln dust, which kills the pathogens in the sludge.

The $6.9 million plant in Truman, Minnesota, will be the first quick MSW-composting plant in the United States using the technology of the French firm OTVD. It is scheduled to start operation in 1991 processing 100 tons of MSW per day. Steps

Table 9.13

MSW COMPOSTING PLANTS IN THE U.S. (99)

Location	Company	Process	Capacity ton/day	Type Waste	Began Operating	Status[a]
Altoona, Pennsylvania	Altoona FAM, Inc.	Fairfield-Hardy	45	Garbage, paper	1951	Operating
Boulder, Colorado	Harry Gorby	Windrow	100	Mixed refuse	1965	Operating intermittently
Gainesville, Florida	Gainesville Municipal Waste Conversion Authority	Metrowaste Conversion	150	Mixed refuse, digested sludge	1968	Operating
Houston, Texas	Metropolitan Waste Conversion Corp.	Metrowaste Conversion	360	Mixed refuse, raw sludge	1966	Operating
Houston, Texas	United Compost Services, Inc.	Snell	300	Mixed refuse	1966	Closed, 1966
Johnson City, Tennessee	Joint USPHS-TVA	Windrow	52	Mixed refuse, raw sludge	1967	Operating
Largo, Florida	Peninsular Organics, Inc.	Metrowaste Conversion	50	Mixed refuse, digested sludge	1963	Closed, 1967
Norman, Oklahoma	International Disposal Corp.	Naturizer	35	Mixed refuse	1959	Closed, 1964
Mobile, Alabama	City of Mobile	Windrow	300	Mixed refuse, digested sludge	1966	Operating intermittently
New York, New York	Ecology, Inc.	Varro	150	Mixed refuse	1970	Under construction
Phoenix, Arizona	Arizona Biochemical Co.	Dano	300	Mixed refuse	1963	Closed, 1965
Sacramento Co., California	Dano of America, Inc.	Dano	40	Mixed refuse	1956	Closed, 1963
San Fernando, California	International Disposal Corp.	Naturizer	70	Mixed refuse	1963	Closed, 1964
San Juan, Puerto Rico	Fairfield Engineering Co.	Fairfield-Hardy	150	Mixed refuse	1969	Operating
Springfield, Massachusetts	Springfield Organic Fertilizer Co.	Frazer-Eweson	20	Garbage	1954 1961	Closed, 1962
St. Petersburg, Florida	Westinghouse Corp.	Naturizer	105	Mixed refuse	1966	Operating intermittently
Williamston, Michigan	City of Williamston	Riker	4	Garbage, raw sludge, corn cobs	1955	Closed, 1962
Wilmington, Ohio	Good Riddance, Inc.	Windrow	20	Mixed refuse	1963	Closed, 1965

[a] Status as of 1970

347

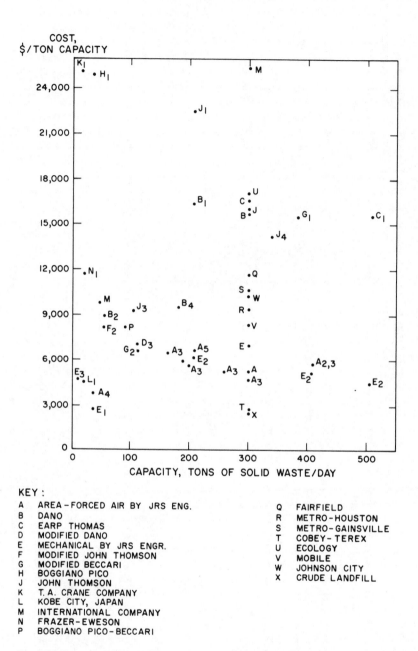

COST, $/TON CAPACITY

KEY :

A	AREA-FORCED AIR BY JRS ENG.	Q	FAIRFIELD
B	DANO	R	METRO-HOUSTON
C	EARP THOMAS	S	METRO-GAINSVILLE
D	MODIFIED DANO	T	COBEY-TEREX
E	MECHANICAL BY JRS ENGR.	U	ECOLOGY
F	MODIFIED JOHN THOMSON	V	MOBILE
G	MODIFIED BECCARI	W	JOHNSON CITY
H	BOGGIANO PICO	X	CRUDE LANDFILL
J	JOHN THOMSON		
K	T.A. CRANE COMPANY		
L	KOBE CITY, JAPAN		
M	INTERNATIONAL COMPANY		
N	FRAZER-EWESON		
P	BOGGIANO PICO-BECCARI		

Fig. 9.22 Cost comparison of compost plants; costs in 1970 dollars (99).

include pulverization, screening to remove noncompostables, and soaking the remaining material with water and microorganisms to accelerate the decaying process. These steps are followed by turning and aerating the compost under controlled temperature and humidity conditions for 28 days (191).

The preferred site for a composting plant is in a rural agricultural area but close to a city and main highway and at least half a mile from built-up areas. The plant must accommodate the precomposting operations of receiving, storage, flow balancing, hand picking, mechanical separation, and shredding. Plastics in the MSW can be removed after the composting process is completed and the material is dry and friable. (Yard wastes collected in biodegradable plastic bags will degrade in the same time frame as tree leaves and can be left in the compost.)

The simplest composting plants in order of sophistication are aerated landfills, windrow-type plants, and the more expensive batch- or bin-type plants. The most complex is the continuous-flow digesters, such as the Hardy digester in Figure 9.23, where the ground MSW is agitated by vertical screws mounted on a rotating radial arm in a circular tank. Aerobic conditions are maintained by admitting air at the bottom of the tank. This process requires a digestion time of about six days. In continuous digesters, seeding is unnecessary because material from previous days is thoroughly mixed in.

Aerobic composting generates heat and water and carbon dioxide formation. Ventilation removes the carbon dioxide to prevent anaerobic conditions that could cause odor problems. Batch-type compost reactors (Fig. 9.24) have not performed well (87). In the batch process, a small amount (1%–5%) of seed material must be added to speed the growth of bacteria. Low-cost, high-capacity turning machines (Figs. 9.25 and 9.26) are used more widely than batch digesters.

Another, rather simple composting method is the "Jersey" process, where the preshredded and salvaged MSW is spread on top of a six-floor concrete digester building. Each day steel trapdoors are opened and the material is dropped onto the floor below. At the end of six days, the material is taken to a warehouse for a six-week maturation process. To eliminate the breeding of flies, the compost pile must be sprayed several times a day. Odor and dust can be controlled but not completely eliminated. Dust in the shredder area can be controlled by water spraying. Occasionally some pockets of anaerobic decompostion will remain, and the only security against complaints is to isolate the plant in a rural area.

Recycling Incinerator Ash

If all the MSW of New York City were incinerated, the residue would amount to 6,000 to 7,000 tons/day, representing a giant disposal problem. About 10% by weight of the incinerator residue is fly ash collected in electrostatic precipitators, scrubbers, or bag filters; the remaining 90% is bottom ash from the primary and secondary combustion chambers. This residue is a soaking-wet complex of metals, glass, slag, charred and unburned paper, and ash containing various mineral oxides. A Bureau of Mines test found that 1,000 pounds of incinerator residue yielded 166 pounds of larger-size ferrous metals, such as wire, iron items, and shredded cans.

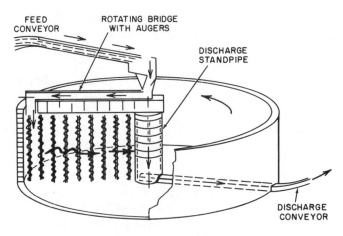

Fig. 9.23 Schematic of Hardy digester (99).

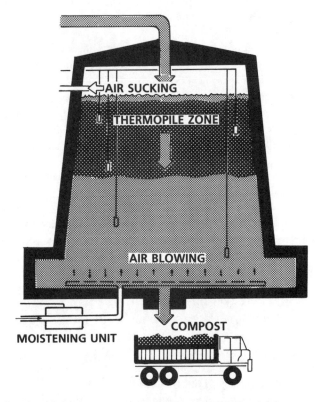

Fig. 9.24 Compost reactor of the Kneer type in Landskrona, Sweden (87).

350

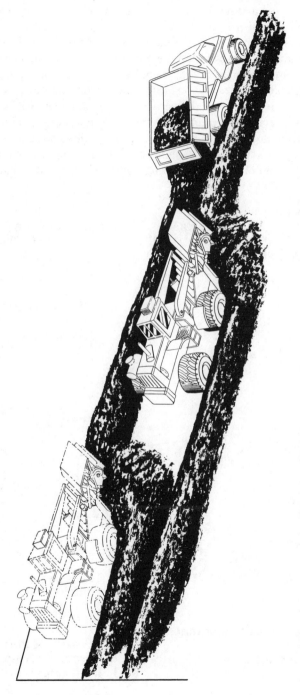

Fig. 9.25 Cobey-Terex turning mechanism. The Cobey-Terex Composter processes row 1 from the left side of the field, moving it one tractor width to the right. Following this operation, the composter moves the second row of fresh material one tractor width to the right (99).

351

The total ferrous fraction was found to be 30.5% by weight; glass represented 50% of the total residue by weight (83).

Common practice in the U.S. is to recover some 75% of the ferrous metals through magnetic separation and to landfill the remaining residue. Incinerator residue has also been used as landfill cover, landfill road base, aggregate in cement and road building applications, and as aggregate substitute in paving materials.

Incinerator residue is processed to recover and reuse some of its constituents and thereby reduce the amount requiring disposal. Processing techniques include the recovery of ferrous materials through magnetic separation, screening the residue to produce aggregate for construction-related uses, stabilization through the addition of lime (which tends to minimize metal leaching), and solidification or encapsulation of the residue into asphaltic mixtures (96).

Table 9.14
ANNUAL AVERAGE SUMMARY OF
MOISTURE, ORGANIC AND ASH CONTENT
OF FLY ASH[a] (96)

	% Moisture	% Organics	% Ash
Mean	4.3	6.2	89.5
Standard deviation	3.1	2.2	4.1
Maximum value	12.4	11.0	94.7
Minimum value	1.7	2.2	80.0
Number of samples	30	30	30

[a] Data reported as percent wet weight.

Table 9.15
INCINERATOR ASH
COMPOSITION (86)

Component	Percentage
Ferrous metal	35
Glass	28
Minerals and Ash	16
Ceramics	8
Combustibles	9
Nonferrous Metal	4

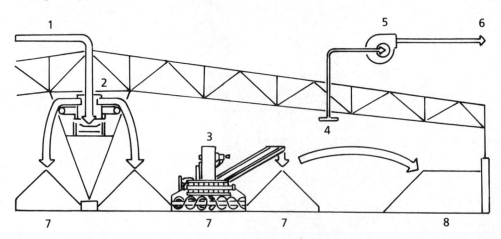

1. Incoming pretreated compost material
2. Conveyor
3. Automatic turning machine
4. Evacuating
5. Fan
6. To compost filter
7. Compost string
8. Maturing

Fig. 9.26 Composting on unventilated plate with the help of a turning machine (87).

Table 9.16
COMPOSITION OF
INCINERATOR ASH
(Under 2″ Fraction)

Component	Percentage
Glass	37
Minerals and Ash	21
Ferrous Metals	19
Ceramics	9
Combustibles	8
Nonferrous Metals	6

An incinerator-residue processing plant might consist of the following operations: (1) fly ash and bottom ash are collected separately, with lime mixed only with the fly ash; (2) ferrous materials are removed from the bottom ash; (3) the ferrous-free residue is screened to separate out the proper particle sizes for use as aggregate; and (4) the remaining oversized items and stabilized fly ash are landfilled. In a more sophisticated ash-processing plant, the ferrous removal and shredding (or oversize removal) are followed by melting of the ash (fusion), resulting in a glassy end-product. This high-tech process has some substantial advantages: It burns all the combustible materials, including dioxins and other trace organics, and encapsulates the metals, thereby preventing their leaching out. The resulting fused product is a glazed, non-abrasive, lightweight black, aggregate. The fusion of combined incinerator ash and sewage sludge is currently practiced in Japan (96).

The first U.S. building to be built from recycled incinerator ash blocks is an 8,000-square-foot boathouse on the campus of the State University of New York at Stony Brook, Long Island. The ash comes from an incinerator in Peekskill and is mixed with sand and cement to form blocks that are as durable as standard cinder blocks (201). This technology has already been used in Europe. The blocks can be used to build seawalls, highway dividers, and sound barriers, in addition to regular buildings. It is the bottom ash (not the fly ash) portion that is considered safe for such applications.

The ash produced by one New York City incinerator has been extensively sampled and evaluated (96). As shown in Table 9.14, fly ash contains substantial quantities of organic materials. About 20% by weight is larger than 2″ (50.8 mm); the metal content of this fraction is over 80% by weight. The overall composition of all the incinerator residue (Table 9.15) differed substantially from the composition of the under-2″ (50.8 mm) fraction (Table 9.16). The test also concluded that the New York State Department of Transportation specifications for Type 3 asphalt binder can be met if 10% combined incinerator ash is mixed in with 90% natural aggregate.

Chapter 10
Recycling Equipment and Materials Recovery Facilities (MRFs)

□
■■

CHAPTER 9 DISCUSSED various recycling strategies, their status, cost, and political implications. This chapter will describe the equipment used in recycling and how recycling plants themselves, or material recovery facilities (MRFs), operate. The first part of this chapter will discuss the various types of devices used in the separation of MSW, including screens, air classifiers, magnetic separators, and shredders. The second part will describe the design, operation, and performance of MRFs, and the operating experience with refuse-derived fuel (RDF) plants.

MSW Separation Equipment

MSW is composed of many types of materials, each having its own characteristics, and therefore clogs, bridges, and binds the processing equipment. This means the design of MSW processing equipment is necessarily more complex than that of equipment used on more uniform substances such as coal or wood. While the early equipment used for MSW separation was modeled on the mining and chemical industries, it had to be modified over time to handle unusual feed items, such as large nonmagnetic pieces of high-manganese steel, large plastic bottles containing gasoline, and long ¼-inch steel cable. Only a well-designed plant can handle these items without jams, damage, explosion, or fires.

The Shredding Process

Shredding has many applications in MSW processing plants (Fig. 10.1). Two of the main benefits of solid-waste shredding are the resulting increase in MSW density and a decrease in volume. Shredded refuse therefore can be used to produce

354

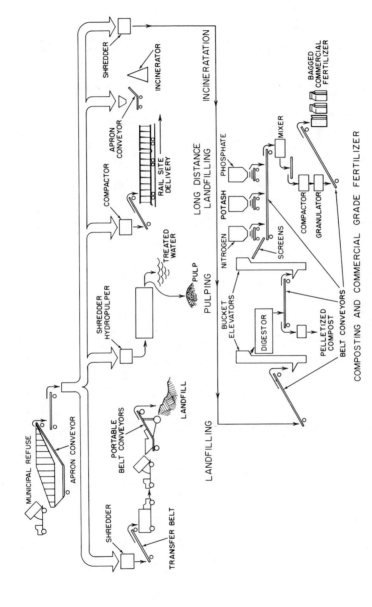

Fig. 10.1 Potential shredder applications in solid-waste processing (81).

355

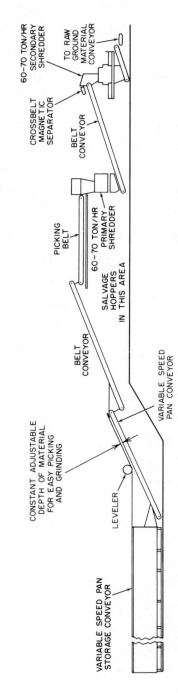

Fig. 10.2 Partial elevation of conveyors and shredders in a compost plant in Houston, Texas (81).

more compact landfills. If a bulldozer can compact a 6-foot depth of ordinary refuse, the resulting density will be about 500 pounds per cubic yard (about 300 kilograms per cubic meter). If shredded refuse were compacted the same way, its density would be about 1,000 pounds per cubic yard (nearly 600 kilograms per cubic meter (81). This 50% reduction in MSW volume extends the life of landfills, reduces transportation costs and eliminates the need for daily earth covering because the more compacted the landfill, the less likely it is that rodents or flies can survive on it (81). In addition to reducing the volume of the solid waste, shredding also improves the performance of magnetic separators (see Fig. 2.2), is a prerequisite for composting (Fig. 10.2), pulp recovery or pyrolysis, and makes incineration more controllable and efficient.

One problem associated with shredding is explosion. In a five-year period, a shredding plant in Albany, New York, reported twenty-three explosions at a cost of over $250,000 (93). Explosions occur when volatile organic chemicals accumulate in the shredder cavities and are ignited by sparks from the rotating cutting elements. Explosions can be reduced by hand-picking in front of the shredder, by ventilating the shredder cavities, and by installing explosion suppression systems. In addition to the potential for explosions, shredding can grind the glass into the paper fibers, which causes wear on the incinerator screw feeders and also results in excessive slagging in the bottom ash. The large pieces of ferrous metals that the shredder cannot break down also cause feeder jams.

The cost of shredding is fairly high. About 6 kWh of electricity per ton of MSW are needed to shred the MSW so that 90% of it will pass a 4-inch (10cm) screen. If the desired size is 0.4 inches (1cm), the energy cost rises to 50 kWh per ton (89). Figure 10.3 shows the ratio of energy cost to size reduction achieved; Figure 10.4 gives data on the power requirements of size-reduction devices at various flowrates for shredding different materials. The total operating cost of shredders includes not only the cost of the electric power used but also the replacement cost of worn shredding hammers. The cost amounts to $1.20/ton of MSW, not including labor. Hammer wear also relates to product size and hammer hardfacing, as shown in Figure 10.5.

Depending on their applications, shredders are also referred to as crushers, pulverizers, hammer mills, hoggers, grinders, and ball mills. They can be vertical or horizontal, single or two stage, low- or high-speed devices. Vertical shredders can handle up to 20 tons of MSW per hour and horizontal ones about 100 tons.

Because MSW is a heterogeneous material, it is not possible to design a shredder that will efficiently reduce the size of *all* components. Brittle material can be shredded efficiently by impact hammers, rock can be reduced to powder in ball mills, and wood and branches can be chipped into shavings by sharp-edged hogs, but the shredder designed for MSW service must handle everything from stones to tires, from soft food particles to mattresses, from documents to rolled carpets or furniture. Tires, appliances, construction debris, and furniture are removed before shredding and processed during off hours.

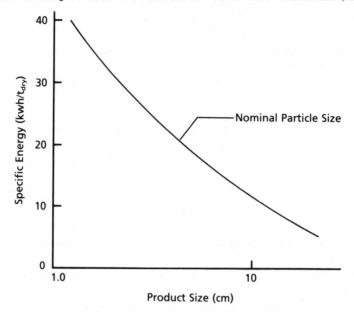

Fig. 10.3 Specific energy requirements for size reduction of MSW (89).

In a typical MSW application, shredding is performed in two steps (Figs. 2.2 and 10.2). The primary shredder might have large hammers and large grill openings capable of shredding material to a range of 6 to 10 inches (150 to 250 mm). If objects are too tough or too large to be crushed on first impact, the hammers, which are held in position by centrifugal force, bend back and pass the object. Eventually the hard object is thrown out through a reject opening (Fig. 10.6). The secondary shredder further reduces the particle size to 1 to 2 inches (25 to 50 mm). In some shredder designs the two stages of size reduction are combined in a single machine (Fig. 10.7), while in others two machines are used in series (Fig. 10.8).

When a shredding rate of fewer than 20 tons per hour is required, a two-stage vertical shredder is best based on flexibility and cost. For larger capacities, the cost of installing two horizontal shredders in series should be compared with the cost of installing several two-stage vertical shredders in parallel. The larger the capacity, the more likely it is that the horizontal shredder installation will be less expensive.

LOW-SPEED SHREDDERS

While the most common shredders for MSW applications are swing-type hammer mills, low-speed shredders have been used for MSW service. Some units pass through difficult-to-shred objects (91). The advantages of low-speed shredders include reduced power consumption and lower maintenance, their ability to open

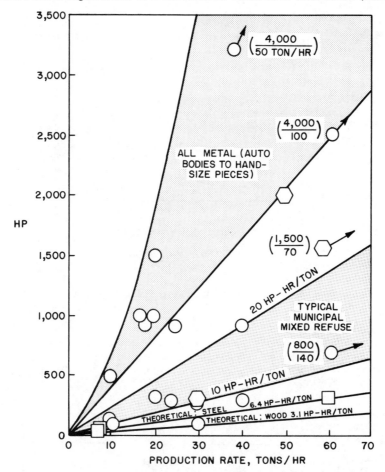

Fig. 10.4 Size reduction power requirements.[a]
[a]Performance data were collected on a total of 27 size-reduction machines manufactured by 21 different companies. The data presented are from manufacturers' ratings, from theoretical calculations, and, to a limited extent, from existing installations. Squares denote drum pulverizers, triangles denote mills, and hexagons denote shredders (81).

plastic bags and expel their contents, a lower risk of explosions (due to low dust generation and to the use of a fine mist spray at the head of the grinder), their ability to provide quick access to the cutter area, and their capability of producing a rough cut (8 to 10 inches) in which glass or ceramics can be separated efficiently without being pulverized. An added advantage is that these shredders require a small foundation.

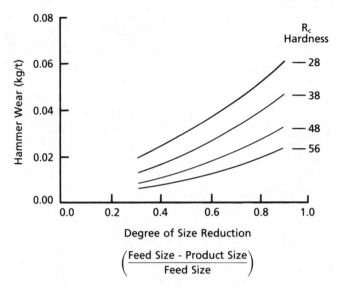

Fig. 10.5 Hammer wear as a consequence of shredding solid waste (89).

Hammer mill without grids and with loose beaters

Hammer mill with grids and fixed beaters

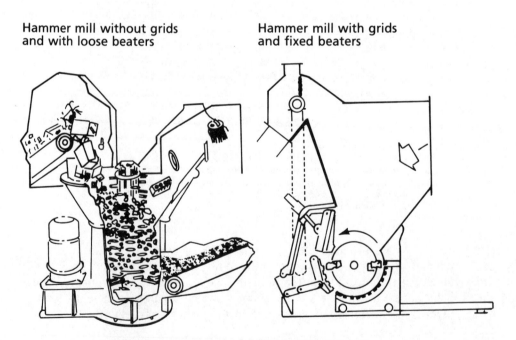

Fig. 10.6 Sketch of two different mills for the shredding of waste (87).

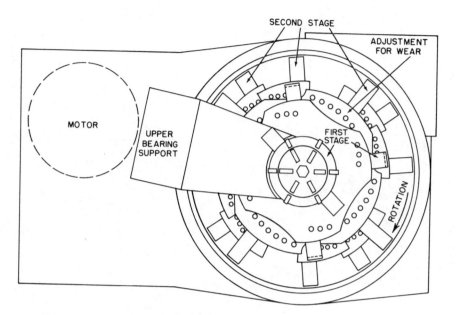

Fig. 10.7 Two-stage vertical grinder, plan view (81).

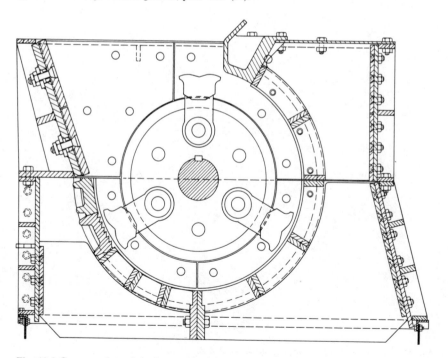

Fig. 10.8 Cross-section of single-stage horizontal hammer mill (81).

361

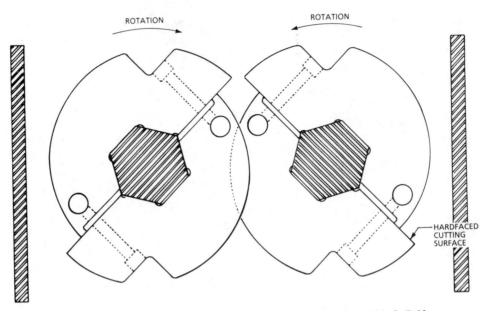

Fig. 10.9 Type II MSW shredder. Input torque specifications: 1 @ 32,000 ft-lb/shaft @ 32 rpm (43,400 N·m/shaft @ 32 rpm); 1 @ 57,000 ft-lb/shaft @ 16 rpm (77,306 N·m/shaft @ 16 rpm) (91).

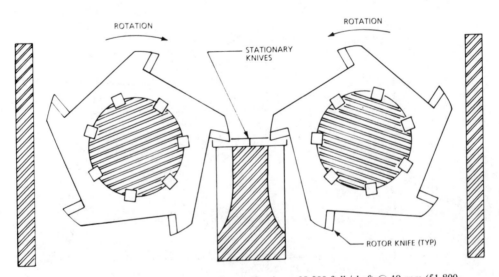

Fig. 10.10 Type I wire shredder. Input torque specifications: 38,200 ft-lb/shaft @ 19 rpm (51,800 N·m/shaft @ 19 rpm) (91).

362

A low-speed MSW shredder is illustrated in Figure 10.9. Because of the deep nip angle of the cutters, it can handle bulky, oversized material. It uses two counter-rotating, intermeshing shafts to draw in the MSW at the pinch point and shred or tear it apart. The two shafts rotate at different speeds and with different torque outputs to give better shearing action. The units can be driven by electric, diesel, or hydraulic drives of from 7.5 to 600 HP.

Low-speed shredders are available for more specialized size-reduction tasks. Figure 10.10 illustrates a wire and cable shredder with a replaceable stationary knife and two counter-rotating shafts with moving knifes that cut against a stationary anvil; the wires are cut but not stretched or torn. Figure 10.11 illustrates a tire shredder, incorporating a set of star-shaped feed rolls that force the individual tires into the cutting zone. The feed rate is kept below the peripheral speed of the cutters, thereby ensuring that the tires are cut in uniform pieces and not extruded through the cutters. Another specialized application for shredders is in automobile shredder plants (Fig. 10.12).

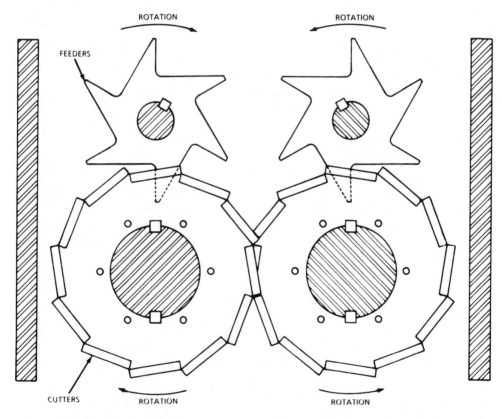

Fig. 10.11 Type III tire shredder. Input torque specifications: 26,500 ft-lb/shaft @ 19 rpm (35,900 N·m/shaft @ 19 rpm) (91).

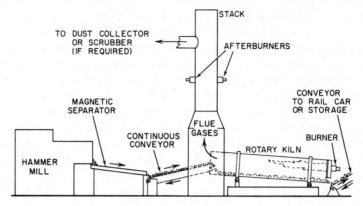

Fig. 10.12 Hammer mill/rotary kiln auto shredder plant with emission controls (171).

Screening Equipment

Screening separates shredded or unshredded MSW into two or more fractions according to size. "Wet-screening" uses water prior to screening; other forms of screening are called "dry screening." Screen designs include rotating or trommel designs (Fig. 10.13), vibrating or ballistic separators (Fig. 10.14), and disk designs. Trommel designs are the most popular for MSW applications. When the screen is located before the shredder, it is called a "pre-trommel"; when it is downstream of the shredder, it is called a "post-trommel" installation.

Screening removes the noncombustible fraction from the shredded MSW or from refuse-derived incinerator fuels. This usually means removing high-density heavy particles such as glass, dirt, or ceramics. When the purpose of screening is to improve the appearance of compost, the goal is to remove particles with high area-to-volume ratios, such as plastics, aluminum, or paper. The effectiveness of screening depends on the design of the screen and the number of screenings. Ballistic

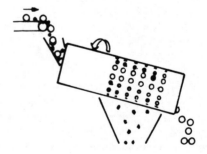

Fig. 10.13 Rotating or trommel screen (87).

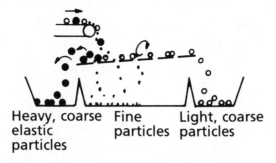

Heavy, coarse   Fine   Light, coarse
elastic   particles   particles
particles

Fig. 10.14 Ballistic or vibrating screen separator (87).

separators give a somewhat higher degree of separation than do rotating trommel screens (87), which have a separation efficiency in the range of 75% to 95% (89). The design parameters that influence their performance include screen loading, aperture size, angle of inclination, rotational speed, and their length and diameter. The effectiveness of separation is directly related to the degree of tumbling and mixing achieved. Screening efficiency drops as loading or inclination rises (Fig. 10.15), but it increases as rotational velocity increases (expressed as a percentage of the critical angular velocity).

Figure 10.16 shows the time required to achieve separation for heavier versus lighter substances. Dryness also increases the efficiency of separation but can contribute to dust formation. When the desired screen diameter and the required MSW throughput is known, the designer can select the appropriate screen length (Fig. 10.17). It has been found that the smaller the distance between the holes on the screen, the greater the risk of clogging. Therefore, it is important to counteract clogging by either dimensional design or through continuous cleaning.

The *combined* performance of shredding and screening on a compost-separation installation in Sweden, evaluated against the dual criteria of degree of separation and plastic removal, is shown in Figure 10.18. Region *a* is the desirable performance area, where the efficiency of separating out the smaller-size materials is high, while the plastic content of the resulting compost is low.

Air Classifiers

During air classification a variety of particles are suspended in a gas stream and separated according to their *aerodynamic* characteristics. In MSW processing, the air classifiers are located either after the shredder or after the screening equipment. The air classifier segregates the refuse stream into ''lights'' (mostly combustibles) and ''heavies'' (mostly inorganics). Its performance is dependent on the rate and uniformity of feed, the applied air-to-solids ratio, and the load on the classifier. Other features affecting separation include the classifier design (horizontal, vertical,

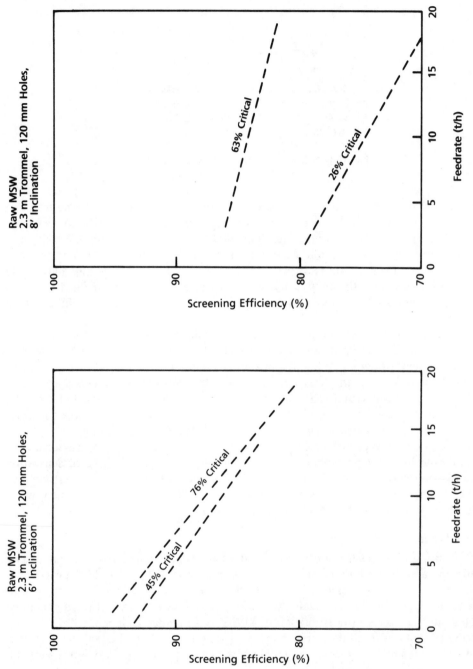

Fig. 10.15 Interrelationship of parameters influencing the screening efficiency of trommel screens (89).

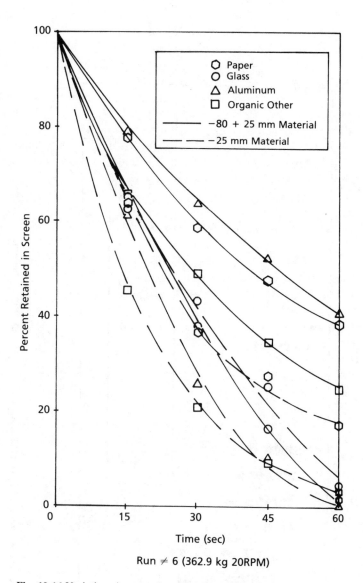

Time (sec)

Run ≠ 6 (362.9 kg 20RPM)

Fig. 10.16 Variations in screening efficiency among solid waste components for a pre-trommel screen operating at zero degrees inclination (89).

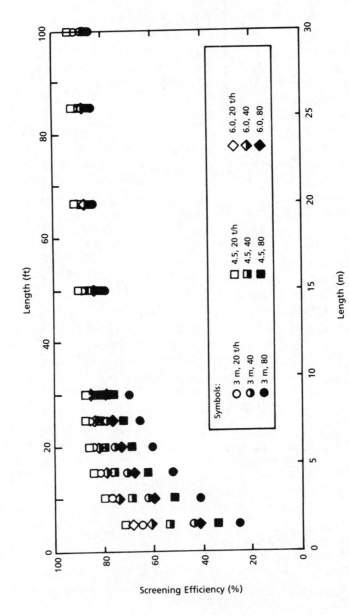

Fig. 10.17 Calculated trommel screening efficiency under various operating conditions (89).

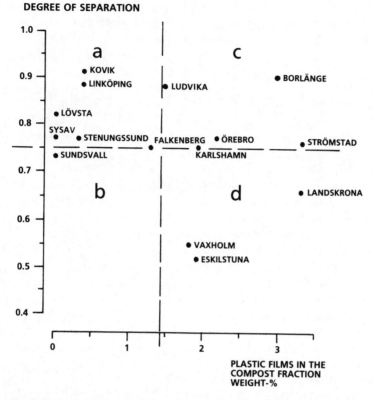

Fig. 10.18 Performance of MSW shredding and screening installations in Sweden (87).

or inclined); the flowrate, velocity, and density of the gas; and the nature of the MSW feed (density, moisture content, particle shape, and size distribution). Typical air-classifier performance ranges are given in Table 10.1.

The energy consumption of air classifiers ranges from 1 to 11 kWh per ton of MSW throughput; their capacities are up to 70 tons per hour. The early air classifiers were designed according to pneumatic transport theory, which assumes homogeneous and uniform-size materials. Neither of these characteristics is true for MSW applications and therefore the air/solid ratios of 2:1 or 3:1 were found to be insufficient. It was found that on MSW services the classifiers could handle only 25% to 50% of their rated capacities because the airflows were insufficient to provide thorough mixing, separation, and transportation. The air/solids ratio on MSW service should exceed 5:1; at lower ratios the feed stream tends to settle in the air classifier column, causing choking.

Table 10.1
TYPICAL RANGES OF AIR CLASSIFIER
PERFORMANCE (89)

	Typical Range
Critical air/solids ratio	2–7
Light fraction composition (%)	
ferrous metals	0.1–1.0
nonferrous metals	0.2–1.0
fines	15–30
paper and plastic	55–80
ash	10–35
Component retained in light fraction (%)	
ferrous metals	2–20
nonferrous metals	45–65
− 14 mesh fines	80–99
paper and plastic	85–99
ash	45–85
Recovered energy (%)	73–99
Specific energy (kwh/t)	1–11
Column loading (t/h)/m²	5–40

The critical air/solids ratio is defined as the point where the light fraction becomes constant and does not rise with an increase in airflow (Fig. 10.19). Airflow should be selected so that the operating air/solids ratio exceeds the critical one. In Figure 10.19 airflow "A" can handle a feed rate of 30 tons/hr; airflow "B" can handle only 20 tons/hr and "C" only 10 tons/hr.

ROTARY-DRUM AIR CLASSIFIERS

Figure 10.20 describes the components of a rotary-drum air-classifier (RDAC) installation for MSW service in Albany, New York (93). The shredded MSW is transported on a 3-foot-wide (0.9 m) rubber belt (1) to a 4-foot-diameter (1.2 m) Ram-Tube conveyor (2), which is a tube with a piston inside. The conveyor protrudes through the large settling chamber (3) into the tapered rotary drum (4), which rotates at about 15 rpms. (The Ram-Tube conveyor can operate without jams). The double-tapered drum (4) is set at an angle of 7.5°. The unit is 9 feet in diameter (2.7 m) and 20 feet long (6 m). The smaller, upper end of the drum protrudes into the settling chamber, separated by only an air seal. The primary fan (7) recirculates the air from the settling chamber (3) and feeds it into the rotary drum (4) at a rate of 85,000 cfm, while the discharge fan, located downstream of a cyclone separator (8), exhausts 31,000 cfm of the total airflow. The airflow moves the "lights" into the settling chamber (3), while the "heavies" fall on a 4-foot-wide (1.3 m) discharge conveyor (6).

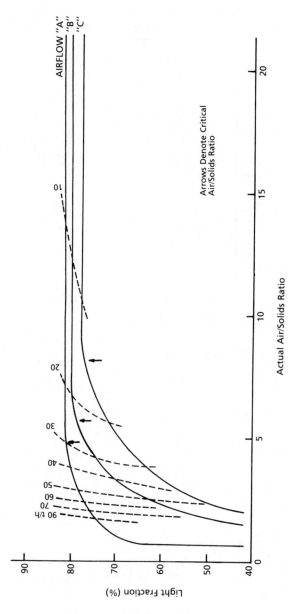

Fig. 10.19 Influence of air/solids ratio upon light fraction split for different airflow settings (89).

371

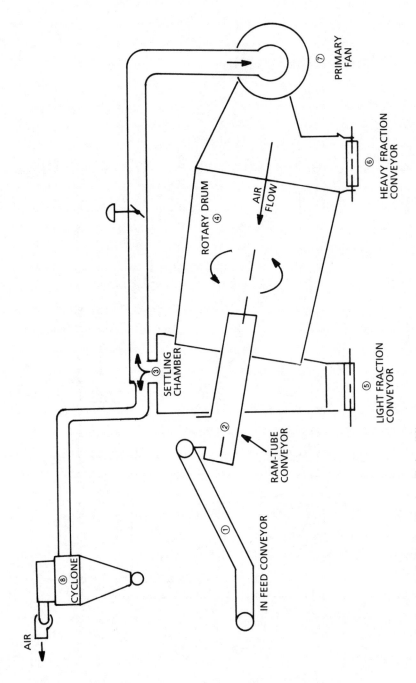

Fig. 10.20 Rotary-drum classifier installation (93).

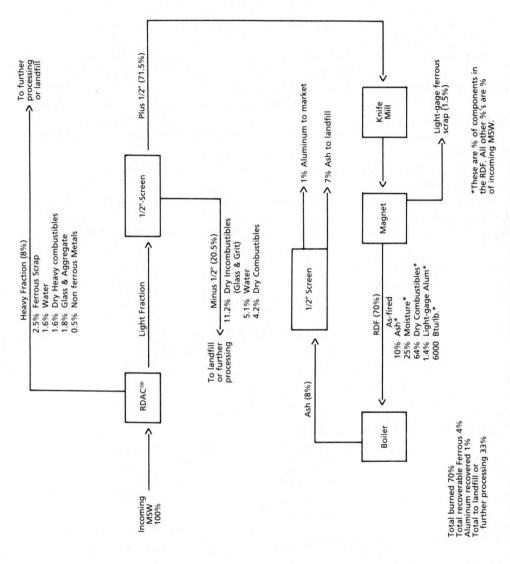

Fig. 10.21 Rotary-drum air-classifier material balance in Albany, New York (93).

Incoming MSW 100%

RDAC™

Heavy Fraction (8%)

2.5% Ferrous Scrap
1.6% Water
1.6% Dry Heavy combustibles
1.8% Glass & Aggregate
0.5% Non ferrous Metals

To further processing or landfill

Light Fraction

1/2"-Screen

Plus 1/2" (71.5%)

Minus 1/2" (20.5%)

11.2% Dry Incombustibles (Glass & Grit)
5.1% Water
4.2% Dry Combustibles

To landfill or further processing

1/2" Screen

1% Aluminum to market

7% Ash to landfill

RDF (70%)

As-fired
10% Ash*
25% Moisture*
64% Dry Combustibles*
1.4% Light-gage Alum*
6000 Btu/lb.*

Ash (8%)

Boiler

Magnet

Knife Mill

Light-gage ferrous scrap (1.5%)

*These are % of components in the RDF. All other %'s are % of incoming MSW.

Total burned 70%
Total recoverable Ferrous 4%
Aluminum recovered 1%
Total to landfill or further processing 33%

373

The material balance and the various percentage splits obtained in the RDAC installation are given in Figure 10.21 (93). When shredded MSW was charged at rates of 40 tons/hr, the RDAC unit required 3 kWh of electricity per ton of MSW processed and provided a 95%/5% split between light and heavy fractions. The airflow velocity was 18 ft/sec (5.8 m/s). When the light fraction was used as boiler fuel, the boiler performance was better than on "crude" (90% ferrous removed, but unclassified) RDF. The improvements were more uniform combustion and a reduction in slagging. Steam production increased from 3.13 to 3.52 pounds of steam per pound of fuel as the fuel was switched from crude RDF to the light fraction obtained from the RDAC classification process. When the $\frac{1}{2}$ inch screen was added (shown in Fig. 10.21), RDF yield was reduced to 73% by weight, but steam production increased to 4.12 pounds of steam per pound of classified and screened light fraction and ash production dropped to 10% by weight of the fuel. When burning crude RDF, the ash production was 23% of the fuel; when burning classified but unscreened lights, it was 20% of the fuel. Table 10.2 gives an overall summary of the three operating modes that were evaluated.

Other findings of the RDAC installation in Albany were that air classification followed by screening virtually eliminated the possibility of explosions. It also removed from the boiler fuel 60% of the acidic elements (nitrogen and sulfur from

Table 10.2
COMPARISON OF YIELD, ASH CONTENT, AND
FUEL ENERGY LOSS (93)

	Crude RDF	Air-Classified Crude RDF 95% Yield	Air-Classified and Screened Unshredded MSW
Yield of RDF as % of incoming material	93%	89%	73%
Ash content, as produced at 23% moisture	23%	20%	10%
Ash content (dry basis)	30%	26%	12.3%
Lb RDF/mm Btu (dry basis)	163	140	121
Lb Ash/mm Btu (dry basis)	49	36	15
Fuel energy loss (HHV)	2% (est)	5.5%	11.1%
Fuel energy of incoming waste in RDF (HHV)	98% (est)	94.5%	88.9%

kitchen and yard waste and chlorine from plastics and rubber items). This resulted in reducing the SO_2 and HCl emissions from the boiler to 50 ppm. The reduction in chlorine also reduced corrosion in the boiler.

Magnetic Separators

Ferrous materials are separated from the shredded MSW by electromagnets (Fig. 10.22) placed above the conveyor belt or installed in other configurations (Fig. 10.23). The collected ferrous metals are removed from the magnets by elastic wipers, which drop them onto a vibrating pan conveyor. The efficiency of ferrous metal separation is reported to be around 80% (90).

Nonferrous metals, primarily aluminum, are separated by eddy current or electrostatic separators. In a typical MRF, metal separation takes place in at least four locations. In the plant in Figure 10.24 ferrous metals are recovered at locations 10, 13, and 19, while aluminum is recovered at location 17. The primary separator (10) picks up the largest quantity of ferrous metals, about 4% of the total feed by weight, the secondary (13) picks up 1%, and the DRDF (19) only 0.8%.

The total ferrous recovery is increased by placing more than one magnetic separator in series. The separators should not be located immediately after milling or shredding devices. In some plants (102) it was found that Dings hockey-stick magnets were not effective and had to be replaced by stronger in-line belt magnets. The stronger belt magnets can pick up larger ferrous items and tend to increase ferrous recovery, but the collected scrap also contains more impurities (entrapment of nonferrous material) and requires further cleaning.

Cleaning the collected ferrous fraction can be accomplished in several ways. The ferrous fraction can be sent through a fired kiln to burn off nonmetallic impurities and paints or it can be washed. The favored method of cleaning in the Rome plant (94) is shredding by hammer mill, which has its own grate. This process cleans the ferrous fraction through friction and also densifies it. A final magnetic separator downstream of the hammer mill is used to separate the metal from the nonmetallic

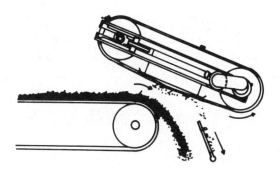

Fig. 10.22 One-stage magnet placed above a conveyor belt (87).

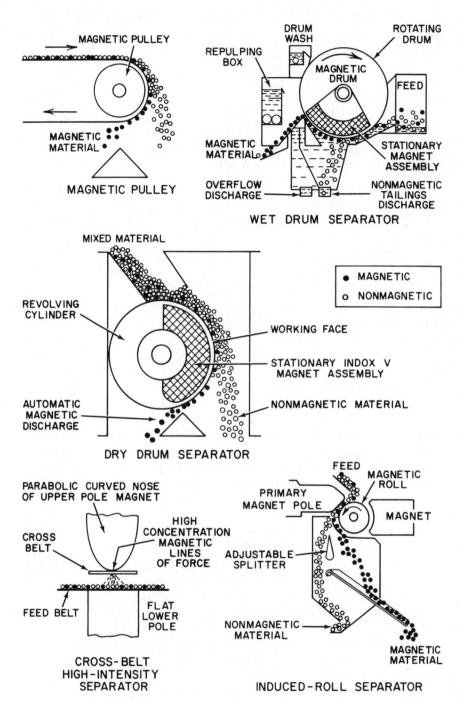

MAGNETIC PULLEY

MAGNETIC MATERIAL

MAGNETIC PULLEY

DRUM WASH

REPULPING BOX

ROTATING DRUM

MAGNETIC DRUM

FEED

MAGNETIC MATERIAL

STATIONARY MAGNET ASSEMBLY

OVERFLOW DISCHARGE

NONMAGNETIC TAILINGS DISCHARGE

WET DRUM SEPARATOR

MIXED MATERIAL

● MAGNETIC
○ NONMAGNETIC

REVOLVING CYLINDER

WORKING FACE

STATIONARY INDOX V MAGNET ASSEMBLY

AUTOMATIC MAGNETIC DISCHARGE

NONMAGNETIC MATERIAL

DRY DRUM SEPARATOR

PARABOLIC CURVED NOSE OF UPPER POLE MAGNET

HIGH CONCENTRATION MAGNETIC LINES OF FORCE

CROSS BELT

FEED BELT

FLAT LOWER POLE

CROSS-BELT HIGH-INTENSITY SEPARATOR

FEED

MAGNETIC ROLL

PRIMARY MAGNET POLE

MAGNET

ADJUSTABLE SPLITTER

NONMAGNETIC MATERIAL

MAGNETIC MATERIAL

INDUCED-ROLL SEPARATOR

Fig. 10.23 Magnetic separator designs (82).

376

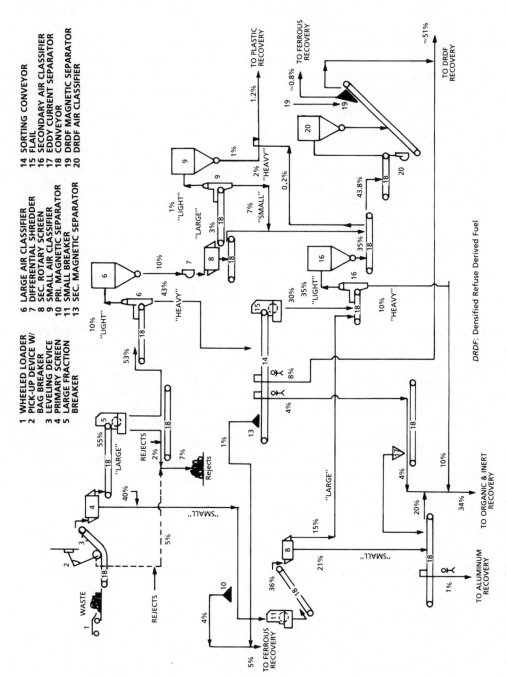

Fig. 10.24 Resource-recovery plant in Rome, Italy (94).

1 WHEELED LOADER
2 PICK-UP DEVICE W/ BAG BREAKER
3 LEVELING DEVICE
4 PRIMARY SCREEN
5 LARGE FRACTION BREAKER

6 LARGE AIR CLASSIFIER
7 DIFFERENTIAL SHREDDER
8 SEC. ROTARY SCREEN
9 SMALL AIR CLASSIFIER
10 PRI. MAGNETIC SEPARATOR
11 SMALL BREAKER
13 SEC. MAGNETIC SEPARATOR

14 SORTING CONVEYOR
15 FLAIL
16 SECONDARY AIR CLASSIFIER
17 EDDY CURRENT SEPARATOR
18 CONVEYOR
19 DRDF MAGNETIC SEPARATOR
20 DRDF AIR CLASSIFIER

DRDF: Densified Refuse Derived Fuel

377

impurities loosened by the mill. The densified ferrous fraction has the appearance of a metal gravel with a specific gravity of about 3.0 or 200 lb/ft^3. This metal fraction can be fired for further cleaning and compressed into blocks or sold as gravel.

In 1985 in a Tennessee plant, the MSW contained 5% ferrous metals and 80% of that was recovered through magnetic separation. At a processing cost of $3/ton and a ferrous revenue of $27/ton, the ferrous income amounted to about $1/ton of MSW processed (90).

ALUMINUM SEPARATION

In the MRF plant described in Figure 10.24, aluminum is separated out by hand-picking enhanced by eddy-current separation (item 17). Eddy currents are induced in metals whenever they are brought into an AC magnetic field. In conductors such as aluminum, these currents create a secondary magnetic field, which opposes the inducing magnetic field and therefore can serve to separate conductors such as aluminum from the nonconductive MSW constituents. Once the aluminum is separated out, it is sent through a crusher, where it is densified to a specific gravity of about 1.0 or 62 lb/ft^3.

Another method of aluminum separation uses metal detectors and air jets (90). Metal detectors actuate the appropriate air jet when it detects a metallic object. The air jet then "blows" that object onto another conveyor. This technique results in an aluminum content of 8% to 10%, compared to the less than 0.5% aluminum by weight in the MSW. The concentrator system removes 70% to 80% of the aluminum containers but only 50% to 60% of foils, pie tins, and similar objects.

The aluminum concentrate can be further separated by an Eddy Current Separator. Aluminum-container recovery by the eddy-current system is in the range of 90% to 95%, but it removes only 30% of other forms of aluminum (other than containers) in the raw waste (90). Thus the overall aluminum recovery by the eddy-current technique is around 75%. Some eddy-current magnetic separators are also turned on and off by metal detectors to conserve power, but rigid supports are necessary to reduce the resulting vibration, which can damage the power-supply wires.

In Tennessee, between 1983 and 1985, the price obtained for a pound of container aluminum ranged from 28 to 52 cents (90). Other forms of aluminum could be sold for about 13 cents per pound. If aluminum beverage containers comprise 0.4% of the MSW and 75% of them are recovered at a selling price of $700/ton, this results in an aluminum revenue of $2.10/ton of MSW (90). For 1989 prices of scrap aluminum, see Table 9.3.

Electrostatic Separators

Electrostatic separators are not widely used in MSW separation applications because they are effective only on finely ground and completely dry materials. Figure 10.25 shows how this separation method could be applied to the dried and ground MSW fraction containing heavy solids, such as glass, ceramics, and metals.

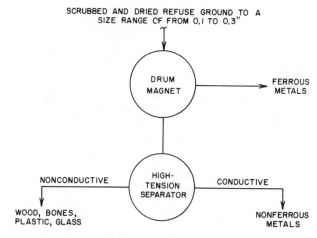

Fig. 10.25 Separation of municipal refuse (82).

Electrostatic separation is achieved by inducing a surface charge onto some components in a granular mixture. One method of inducing a surface charge is by conductive induction (Fig. 10.26). This can be done by a grounded rotor feeding a mixture of conductive and nonconductive particles into an electric field (Fig. 10.26). The nonconductive (dielectric) particles become polarized and thus are repelled by the active (negative) electrodes, while the conductive metallic particles become equipotential with the grounded rotor and are attracted to the active electrode.

Ion bombardment or "high tension" charging is another method of inducing a surface charge onto nonconductive particles. Here an ion beam, generated by an ionic electrode, imparts a surface charge onto the particles in its path. Conductors redistribute that charge, while nonconductors remain polarized by it. When they are both placed on a grounded rotor, the conductors can leave that surface, while the polarized nonconductors cannot (Fig. 10.27).

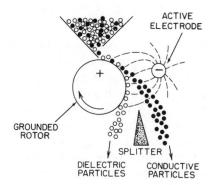

Fig. 10.26 Separation by conductive induction (82).

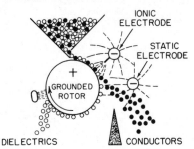

Fig. 10.27 Separation by ion bombardment (82).

While electrostatic separation is seldom used on general MSW, it is an attractive alternative to burning when the goal is to separate plastic insulation from wire fragments (if this waste has already been granulated). Another application for electrostatic separation is the recovery of metals (including tantalum) from granulated electronic scrap.

Materials Recovery Facilities (MRFs)

Some MRFs achieve full recycling by first separating the glass, paper, metal, and plastic constituents from the MSW and then producing compost from the rest. Other plants produce refuse-derived fuel (RDF) for incinerators by removing only the noncombustibles from the MSW. As such, RDF plants do not completely recycle the MSW, since the incinerator residue might still require disposal by landfilling (although the incinerator ash may be reused as filler in road construction).

The operating experience with RDF plants is extensive, while MRFs are relatively new. Yet because the interest in MRF plants is rising (construction proposals are being issued at better than one per month), equal emphasis will be given to both types. In August 1989 alone, (172): Huntington, New York, awarded a contract to Ogden Martin Systems to build a 200 TPD MRF plant; Howard and Prince George counties in Maryland received proposals from Browning-Ferris Industries to build two MRF plants; and Cape May County, New Jersey, accepted bids on a 300 TPD MRF plant.

MSW-Processing Concepts

The ultimate goal of processing MSW is to recycle all the material and energy values in the refuse and thereby completely eliminate the need for landfilling. Figure 10.28 illustrates the concept for a regionwide reclamation plant for processing both MSW and sewage sludge. In this concept the size-reduction, separation, and classification processes recover fibers, metals, and glass. A pyrolysis unit can generate both fuels for energy and activated carbon for the tertiary treatment of wastewater. The organic wastes are converted in a digestion-composting process into a pathogen-free product for soil conditioning. This type of total recycling of MSW has not been successful for two reasons: The various biogasification and thermal gasification pro-

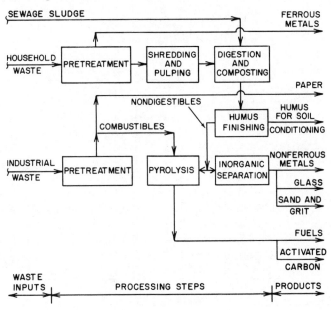

Fig. 10.28 Waste recycling plant (85).

cesses did not perform well (89), and markets for the compost were not always available.

Other processing concepts, which instead of pyrolysis and composting separate the constituents of the MSW and send the combustible fraction either to paper pulp and/or heat recovery (Fig. 10.29), were more successful. In these plants up to 40% of the materials are recovered (89) and the remaining fraction is used to generate steam and/or electricity. Naturally there is a built-in conflict between the recovery and reuse of combustible materials and the recovery of heat energy because the heating value of the incinerator fuel is reduced as combustibles are removed from it.

The early concepts of MSW processing were modified as experience accumulated in the field. These newer, successfully operating MRFs are discussed in the following pages. A few of these plants aim at the complete recovery of all MSW constituent materials, while in the majority the recovery of materials is only a "pretreatment" step in the generation of RDFs for incineration.

The most successful MSW processing plants are flexible enough to respond to changes in the market conditions for both the recovered materials and energy. Optimized MRFs of the future will likely have the flexibility to switch operating modes as a function of MSW load and market conditions. In such plants, paper fraction, for example, can be sent to paper companies, to composting, or to incineration, depending on which choice is more profitable.

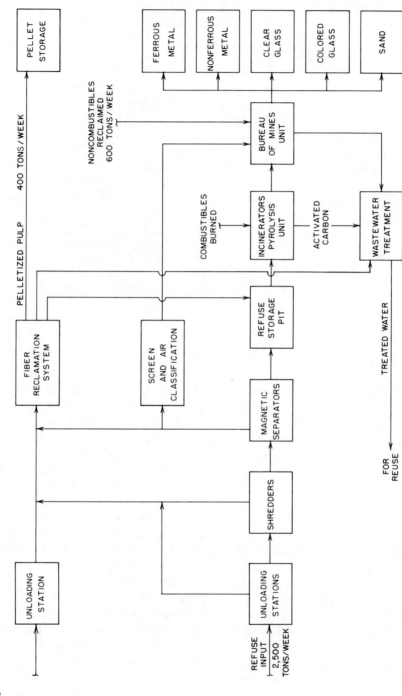

Fig. 10.29 Municipal refuse recycling plant (85).

382

A Brief History of RDF Plants in the United States

In the United States the market for MSW-based compost is limited and the majority of waste-processing plants convert the combustible fraction into RDF for incineration. RDF is derived from MSW by removing the noncombustibles and converting the combustible fraction into pellets, briquets, or other forms of easily handled and burned solid fuel. RDFs are classified according to the type and degree of processing and the form of fuel produced (Table 10.3).

Preprocessing solid waste into RDF has been practiced since the 1930s. The first commercial-scale RDF facility was built in 1972 in St. Louis, Missouri. The MSW was processed by shredding, magnetic separation, and air classification. Before 1982 twenty-three preprocessing plants were built in the United States. In eleven of these plants the RDF was burned in boilers designed for conventional fuels such as coal and not for RDF (Table 8.3.) Eight of these installations have since closed. In twelve of the twenty-three projects, the boilers were specifically designed to burn RDF (Table 10.4). With one exception, all of these installations are still in operation.

Between 1982 and 1984, many "mass burning" plants were contracted, and between 1985 and 1987, eleven new RDF-preprocessing facilities were financed, at a cost of $1.5 billion (Table 10.5) (104). All but one (Eden Prairie, Minnesota) use boilers designed for RDF service. Most use bag breakers and particle-size classification prior to shredding. A number of these plants have selected the simple, labor-intensive hand-picking methods of recovering aluminum and plastics instead of the

Table 10.3
ASTM CLASSIFICATION OF RDFs (105)

Class	Form	Description
RDF-1 (MSW)	Raw	Municipal solid waste with minimal processing to remove oversize bulky waste
RDF-2 (C-RDF)	Coarse	MSW processed to coarse particle size with or without ferrous metal separation such that 95% by weight passes through a 6 in. square mesh screen
RDF-3 (f-RDF)	Fluff	Shredded fuel derived from MSW processed for the removal of metal, glass, and other entrained inorganics; particle size of this material is such that 95% by weight passes through a 2 in. square mesh screen
RDF-4 (p-RDF)	Powder	Combustible waste fraction processed into powdered form such that 95% by weight passes through a 10 mesh screen (0.035 in. square)
RDF-5 (d-RDF)	Densified	Combustible waste fraction densified (compressed) into pellets, slugs, cubettes, briquettes, or similar forms
RDF-6	Liquid	Combustible waste fraction processed into a liquid fuel
RDF-7	Gas	Combustible waste fraction processed into a gaseous fuel

Table 10.4
PRE-1982 RDF PLANTS WITH BOILERS
SPECIFICALLY DESIGNED FOR
RDF SERVICE (104)

Facility Location	Capacity (tons/day)	Year Started	1987 Status
Akron, Ohio	1000	1979	Operating
Albany, New York	750	1981	Operating
Columbus, Ohio	2000	1983[a]	Operating
Duluth, Minnesota	400	1980	Operating
Hamilton, Ontario	500	1974	Operating[b]
Haverhill/Lawrence, Massachusetts	1300	1984[a]	Operating[c]
Niagara Falls, New York	2000	1981	Operating
Miami, Florida	3000	1982	Operating
Hempstead, New York	2000	1980	Shut Down
Rochester, New York (Kodak)	120	1974	Operating
Toronto, Ontario	220	1978	Operating
Wilmington, Delaware	1000	1982	Operating

[a] Financed prior to 1982.
[b] Undergoing rehabilitation work.
[c] Undergoing expansion.

Table 10.5
NEW PREPROCESSING FACILITIES IN THE
UNITED STATES POST-1982 (104)

Facility Location	Capacity (tons/day)	1987 Status
Biddeford, Maine	800	Operational[a]
Detroit, Michigan	4000	Construction
Eden Prairie, Minnesota	400	Operational[b]
Hartford, Conn.	2000	Start-up
Honolulu, Hawaii	2160	Financing
Orrington, Maine	800	Start-up[a]
Portsmouth, Virginia	2000	Operational[b]
Rochester, Mass.	1500	Construction
San Marcos, California	1700	Financing
Newport, Minnesota	1000	Construction[c]
West Palm Beach, Florida	2000	Financing

[a] Designed for co-firing with wood chips.
[b] Designed for co-firing with coal.
[c] Existing coal-fired boilers extensively modified.

capital-intensive mechanical separation processes. Other RDF plants, like the Ramsey/Washington Resource Recovery Facility of Northern States Power Co. in Newport, Minnesota, are fully automated and computerized, converting 70% of 1,400 TPD of MSW into RDF. Other highly automated RDF plants include one in Elk River, Minnesota, and another in West Palm Beach, Florida.

Two advantages of converting MSW into RDF (and in the process recycling all the noncombustibles) include the income from material recovery and the reduction in the quantity of ash residue. In the past, the ability to co-fire with other fuels was considered to be another advantage, but RDF-coal co-firing installations turned out to be troublesome. Ash production in RDF plants is about 8% to 10% by weight versus 20% to 25% in mass-burning plants. Some argue that when metals and plastics are removed during processing, the ash can be landfilled, while the residue from mass-burning plants cannot (104). This debate is not over yet, but a mounting body of opinion favors disposal from mass-burn incinerators in landfills built for hazardous waste.

RDF Co-Firing in Electric Power Plants

During the last fifteen years, nine American utilities have co-fired some 800,000 tons of RDF with coal or oil. As of 1988 only four utilities were still continuing that practice (Table 10.6). The reason for this reduced interest in RDF co-firing is economical, since the RDF value does not equal the value of the fuel it replaces; burning RDF not only increases capital costs but also the cost of operation and maintenance. While the operation of RDF-burning incinerators is superior to mass-burning ones because of the faster response to load variations and increased efficiency, their performance is inferior to the efficiencies of 100% coal-firing power plants.

According to the Electric Power Research Institute (EPRI), the co-firing of coal with 15% RDF (on a heat-input basis) reduces boiler efficiency by 1.5% to 3.5% relative to 100% coal firing. The loss is higher in retrofit units and less in boilers designed originally for co-firing. EPRI recommends that RDF fuel for co-firing have a maximum pellet size of 2.5 inches (standard RDF pellets are 4 to 6 inches), an ash content of 12%, a moisture content of 24%, and a heating value of 5,900 BTU/lb. These specifications are not easily met, as both ash and the moisture content tend to vary with the season and the source of waste.

Operational problems, particularly in retrofit installations, include difficulties in RDF handling, slagging of the upper furnace wall, loss of electrostatic precipitator efficiency, boiler tube corrosion, and accumulation of bottom ash. This experience is similar to that of the 100% RDF-burning HRIs (Tables 8.3 and 10.4) insofar as the performance of retrofit units is inferior to units designed specifically for RDF service.

AN RDF BURNING HRI IN AKRON, OHIO

Akron has a population of 250,000 and collects about 200,000 tons of MSW per year. The Akron plant, the first HRI in the United States, has a dedicated boiler designed for RDF service and provides energy for downtown Akron (Table 10.4).

Table 10.6
ELECTRIC UTILITY RDF CO-FIRING EXPERIENCE (104)

Location	Power Plant	Unit No.	Capacity (MW)	Unit Boiler Mfr.	Firing Method	Co-firing Fuel	Dump Grate	RDF Heat Input (%)		RDF Mass Feed Rate (t/h)		RDF Co-fired Through 1987 (t)	Co-firing Start Date	Commercial Start Date	Co-firing Duration (yrs)	Status
								Max.	Avg.	Max.	Avg.					
Ames, Ia.	Ames	7	35	C-E	Suspension	Coal	Yes	21.9	21.9	13.0[a]	6	367,500	1975	April 1978[b]	10.0	Operating
Ames, Ia.	Ames	8	65	B&W	Suspension	Coal	Yes	21.9	21.9	13.0[a]	6	367,500	1981	April 1978[b]	10.0	Operating
Baltimore, Md.	Crane	2	200	B&W	Cyclone	Coal	No			7.9	5.6	128,740	1980	February 1984	4.1	Operating
Bridgeport, Conn.	Bridgeport, Harbor	1	80	B&W	Cyclone	Oil	No	51				7,900	1979	November 1979	0.2	Shut down 1981
Chicago, Ill.	Crawford	7	240	C-E	Suspension	Coal	No	10				20,000	1978	October 1978	1.3	Shut down 1979
Chicago, Ill.	Crawford	8	358	C-E	Suspension	Coal	No									
Lakeland, Fla.	McIntosh	3	364	B&W	Suspension	Coal	Yes	10				58,176	1983	February 1983	4.1	Operating
Madison, Wis.	Blount	8&9	50 ea	B&W	Suspension	Coal	Yes	26	11	13.0	5.4	81,665	1975	January 1979	9.3	Operating
Milwaukee, Wis.	Oak Creek	7&8	310 ea	C-E	Suspension	Coal	No	20	15	30	25	100,000	1977	March 1977	3.5	Shut down 1980
Rochester, N.Y.	Russell[c]	1	42	C-E	Suspension	Coal	Yes									
Rochester, N.Y.	Russell	2	63	C-E	Suspension	Coal	No					47,900	1981	September 1981	3 ±	Shut down 1984
Rochester, N.Y.	Russell	3	63	C-E	Suspension	Coal	Yes									
Rochester, N.Y.	Russell	4	75	C-E	Suspension	Coal	Yes	15	10							
St. Louis, Mo.	Meramec	1&2	125 ea	C-E	Suspension	Coal	No	27	10		9.1	48,972	1972	April 1972	3.7	Shut down 1975
												Total 861,000				

Source: Electric Power Research Institute.
Notes: C-E, Combustion Engineering; B&W, Babcock & Wilcox. Heat input and mass feed rate from either yearly averages or specific tests. Best available measured values are shown but actual current usage may differ. Madison 5.4 t/h is 1961 average and St. Louis 9.1 t/h is from plant records.
[a] Maximum during unit 8 3-h boiler test in 1982 was 14.7 t/h.
[b] Trial operation in 1975 and 1976—commercial since 1978 grate installation.
[c] Induced draft fans limit cofiring to about 90% of capacity.

Built in 1979 at a capital investment of $54.5 million, its initial purpose was to burn 1,000 TPD of MSW and shredded tire chips to generate steam while recovering the ferrous metals from the MSW feed. The main components of the plant are shown in Figure 10.30. The MSW is weighed and dumped (at a truck turnaround time of about five minutes) in a storage pit sized for 2–3 days of supply. Hydraulic rams feed the waste from the storage pit into two 1,500 horsepower hammer-mill shredders, each having a 60–75 TPH capacity. After the MSW is reduced to a particle size of 4–6 inches, it is sent through a double-drum-type magnetic separator. The ferrous-metal fraction is air-cleaned to remove paper and other contaminants. The ferrous metals are sold at $2 to $4/ton.

In 1985–86 the recovered energy was used in three different forms. An 18-mile-long pipe loop distributed 560 PSIG steam to more than 100 municipal and commercial users in downtown Akron. The steam was sold on a continuous basis for $9 to $10.25 per thousand pounds depending on the quantity purchased. Because the steam supply cannot be interrupted, a pay fuel (natural gas) was used to supplement the RDF so that the amount of steam generated always matched the load. On average, 5 pounds of steam were produced from each pound of RDF. A 2-mile pipe loop distributed 210° F hot water (produced from turbine generator exhaust) for district heating purposes. Two 2-MW turbine generators produced electricity for in-plant use.

A serious accident at the Akron plant (including fatalities) was caused by the unauthorized disposal of sawdust saturated with highly volatile solvents. This resulted in several severe explosions in the shredders. After the explosions, plant procedures were substantially revised. Personnel are no longer allowed in the shredder building during processing, and some sources of solid waste—from paint stores, cleaning establishments, and hardware stores—are no longer accepted. Pressure-relief systems were added to minimize explosion damage in "hardened" areas and solid vent deflectors were replaced with blast mats. In addition, flammable vapor detection and ventilation and sprinkler systems were added to the pit area and in the shredder building.

Although the economics of operation in the Akron plant have improved (Table 10.7), the plant is not yet profitable. In 1986 total revenues amounted to $6,721,000. Of this sum $4,803,000 was from the sale of the steam and $1,582,000 was from $9/ton tipping fees. The rest of the income was for the sale of hot water and ferrous materials. In 1987 the revenue was projected to be $42.54/ton. The operating costs for that same year included $46.75/ton for plant operation (Fig. 10.31), $30.57/ton for debt service, and $2.28/ton for the disposal costs of the 23% ash and residue. To carry the plant's debt service and still break even in 1987, the tipping fee would have to be increased from $9 to $46.06/ton (103).

An RDF Facility in Lawrence, Massachusetts

The Lawrence plant in Haverhill has been operating successfully since 1984, feeding 100% RDF to a 250,000-lb/hr boiler designed for RDF service (Fig. 10.32). In this facility 1,300 TPD of MSW is separated into 983 TPD of RDF fuel, 260 TPD of glassy residue, which is landfilled, and 57 TPD of ferrous metals, which are sold.

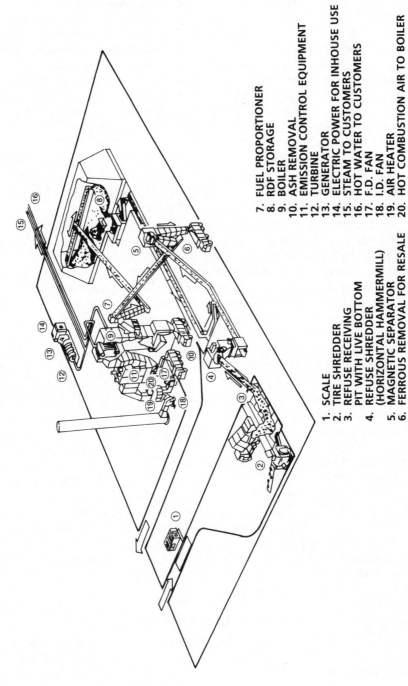

1. SCALE
2. TIRE SHREDDER
3. REFUSE RECEIVING
   PIT WITH LIVE BOTTOM
4. REFUSE SHREDDER
   (HORIZONTAL HAMMERMILL)
5. MAGNETIC SEPARATOR
6. FERROUS REMOVAL FOR RESALE

7. FUEL PROPORTIONER
8. RDF STORAGE
9. BOILER
10. ASH REMOVAL
11. EMISSION CONTROL EQUIPMENT
12. TURBINE
13. GENERATOR
14. ELECTRIC POWER FOR INHOUSE USE
15. STEAM TO CUSTOMERS
16. HOT WATER TO CUSTOMERS
17. F.D. FAN
18. I.D. FAN
19. AIR HEATER
20. HOT COMBUSTION AIR TO BOILER

Fig. 10.30 Process flow diagram. (Courtesy of Babcock & Wilcox [103].)

Table 10.7
FINANCIAL PERFORMANCE OF AKRON, OHIO, PLANT (103)

	Actual *Oct 85–Dec 86*	*Actual* *1986*	*1987* *Budget*	*Actual* *1st Qtr 87*
Revenues	$8,206,000	$6,721,000	$8,890,000	$2,298,000
Less cost	$12,562,000	$10,573,000	$9,769,000	$2,263,000
Net income (loss)	($4,356,000)	($3,851,000)	($879,000)	$35,000
Tons processed	211,674	186,243	208,960	58,630
Net loss per ton	− 20.58	− 20.68	− 4.21	+ 0.60

The MSW passes through two parallel 70 TPH Heil shredders producing an output particle size of 90% under 4 inches (101.6 mm). The ferrous metals are removed by Dings and head pulley magnets.

The shredded refuse is passed through a two-stage, 12.5-foot-diameter, 60-feet-long (3.8m × 18.24m) trommel screen. The first stage is provided with 1-inch (25.4mm) holes to remove the glassy residue. The second stage is provided with 6-inch (152.4mm) holes that separate the oversize material for further shredding and send the under-6-inch fraction to RDF storage. The RDF produced has a heating value of over 6,000 BTU/lb (13.96 mJ/kg); the ash content is less than 15% and its particle size is 97% under 4 inches (101.2mm).

Figure 10.33 shows the boiler plant, which contains two redundant fuel-delivery systems consisting of horizontal, inclined return-feed conveyors. The feeding rate exceeds the boiler requirements, resulting in the return of some excess RDF to the feed area. The boiler is provided with six Detroit feeders. Two Heil distribution augers are mounted on top of the feeders. A slide gate is provided between each feeder and auger so that any of the feeders can be isolated for maintenance. Each inclined conveyor can feed either distribution auger.

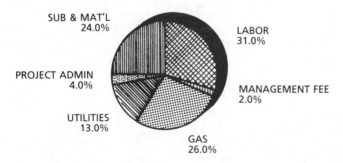

Fig. 10.31 Annual distribution of costs at an RDF plant in Akron, Ohio (103).

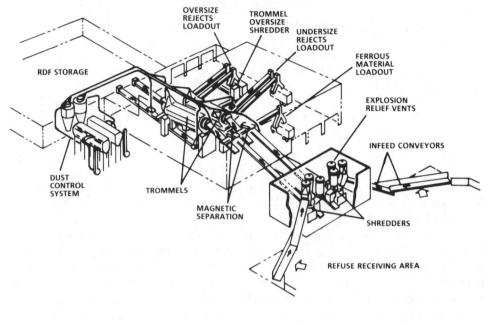

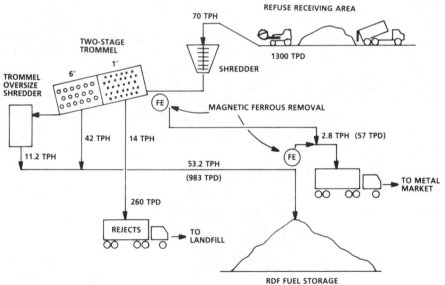

Fig. 10.32 RDF preparation plant in Haverhill, Massachusetts (95).

390

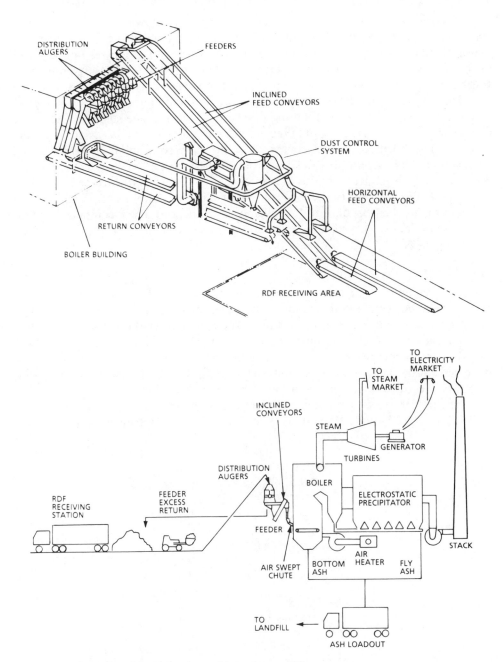

Fig. 10.33 RDF boiler plant in Lawrence, Massachusetts (95).

391

Each feeder contains two hoppers—a surge and a lower RDF hopper. The surge hopper is kept full all the time by the over-running distribution augers. A hydraulic ram moves the RDF from the surge hopper into the lower hopper. The lower hopper is provided with three level switches. The ram in the surge hopper is started and stopped to keep the level in the lower hopper within the desired operating gap. The RDF from the lower hopper is charged into the boiler by a variable-speed inclined conveyor, controlled by firing-rate boiler controls. The steep incline of the conveyors causes a mixing and fluffing action in the hopper, while the air sweeping of the spouts (located on the front wall of the furnace) minimizes bridging and compaction. The system provides a uniform distribution of RDF over the moving grate and also gives fast response to load changes. System performance is satisfactory. The RDF bulk density is consistent and its flow is smooth. The variable-speed feeders provide fast response to load changes. The almost 100% equipment redundancy guarantees a very reliable system, with the availability of the entire plant exceeding 98%.

Modeling RDF Performance

As community recycling increases, the feed to the RDF plant will change. Studies have determined the heating value and composition of the RDF that might result from different degrees of recycling. Table 10.8 gives the MSW composition that was assumed for the purposes of such a study. In this model it was assumed that the pretreatment steps include size reduction (to about 5 cm), screening, magnetic separation, and air classification. Table 10.9 gives the properties of the MSW and of the recovered RDF after various degrees of recycling.

The model shows that the ash content drops and the heating value rises as the MSW is processed into RDF, and the type and degree of recycling has only a limited

Table 10.8
COMPOSITION OF WASTE
FOR BASE CASE (89)

Component	Percent As-Received
Ferrous	5.5
Aluminum	0.9
Glass	9.5
Mixed paper	22.6
Newsprint	11.8
Corrugated	12.2
Non-PVC plastic	2.9
PVC plastic	0.3
Yard waste	12.5
Food waste	2.5
Other noncombustible	9.5
Other combustible	9.8

Table 10.9
CALCULATED MSW AND RDF PROPERTIES AND COMPOSITIONS RESULTING FROM DIFFERENT DEGREES OF RECYCLING (89)

Scenario	Heating Value (Btu/lb Wet)	Percent Ash (Dry)	Ultimate Analysis (Percent)						Heavy Metal Analysis (mg/kg)									
			C	H	O	N	S	Cl	Sb	As	Ba	Cd	Cr	Cu	Pb	Hg	Ni	Zn
Baseline Case																		
MSW	3970	36.6	32.1	4.3	25.8	0.58	0.17	0.33	53	4.9	2160	14.4	210	720	630	18	220	290
RDF	5670	11.0	44.3	5.9	37.7	0.44	0.16	0.49	68	5.4	2620	14.0	200	170	500	23	40	160
— 30 Percent Fe, Al, and Glass																		
MSW	4200	32.0	34.0	4.6	27.7	0.62	0.18	0.36	55	5.1	2220	15.0	200	570	600	18	160	270
RDF	5740	9.7	44.9	6.0	38.3	0.45	0.16	0.51	65	5.2	2510	13.4	190	140	470	22	30	130
— 30 Percent Fe, Al, Glass, Newsprint, and Corrugated																		
MSW	4070	35.0	33.3	4.4	26.1	0.70	0.19	0.37	62	4.8	2500	16.5	210	600	550	21	170	300
RDF	5710	11.0	44.5	6.0	37.3	0.52	0.17	0.56	82	5.3	3200	16.2	210	160	440	28	30	160
— 30 Percent Newsprint																		
MSW	3905	37.9	31.6	4.2	25.2	0.61	0.17	0.34	55	5.1	2250	15.0	210	750	590	18	230	300
RDF	5635	11.6	44.0	5.9	37.4	0.47	0.16	0.52	74	5.9	2860	15.1	200	180	450	25	40	170
— 50 Percent PVC																		
MSW	3950	36.8	32.1	4.3	25.8	0.59	0.17	0.24	52	4.8	2190	14.4	210	730	640	18	220	290
RDF	5655	11.1	44.3	5.9	37.8	0.45	0.16	0.34	67	5.3	2680	14.0	200	170	500	23	40	160
— 50 Percent Yard Waste																		
MSW	4055	37.6	31.7	4.2	25.5	0.50	0.16	0.33	55	5.1	2230	15.0	220	750	660	18	230	300
RDF	5720	11.0	44.2	5.9	37.8	0.41	0.16	0.49	69	5.5	2670	14.1	210	170	500	23	40	160
— 50 Percent Food Waste																		
MSW	3990	36.7	32.2	4.3	25.8	0.57	0.17	0.32	53	5.0	2170	14.2	210	700	630	18	220	270
RDF	5680	11.0	44.3	5.9	37.8	0.44	0.16	0.48	68	5.4	2620	13.9	200	160	490	22	40	150

effect on ash content or heating value (89). The nitrogen content of the RDF is consistently lower than that of the MSW and the sulfur content is relatively unaffected by processing, while PVC recycling has a substantial effect on the chlorine content of the RDF. The calculated heavy-metal analysis shows that because of the magnetic separation of ferrous metals, the concentration of lead (Pb) and zinc (Zn) is lower in the RDF than in the MSW.

Modeling is a useful tool in the evaluation of RDF processes. One can estimate the effect of the degree of size reduction, the influence of the opening sizes in screening equipment, and the effect of placing shredders up or downstream of screening or air-separation equipment. Some modeling calculations can also estimate the Base/Acid Ratio, Slagging Index, and Fouling Index values, which give an early indication of the likely maintenance and operating problems associated with the particular process.

THE MRF PLANT IN ROTTERDAM

A typical European MSW-processing MRF is the Recycling Zoetermeer BV in Rotterdam, the Netherlands, which I visited in 1988. The plant has been in operation since 1982 and serves a population of 210,000 people, who generate about 70,000 tons of MSW per year, or about one kilogram (2.2 pounds) per person per day. The MSW is separated into organics (53% by weight), RDF (31%), glass/inerts (13.5%), and metals (2.5%). In 1988 the plant charged a tipping fee of 76 guilders ($43) per ton and sold the recovered steel at 60 to 100 guilders ($34 to $57) per ton and the RDF at 20 guilders ($11) per ton. With these operating parameters, the plant is profitable (109).

The plant is located in the industrial section of the city. Trucks and railroad cars wait in front of the building. Inside, the noise level is high enough so that one must raise one's voice to be understood. The tipping floor is big enough for two days' collection of MSW. The oversize items in the solid waste are separated and the rest (mostly in plastic bags) is placed on the shredder feed conveyor.

The rotating hammers in the 350-horsepower shredder break down the MSW to a particle size of 4 to 8 inches (10 to 20 cm). During my two-hour visit, there was one jam in the shredder, (caused by some steel cable being fed to it), which took about fifteen minutes to clear. The shredded MSW is conveyed through steps of magnetic separation, air classification, and screening. The plant is automatically and remotely operated from a central glass-paneled control room so located that the whole plant is visible from it. A full-size graphic-display panel (Fig. 10.34) shows the operating status of all equipment, while the instrumentation below the graphic displays controls the more important process variables, such as airflows.

The organic and inert fractions are collected in railroad cars and disposed of as soil conditioners and filling materials, respectively, and generate no revenue. The metallic fraction, collected separately (Fig. 10.35) is sold to scrap dealers. The RDF, which is compressed into bales (Fig. 10.36), is sold as fuel.

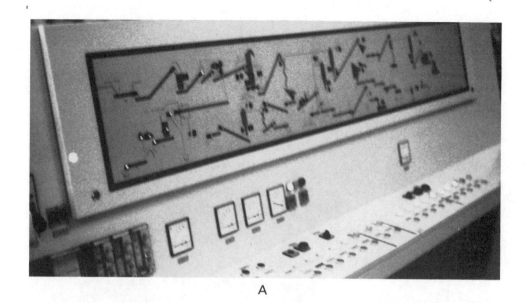

A

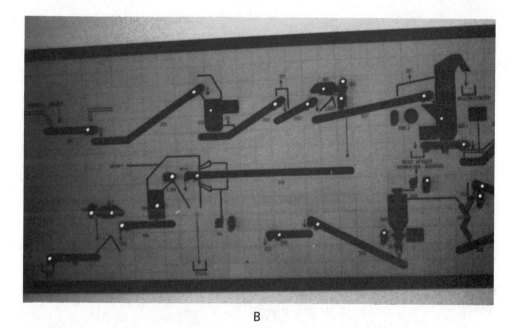

B

Fig. 10.34 The control panel (A) and part of the graphic display section (B) in an MSW processing plant in Rotterdam, the Netherlands (109).

Fig. 10.35 Bottle caps are separated from the metal fraction in an MSW processing plant in Rotterdam, the Netherlands (109).

Fig. 10.36 RDF baling machine in an MSW processing plant in Rotterdam, the Netherlands (109).

396

THE MRF PLANT IN ROME

The Sorain-Cecchini MRF plant in Rome has been in operation for over twenty years. It recovers ferrous metals (6% by weight), aluminum (1%), "organics" (34%), and film plastics (1%), while generating densified RDF (51%) and rejecting 7% over-size items (Table 10.10). The total plant capacity is 1,200 tons per day (94). The main processing steps involve magnetic separation for ferrous-metal removal, eddy-current separation for aluminum recovery, rotary screens for separation by size, and air classifiers for separation by density. The overall process consists of some 80 pieces of equipment, which are flexible enough to give different splits as market conditions change.

The tipping area is under a slight vacuum to prevent the escape of odors. The charging conveyor (item 1 in Fig. 10.24) is provided with a pick-up device (2) that breaks the bags and removes bulky reject items. A leveler device (3) meters the waste-flow rate and also removes rejects. The primary screen (4) separates the "large" (over 8 inch) fraction from the smaller, heavier fraction. The approximately 55% "large" fraction of paper, wood, and plastic is fed to the 10–20 rpm large breaker (5), which reduces particle size and breaks the plastic bags. The large breaker automatically rejects any items it cannot break (about 2%). The output (53%) is sent to the large air classifier (6), where the lighter (10%) sheet paper and plastic fraction is separated from the heavier (43%) cardboard, wood, and rags. The reject fraction consists mostly of "white goods" (appliances) but includes bulky items such as bedsprings. This fraction is hauled away by subcontractors.

The light fraction (10%) passes through a differential shredder (7), which breaks up only the paper. It is followed by a rotary screen (8), which separates out the lighter 3% of the stream, containing the plastic film. This stream is sent to the small air classifier (9), where the 1% light fraction is taken to plastics recovery, while the remaining paper and rag fragments are included with the densified RDF. The re-covered plastic film (mostly polyethylene) is shredded into square-inch flakes, cleaned by washing, and air-dried. The dry flakes are melted and fed into an extruder and the pellets are shipped to plastic film manufacturers.

To separate metal, the 40% heavy fraction from the primary screen (4) is passed through the primary magnetic separator (10), which removes 4% and sends that fraction to ferrous recovery. The remaining fraction is further homogenized in the small breaker (11) and separated in the secondary rotary screen (8) into the 15% large fraction (over 4 inches), consisting mainly of paper, wood, and plastics, and is sent to DRDF recovery. The 21% small fraction (under 4 inch), consisting of organics, glass, sand, ashes, and aluminum, is sent to a conveyor (18); aluminum (1%) is removed by hand and the rest (20%) sent to organic recovery.

The 43% heavy fraction from the large air classifier (6) is passed through the secondary magnetic separator (13), which removes 1% and sends that fraction to ferrous recovery. The remaining fraction travels on the sorting conveyor (14), where semiautomatic devices and inspectors remove the cardboard (8%), which is sent to

Table 10.10
ROME-EAST—DECADE 1 JANUARY 1975 TO 31 DECEMBER 1984
RECOVERY OF MATERIAL FRACTIONS—ANNUAL (94)
(in thousands of metric tons)

	1975	1976	1977	1978	1979	1980	1981	1982	1983	1984
1. Input waste fractions	327.0	338.0	347.0	344.0	297.0	286.0	270.0	275.0	280.0	285.0
2. Paper (pulp)	19.6	20.2	20.0	20.5	20.8	19.2	18.6	18.4	18.5	18.2
3. Metal (ferrous)	11.2	11.4	11.5	11.1	9.1	8.4	7.8	8.1	8.1	8.4
4. Plastic (film)	—	—	.4	1.4	2.1	1.7	1.6	1.9	2.1	2.0
5. Organic (feed)	82.0	85.0	87.0	86.0	68.0	51.0	—	—	—	—
6. Organic (compost I)	65.0	63.0	66.0	65.0	67.0	75.0	79.0	78.0	78.0	77.0
7. Organic (compost II)	81.0	85.0	88.0	82.0	107.0	109.0	123.0	127.0	130.0	132.0
8. Glass	Test	Test	Test	Test	Test	—	—	—	—	—
9. Drdf	—	—	—	Test	Test	—	1.2	1.1	1.6	1.8
10. Materials (recovered)	185.0	188.1	194.6	188.6	212.8	218.4	231.2	234.5	238.3	239.4
11. Incineration fraction	140.0	140.0	145.0	150.0	160.0	160.0	160.0	160.0	160.0	160.0
12. Steam sold	50.0	50.0	50.0	48.0	46.0	45.0	60.0	58.0	60.0	60.0
13. Landfilled residue	47.0	48.0	45.0	46.0	52.0	54.0	53.0	54.0	53.0	54.0
13.1 from 11	42.0	42.0	43.5	45.0	48.0	48.0	48.0	48.0	48.0	48.0
13.2 from 2 and 9	5.0	6.0	1.5	1.0	4.0	6.0	5.0	6.0	5.0	6.0

DRDF. Any missed recovery items and rejects (4%) are sent to the eddy-current separator (17) for aluminum removal and then to organic recovery. The removed aluminum is crushed and densified to a specific gravity of 1.0 (62 lb/ft³) before being placed in storage. The 30% fraction remaining on the sorting conveyor (14) is mostly paper and is sent to the flail (15), where it is broken down before being sent to the secondary air classifier (16). The three magnetic separators (10), (13), and (19) send 5.8% to ferrous recovery, where it is shredded by the abrader hammer mill. The shredding step is followed by cleaning through firing or washing and a final magnetic-separation step to remove the nonmetals that were loosened by the abrader.

The "organics" fraction, left over from the plant feed after the removal of metals, plastic film, and paper is essentially a heavy fraction of small-size particles containing organics, glass, ceramics, sand, ashes, hard plastics, and small pieces of wood. This fraction is placed in an aerobic digester and broken down into raw compost. After the removal of glass, ceramics, and other inorganic rejects, the raw compost is subcontracted for further processing. This processing splits the organic fraction into a "feed" fraction, a high-quality compost fraction, and a residue, which is usually landfilled.

The 15% large fraction from the secondary rotary screen (8) and the 30% from the flail (15) are sent to the secondary air classifier (16), which removes all paper (35%) and sends it to the DRDF air classifier (20), together with the "small" fraction (7%) from the secondary rotary screen (8) and the "heavy" fraction (2%) from the small air classifier (9). After the DRDF magnetic separator (19) removes the remaining ferrous metals, the DRDF is densified into flakes in the recovery line. The DRDF is stored at a specific gravity of 0.6 (38 lb/ft³). The heavy fraction (10%) from the secondary air classifier (16) is sent to organic recovery.

The DRDF obtained is relatively clean and its sulfur and chlorine content is low, as most metals, hard plastics (PVC, PET), and other impurities have been removed from it. The Sorain process can switch from producing DRDF to generating paper pulp, depending on market conditions.

THE MRF PLANT AT GELLATIN, TENNESSEE

National Recovery Technologies Inc. is the owner and operator of the Gellatin MRF, which has been in operation since 1982. It produces RDF by removing from the MSW 80% of the glass, ferrous metals, and grit and 75% of the aluminum containers. The capacity of the plant is 10 tons/hr. Relative to mass burning, the use of RDF fuel at the boilers increases the combined boiler efficiency and availability by 20% (90). The corresponding increase in revenue is about $7/ton of MSW. Therefore, in order for the Gellatin plant to be profitable, the net preprocessing costs must be under $7/ton.

The MRF (Fig. 10.37) consists of a rotary separator, which is a cylinder 9.5 feet (2.9m) in diameter and 16 feet (5.9m) long, rotating at 12.5 rpm, and has three sections. The input section contains cutters and breakers that open bags and break glass. The ferrous-removal section contains permanent magnets. The ferrous metal

400

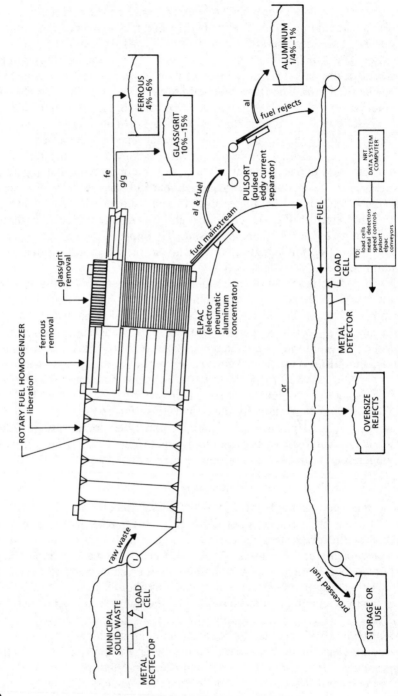

Fig. 10.37 The process of materials recovery (90).

is wiped off the magnets and a vibrating conveyor takes it to storage. The exit section contains longitudinal lifters that collect glass and grit and drop this heavy fraction on a vibrating pan conveyor, which carries it to storage. The remaining material (primarily combustibles and aluminum) exits the rotary separator onto an array of ten metal detectors that identify the aluminum containers. The aluminum is then blown onto the collection conveyor by air jets and removed by an eddy-current separator.

A two-man crew (per shift) performs all repair and maintenance on the system. Equipment availability [operating time/(operating time + downtime)] is not as good as some other processes but still is over 80% (90). The recovery efficiencies of this process are given in Table 10.11.

Table 10.11
ROTARY FUEL HOMOGENIZER RECOVERY EFFICIENCIES
TESTING (90)

DATE: *14–15 November 1985*
SAMPLE: *Residential refuse, hand-sorted*
TEST SAMPLE PROFILES

	Input Waste Composition					
Feedrate (tph)	*Paper, Plastics Textiles, etc.*	*Wet Organics*	*Glass*	*Fines*	*Ferrous and Al*	*Sample Wt. (lbs)*
3.6	59.9%	10.0%	14.6%	9.4%	6.1%	288
7.7	68.5%	9.0%	9.1%	6.8%	6.3%	1084[a]
12.3	80.7%	5.2%	6.4%	2.6%	5.0%	763
15.7	71.7%	4.2%	7.7%	7.8%	5.9%	1000[b]

[a] Includes 1.5 lb oversize bulky construction debris.
[b] Includes 26 lb of oversize bulky construction debris.

NONCOMBUSTIBLE RECOVERY EFFICIENCIES

Feedrate (tph)	*Ferrous Recovery*	*Glass Recovery*	*Fines Recovery*[a]
3.6	95.8%	89.0%	71.3%
7.7	89.5%	80.7%	48.8%
12.3	77.6%	87.0%	85.1%
15.7	63.4%	77.4%	35.2%

[a] Fines stream processed through glass clean-up air knife. Modifications underway to bypass air knife for fines stream. Analyses show fines 80%+ noncombustible.

COMBUSTIBLES REMOVED IN PROCESSING

Feedrate (tph)	*Paper, Plastics Textiles, etc.*	*Wet Organics*[a]
3.6	3.3%	62.2%
7.7	2.9%	58.6%
12.3	2.4%	68.7%
15.7	2.4%	67.8%

[a] Wet organics are high moisture/low BTU food waste present in the glass stream. Further processing of this stream can return these to the fuel fraction if desired.

In 1984–85 operating and maintenance expenses averaged $2.53/ton of MSW. The initial capital cost of the facility was $800,000, which corresponds to about $4,000 per TPD capacity. The debt service at 10% interest for twenty years amounts to $7,700/month or $1.38/ton of MSW. Therefore, the total debt and operating cost is $3.91/ton. Glass and grit disposal costs are additional. The income includes the sale of aluminum at $700/ton and ferrous metal at $24/ton. Assuming 0.4% aluminum in the MSW and 75% recovery efficiency, the income from aluminum recovery is $2.10/ton of MSW; the ferrous revenue, assuming 5% content and 80% recovery, is $0.96/ton of MSW. Therefore, the total value of the recovered materials is $3.06/ton of MSW. This income alone is insufficient to pay for the debt and operating costs.

Table 10.12
PROJECTED MATERIALS RECOVERY FACILITY ECONOMICS (90)
(operation in conjunction with an energy recovery facility)

67,000 tons per year processed 3 shifts, 350 days/yr, 10 tph, 85% availability		
Income		
Aluminum sales (0.4%, 75% recovery, $700/ton)		140,700
Ferrous sales (5.0%, 80% recovery, $24/ton)		64,300
Increased energy production efficiency (9.5%)		183,300
Increased energy production availability (10%)		193,000
	Total annual income	$581,300
Expenses		
Salaries		120,700
Insurance		6,700
Supplies		26,800
Utilities		10,100
Accounting		5,400
Debt service $800,000 @ 10% for 20 years		92,400
	Total annual expenses	$262,100
	Net annual profit	$319,200

Notes
1. Energy production based upon 3 lbs-steam per lb of raw waste, an energy value of $36.00 per ton of raw waste, and 80% of increased steam sales.
2. Credit for energy plant from increased tipping revenue is not included.
3. Combustible loss is included in increased energy production percentage.
4. Capital cost assumes utilization of energy facility storage and feeding capacity (crane or front-end loaders).
5. Ferrous revenue includes $3.00 per ferrous ton shredding costs.
6. Supplies include all maintenance item costs.

To make the plant profitable, some credit must be taken for increased energy production in HRI. Assuming 3 pounds of steam per pound of waste and a steam value of $6/1,000 lb, the total value of the steam generated is $36/ton of MSW. Assuming that preprocessing at the Gellatin facility increases the efficiency and availability of the HRI by 20%, and assuming that 80% of this added steam production is saleable, this would add an income of $5.76/ton of waste and would make the operation profitable. Some of these cost-benefit figures are summarized in Table 10.12.

A New MRF Plant in Rhode Island

In 1989 a new 80 TPD MRF was started up in Rhode Island (Fig. 10.38). Designed and operated by New England CR Inc. in conjunction with Maschinenfabrik Bezner of West Germany, this highly automated plant can sort and recover the recyclables from partially separated MSW containing metallic, glass, and plastic

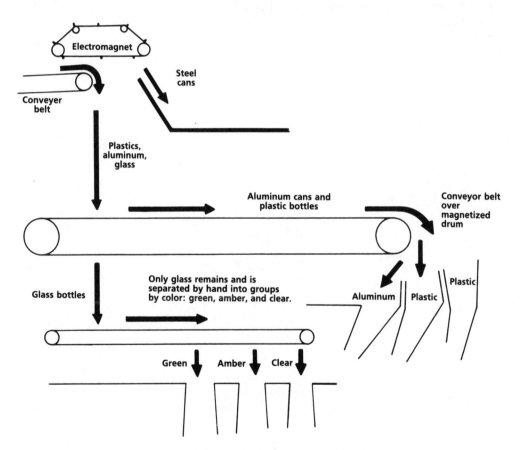

Fig. 10.38 The operation of a solid-waste processing plant in Rhode Island. (New England CRInc. New York Times/Bohdan Osyczka/May 2, 1989 [160].)

cans, bottles, and other containers but not paper and organics. The partially separated MSW enters the plant on a conveyor belt, which first passes under an electromagnet that attracts the tin-plated steel cans and carries them off to be shredded. The remaining waste stream falls down toward another conveyor belt. As it falls, it encounters a rolling curtain of chains. The lighter objects (aluminum and plastic cans) cannot break through and are diverted toward a magnetized drum. The heavier (mostly glass) bottles pass through the curtain and arrive on a hand-separation belt, where they are separated manually by color.

As the plastic and aluminum containers reach the magnetized drum, the aluminum objects are dropped into a separate hopper. The plastic objects continue on the conveyor belt and are later sorted according to weight.

The plant design appears to be simple enough to guarantee reliability. The concept of this type of MRF plant is promising because it simplifies the process of source separation by allowing cans and containers of all types to be placed in the same bin.

Postscript

◻
◼◼

THROUGHOUT THIS BOOK we have seen that the problems associated with MSW disposal are symptoms of an overindustrialized society that values comfort and economic growth above conservation. Today, however, a fundamental change is taking place in our understanding of the world. We now understand just how finite the planet is. One purpose of this book is to help speed that realization. Another goal is to reassure the reader that science and technology and human ingenuity, when applied properly, can solve our problems. We should not be scared by the prophets of doom who see only overpopulation, ozone depletion, acid rain, or the greenhouse effect. They forget that we have also doubled our life span, have nearly eliminated starvation, and that the ideals of disarmament, democracy and human rights are spreading throughout this planet.

The planet's biosphere is of fixed dimensions. It is a ten-mile-thick layer that extends over a surface of 200 million square miles. This thin crust of earth, air, and water sustains life; we mold this environment and are molded by it. By overgrazing we can cause desertification, by excessive lumbering we cause soil erosion and flooding. By planting single crops, we shorten the food chain and transform a complex, stable ecosystem into a simple, unstable one (Fig. 1).

More than ever before, the future depends on the choices we make. In a sense we have become responsible for our own evolution. Therefore, we must develop new values and cultural attitudes to ensure survival. A good starting point would be to tame science and its offspring, technology. Another would be to revise our economy around the concept of a "green" GNP (G-GNP). Science should not be a "value-free" force, but must be subordinated to human goals and values, and technology

405

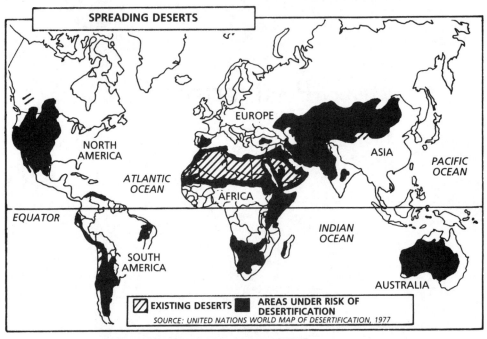

SPREADING DESERTS

EUROPE

NORTH AMERICA

ASIA

PACIFIC OCEAN

ATLANTIC OCEAN

AFRICA

EQUATOR

INDIAN OCEAN

SOUTH AMERICA

AUSTRALIA

EXISTING DESERTS   AREAS UNDER RISK OF DESERTIFICATION

SOURCE: UNITED NATIONS WORLD MAP OF DESERTIFICATION, 1977

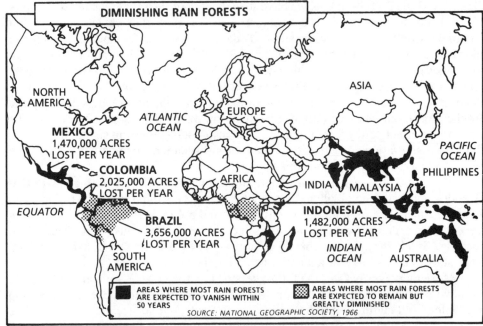

DIMINISHING RAIN FORESTS

NORTH AMERICA

ATLANTIC OCEAN

EUROPE

ASIA

PACIFIC OCEAN

MEXICO
1,470,000 ACRES
LOST PER YEAR

PHILIPPINES

COLOMBIA
2,025,000 ACRES
LOST PER YEAR

AFRICA

INDIA   MALAYSIA

EQUATOR

BRAZIL
3,656,000 ACRES
LOST PER YEAR

INDONESIA
1,482,000 ACRES
LOST PER YEAR

INDIAN OCEAN

SOUTH AMERICA

AUSTRALIA

AREAS WHERE MOST RAIN FORESTS ARE EXPECTED TO VANISH WITHIN 50 YEARS

AREAS WHERE MOST RAIN FORESTS ARE EXPECTED TO REMAIN BUT GREATLY DIMINISHED

SOURCE: NATIONAL GEOGRAPHIC SOCIETY, 1966

Fig. 1 Areas of diminishing rain forests and spreading deserts (Bernard Bennell/The Globe and Mail [127].)

406

should not attempt to "govern" or "conquer" nature. Nature's processes operate on a higher level of complexity and on a longer time frame than what human science is capable of comprehending. We do not know what the consequences might be of our toxic discharges into the oceans by the year 3000 or what awaits the generations that will inherit our radioactive wastes. The environment is a complex web, and straining some of the strands affects the entire web.

Scientists do not yet know how much acidic fumes, fluorinated hydrocarbons, or carbon dioxide gases can be released before causing irreversible climatic changes. We have only disconnected bits of information, such as that each car releases five tons of carbon dioxide in a year, that 6,000 oil production platforms are already operating in the ocean, that food in the United States travels an average of 1,300 miles before being consumed, and that our population increases by one billion every twelve years (191). Similarly, human science does not yet know how much dioxin, radioactivity, heavy metals, PCBs, or pesticides can be discharged before the food chain itself is poisoned. Science is ignorant of the time lag between discharging pollutants and feeling their accumulated effects. The time scale of technology is based on the human lifespan, while a single displacement of the waters in the deep oceans takes 1,000 years and the half-life of plutonium is 24,000 years.

Today economists determine the net output of a society by subtracting capital depreciation—the loss of machines, vehicles, and buildings—from the goods and services produced. In this accounting system, cutting down trees to build a parking lot increases the GNP. Traditional economists assume that all wasted assets are replaceable, that none is essential to economic growth and human welfare, including oil or the ozone layer. This balance sheet disregards changes taking place in our environment. "Green-accounting" would subtract from the GNP the damage done to the air, water, soil, and animal and plant life and take into account the cost of restoring them. The goal of environmental accounting should be to maintain the life-support systems of the planet so that future generations will have as much productive capacity as ours. Therefore, the G-GNP can be calculated by subtracting from the GNP the cost of the measures needed to guarantee a stable and sustainable use of the environment.

Such accounting is needed, for example, to know what percentage of the GNP should be spent on nature restoration or nature preservation. What the Japanese have already proved and what the G-GNP could quantify is that income alone is not a satisfactory measure of the quality of life, and that many developing nations that have not yet squandered their natural assets can still save the expense of restoration (Fig. 2).

Strict environmental standards do not only increase the cost of production, they also foster innovation, upgrading and competitiveness. When processes are reengineered to meet new regulations, such as the Clean Air Act of 1990 (202), the updated process will not only be less polluting, but it will also be more efficient and will produce a less expensive and better-quality product. If regulations aim to prevent pollution, instead of legislating cleanup, the resulting environmental technology can

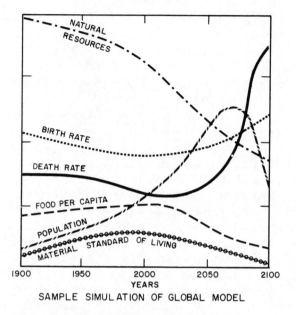

SAMPLE SIMULATION OF GLOBAL MODEL

Fig. 2 Computer simulation of world trends. (Courtesy of the World Future Society, Washington, DC [125].)

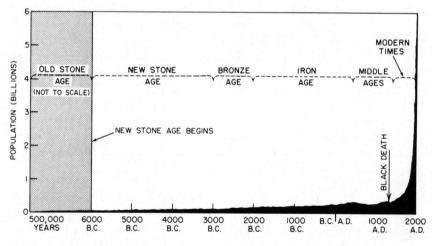

Fig. 3 Growth of human population. (From redrawing by AEC of figure from *Population Bulletin*, Population Reference Bureau, Inc., vol. 18, no. 1, February 1962 [125].)

provide a competitive advantage in the international marketplace. That marketplace is large, the yearly environmental expenditures in Europe alone exceed $50 billion, and American exports are gaining in the very areas where American regulations are the strictest.

Three hundred years ago the world's population doubled every 250 years. Today it doubles in less than a century (Fig. 3). Birthrates have been dropping in the more affluent parts of the world, but the population of poor countries will rise by three billion in the next thirty years. In the last twenty years the population of the third world increased by 1.3 billion people, more than the entire population of the developed countries. While food production has so far increased sufficiently to match population growth, it has done so at the expense of drastically increasing carbon dioxide emissions and the use of fertilizers and pesticides. In fact, to accomplish the 33% worldwide increase in food production, the use of fertilizers was increased by 146% and pesticides by 300% (125). This method of food production requires the expenditure of more and more energy in the form of fertilizers, herbicides, and pesticides even as we deplete the fossil energy sources (Fig. 4). In addition, agriculture uses an excessive amount of water. In the United States 15% of crop land is irrigated and 40% of the irrigation water is pumped from groundwater (191).

The total amount of land that is suitable for agriculture is about 7.86 billion acres, but only about 3.8 billion acres are under cultivation and the availability of

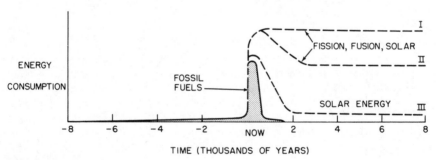

TIME (THOUSANDS OF YEARS)

THREE TRAPS:
   1. POPULATION (FOOD)
   2. ENTROPY (RESOURCES, ENERGY, POLLUTION)
   3. WAR (HUMAN NEEDS AND BEHAVIOR)

THREE ASSUMPTIONS:
   I. SUCCESSFUL CONTROL- AVOIDANCE OF TRAPS
   II. SERIOUS OVERSHOOT OF POPULATION -TEMPORARY STATE OF CHAOS
   III. SERIOUS FOOD, RESOURCES, AND ENERGY SHORTAGES- CONFUSION AND CHAOS- NUCLEAR WAR

Fig. 4 The great transition. (From AEC drawing from "Why Fusion?" by W. C. Gough, June 1970 [125].)

agricultural land is shrinking (125). The amount of land needed to feed one person for one year is about one acre (125). This would suggest that the Earth's population should not exceed 3.8 billion, yet it has already passed 5 billion.

It is widely believed that the emissions of carbon dioxide, methane, and refrigeration chemicals are warming the earth's atmosphere. The average American generates 19 tons of carbon dioxide in a year, while the average Indian generates only one ton. Today, the total yearly global carbon dioxide emission is about 20 billion tons. By 2020 it is projected to rise to 27 billion tons. Rich nations are projected to cut their per capita production of carbon dioxide in the next thirty years by 20% by switching to solar and nuclear energy, while poor nations are projected to double their per capita emissions (198). This suggests that the greenhouse effect will become more critical as the world population rises, and it will be further aggravated as the affluence of the population rises.

A reasonable estimate of the total known fossil deposits on this planet, including Uranium 235, the energy source of "conventional" nuclear power plants, is about 8×10^{18} kg-cal (32×10^{18} BTU) (128). The "potentially discoverable" quantity of all fossil fuel reserves is obviously much more, but its exact value cannot be predicted easily. The world's energy consumption is nearing 0.2×10^{18} BTU a year and could reach 0.3×10^{18} BTU a year by the turn of the century (128). Therefore, for the first time in history, we are faced with the decision of finding new energy sources.

Because of the greenhouse effect, meeting our growing energy needs by burning more fossil fuels would be undesirable even if the supply were inexhaustible. For

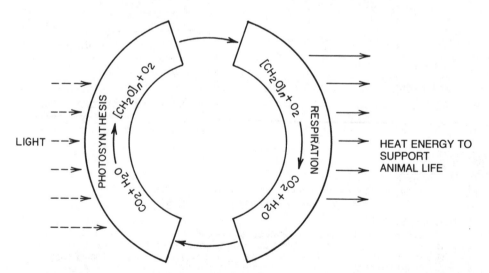

Fig. 5 Photosynthesis in plants consumes carbon dioxide, water and the energy of light. Respiration in plants or animals releases these compounds plus heat energy which result from the combustion of carbohydrates. The largest producers of oxygen are sea algae, an important reason for cleaning polluted oceans (128).

each unit of electricity generated, two units of heat energy are discharged in the form of thermal pollution, in addition to the chemical pollutants released. Burning solid wastes is undesirable for the same reasons.

Between 1945 and 1975 the population in the United States increased 50% and the per capita consumption rose by 25%, while pollution increased by 2,000% (128). Therefore, traditional fossil fuels (including Uranium 235) should be replaced by less polluting and inexhaustible energy sources, such as solar, alcohol, hydrogen, and possibly fusion (thermonuclear) energy.

In one year the total energy received from the sun, a reliable thermonuclear reactor, is about 200 times the total fossil fuel consumption of the world; 30% of this radiation is reflected back into space, 47% is absorbed by the atmosphere, 23% is consumed by the hydrologic cycle, and only 0.02% is used to support life (Fig. 5). If the present electric power generating capacity of the United States (800 billion watts) were to be met by solar energy, we would have to cover a 4,000-square-mile area with solar collectors. Plants and algae are the most cost-effective solar collectors, and if the algae harvest is converted to alcohol, we can obtain an inexhaustible source of clean energy.

What is Waste?

Waste is the wrong material in the wrong place at the wrong time. Organic materials returned to the soil, as nature intended, are not wastes because they fertilize and condition the soil. Organic materials dumped in the ocean, buried in landfills, or burned in incinerators *are* wastes simply because they are disposed of in the wrong place. On the other hand, if organic waste is discharged in a wetland treatment system, as it is done in Arcata, California, the waste can become a nutrient-rich resource for raising salmon.

Nature operates in a closed material system with no wastes. The life-cycle, as illustrated in Figure 5, utilizes sunlight as its energy source and needs oxygen, CO_2, and water to sustain life. Nature produces carbohydrates and oxygen from water and carbon dioxide. Animals consume the carbohydrates and obtain muscle energy while digesting these larger molecules. Nature has developed a remarkably complex and stable food chain both in producing these large molecules and in breaking them down. The food chain of decomposition is just as complex and just as stable as is the food chain of production. The role of the "decomposers"—insects, fungi, and bacteria—is just as important as the role of the "producers." Nature is capable of returning all chemical compounds into the earth's pool of resources, because *it never produces anything (living or by-product) that it cannot decompose.*

We, on the other hand, concentrate only on the production side of the "produce-decompose" cycle. We spend our material and mental energies on producing, without any thought to how a product will be returned to the earth's pool of resources. While nature's attention is distributed 50%–50% between production and decomposition, 99% of our energies are spent on production and almost none on recycling. This is the main cause of the solid waste problem and of the depletion of the earth's re-

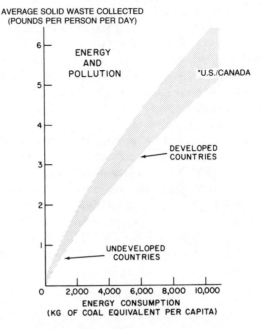

Fig. 6 Relationship of energy to living standards, materials use, and pollution. (From AEC drawings from "Why Fusion?" by W. C. Gough, June 1970 [125].)

sources. This conflict with the natural life-cycle becomes more pronounced as a society becomes more "developed" (Fig. 6). We can only hope that a better understanding of the exhaustible nature of the planet will reverse this trend.

Recycling

We can no longer afford to operate the economy as an "open" pipeline (see Fig. 1.1). We should add environmental costs to production costs and levy a disposal tax on products that are not recyclable. As a consequence, the emphasis would shift to durability and quality. Industries could become more profitable by minimizing the material flow through a "closed" pipeline and by maximizing the quality and useful life of the products. Recycling would not mean scarcity, but a shift from quantity to quality.

A ton of recycled wastepaper produces the same amount of paper as 17 trees (86). If not recycled, wastepaper creates a disposal problem, while the excessive cutting of trees diminishes a valuable and endangered natural resource. The percentage of European forests damaged by acid rain, for example, has increased from 8% to 52% in the last decade (191). This being the case, the recycling efforts should be increased. But in some areas this logical conclusion is in conflict with economic reality. It costs less to make paper from trees because the environmental costs of paper production are not included in the price of the paper but are borne by the community, which pays for the disposal of the wastepaper.

In order to make the economic environment more favorable to recycling, markets for the recycled goods must be established. Today, this market is fragile and unsupported. The industrial users of recycled materials are afraid to plan on the continued availability of such supplies, while the municipalities are equally afraid to intensify their efforts to increase recycling, because this might flood and destroy the already fragile market. Therefore, in this period of transition, it is necessary for the government to legislate the recycling industry into existence by requiring the use of waste paper in newsprint, the use of recycled plastics in containers, by labeling goods "reusable," "recycled" or "recyclable," and by establishing monetary incentives, tax advantages, and price subsidies until the recycling industry matures and stabilizes.

The cost of goods must reflect the total cost of their production. When the social, environmental, and disposal costs are reflected in the price tag, the forces of the free market will naturally convert the GNP-oriented "linear" economy into a quality-oriented "closed-loop" system (see Fig. 1.1). As the costs of power, raw materials, and pollution increase, market forces will penalize resource-intensive products and favor labor-intensive ones designed to last longer. Therefore, an economy that protects the environment also tends to create jobs while it makes polluting products more expensive.

In an economic climate where the cost of a product reflects its disposal cost, market forces will act to reduce excessive packaging, which today accounts for 33% of all household trash.

Education will have to play a major role in this conversion. The next generation must grow up fully understanding such relationships as were described in this book. Until this educational process is completed we will continue to receive messages in the mail, which on the one hand tell us Earth is in peril—and on the other, are printed on chlorine-bleached, chemically treated paper, colored with toxic inks, and held together by glue or plastic. A major step toward environmental understanding would be to make the teaching of environmental sciences compulsory in schools of the world.

When Copernicus discarded the concept of an earth-centered and stationary Universe, the Earth continued to travel undisturbed on its orbit around the Sun, yet the consequences of his discoveries were revolutionary. Today, the concept of Earth is changing again, and this change is just as revolutionary. It took several centuries until the meaning of the Copernican view penetrated the human subconscious. It should not take that long, until our behavior starts reflecting the new understanding that our environment is neither infinite nor inexhaustible.

Appendix

ABBREVIATIONS AND SYMBOLS

AC	Alternating current	cfm	Cubic feet per minute
acf	Actual cubic feet*	CH_4	Methane
acfm	Actual cubic feet per minute*	cm	Centimeter
acm	Actual cubic meter*	CO	Carbon monoxide
acmm	Actual cubic meter per minute*	CO_2	Carbon dioxide
Al	Aluminum	C:N	Carbon-to-nitrogen ratio
ANCI	Annual National Cancer Incidence	Cr	Chromium
APC	Air-pollution control	CRT	Cathode ray tube (like a TV screen)
ASME	American Society of Mechanical Engineers	cu.m.	Cubic meter
		cu.yd.	Cubic yard
atm.	Atmospheric (pressure)	d	Day
av.	Average	D/A	Direct acting
BACT	"Best available control technology"	DC	Direct current
		DCS	Distributed control system
BAT	"Best available technology"	DIF	Dry and inert-free
bbl.	Barrel	DRDF	Densified refuse-derived fuel
BHP	Break horsepower	ds	Dry solids
BTU	British thermal unit	dscf	Dry standard cubic foot (moisture-free at atmospheric conditions)
C	Centigrade		
CEM	Continuous emission monitoring		
		d/p	Differential pressure
cf	Cubic feet	EPA	Environmental Protection Agency
CFC	Chlorofluorocarbons		
		ESP	Electrostatic precipitator (also EP)

*Gas or vapor volume is measured at actual operating pressure and temperature.

414

F	Fahrenheit
fps	Foot per second
FRP	Fiberglass-reinforced plastics
ft	Feet
g	Gram
ga.	Gauge
GC	Gas chromatography
gr	Grain
GNP	Gross national product
HDPE	High-density polyethylene
H_2	Hydrogen
H_2S	Hydrogen sulfide
HC	Hydrocarbon
HCl	Hydrochloric acid
HF	Hydrogen fluoride
HHV	Higher heating value
HP	Horsepower
hr	Hour
HRI	Heat-recovery incinerator
H_2SO_4	Sulfuric acid
K°	Kelvin (degree of temperature)
kcal	Kilocalories
kg	Kilogram
kJ	Kilo Joule
kPa	Kilo Pascal
kW	Kilowatt
kWh	Kilowatt-hour
L	Liter
lb	Pound
lbf	Pound force
lbm	Pound mass
lbs	Pounds
m	Meter
mb	Millibar (pressure)
mg	Milligram
MG	Million gallon
MGD	Million gallons per day
MLR	Maximum lifetime risk
Mo	Molybdenum
mm	Millimeter
MPa	Mega Pascal
mps	Meter per second
MPW	Mixed paper waste
MRF	Materials-recovery facility
MS	Mass spectroscopy
MT	Metric ton
MW	Megawatt

N	Normal (normal conditions of 273 K and 1.0 atm.)
NA	Not available (or not applicable)
ND	Not detectable
Ncm	Normal cubic meter ("normal" in Europe means "standard" in the U.S.)
NFPA	National Fire Protection Association
ng	Nanogram (10^{-8}g)
NH_3	Ammonia
NIMBY	"Not In My Back Yard"
NKP	Nitrogen, phosphorous, and potassium
Nm^3	Normal cubic meter (at atmospheric pressure and at 273 K temperature)
NOAA	National Oceanic and Atmospheric Administration
NOx	Nitrogen oxides (monoxides and dioxides combined)
NR	No result
NTR	National Recovery Technologies
NSR	New Source Review
OD	Outside diameter
OEC	Organization of Economic Cooperation
OFA	Overfire air
Pa	Pascal (unit of pressure)
Pb	Lead
PCB	Polychlorinated biphenyls
PCDD	Polychlorinated dibenzo-p-dioxins
PCDF	Polychlorinated dibenzo-furans
PET	Polyethylene terephthalate
pg	Picogram (10^{-12} g)
pH	A unit of acidity (under 7) or alkalinity (over 7)
ppb	Parts per billion (by weight or volume)
ppm	Parts per million (by weight or volume)
PSD	Prevention of significant deterioration
psi	Pounds per square inch (pressure unit)

PSIG	Pounds per square inch gauge (gauge refers to "above atmospheric")
PTC	Performance Test Code
R°	Rankin degree temperature
R/A	Reverse acting
Rc	Rockwell (hardness)
RCRA	Resource Conservation and Recovery Act
RDAC	Rotary-drum air classifier
RDF	Refuse-derived fuel
s	Second
SARA	Superfund Amendments and Reauthorization Act
scf	Standard cubic foot (standard refers to 60°F and atmospheric pressure)
scfm	Standard cubic foot per minute
scm	Standard cubic meter
SD	Standard and dry (moisture-free and at 60° F and atmospheric pressure)
SDCF	Standard dry cubic foot
SI	System International (International System of Units)
SO_2	Sulfur dioxide
SO_3	Sulfur trioxide
SOx	All forms of sulfur oxides
SP	Set point
sq	Square
sq.ft.	Square feet
t	Ton
TAL	TA Luft (air-quality guidelines of Germany)
TLV	Threshold limit values
TPD	Tons per day
TPH	Tons per hour
TWh	Terawatt hour
wt	Weight
VPPM	Volumetric parts per million
VS	Volatile solids
yd	Yard
YR	Year
Zn	Zinc
′	Minutes
″	Inch
$''H_2O$	Inch of water column (pressure)
μ	Micron
μg	Microgram
μm	Micron (micro meter)

Multiplication Factors		Prefix	*SI* *Symbol*
1 000 000 000 000 000 000	$= 10^{18}$	exa	E
1 000 000 000 000 000	$= 10^{15}$	peta	P
1 000 000 000 000	$= 10^{12}$	tera	T
1 000 000 000	$= 10^{9}$	giga	G
1 000 000	$= 10^{6}$	mega	M
1 000	$= 10^{3}$	kilo	k
100	$= 10^{2}$	hecto	h
10	$= 10^{1}$	deka	da
0.1	$= 10^{-1}$	deci	d
0.01	$= 10^{-2}$	centi	c
0.001	$= 10^{-3}$	milli	m
0.000 001	$= 10^{-6}$	micro	μ
0.000 000 001	$= 10^{-9}$	nano	n
0.000 000 000 001	$= 10^{-12}$	pico	p
0.000 000 000 000 001	$= 10^{-15}$	femto	f
0.000 000 000 000 000 001	$= 10^{-18}$	atto	a

CONVERSION FACTORS

To Convert	Into	Multiply by
	A	
Abcoulomb	Statcoulombs	2.998×10^{10}
Acre	Sq. chain (Gunters)	10
Acre	Rods	160
Acre	Square links (Gunters)	1×10^{5}
Acre	Hectare or sq-hectometer	.4047
acres	sq feet	43,560.0
acres	sq meters	4,047.
acres	sq miles	1.562×10^{-3}
acres	sq yards	4,840.
acre-feet	cu feet	43,560.0
acre-feet	gallons	3.259×10^{5}
amperes/sq cm	amps/sq in.	6.452
amperes/sq cm	amps/sq meter	10^{4}
amperes/sq in.	amps/sq cm	0.1550
amperes/sq in.	amps/sq meter	1,550.0
amperes/sq meter	amps/sq cm	10^{-4}
amperes/sq meter	amps/sq in.	6.452×10^{-4}
ampere-hours	coulombs	3,600.0
ampere-hours	faradays	0.03731
ampere-turns	gilberts	1.257
ampere-turns/cm	amp-turns/in.	2.540
ampere-turns/cm	amp-turns/meter	100.0
ampere-turns/cm	gilberts/cm	1.257
ampere-turns/in.	amp-turns/cm	0.3937
ampere-turns/in.	amp-turns/meter	39.37

To Convert	Into	Multiply by
ampere-turns/in.	gilberts/cm	0.4950
ampere-turns/meter	amp/turns/cm	0.01
ampere-turns/meter	amp-turns/in.	0.0254
ampere-turns/meter	gilberts/cm	0.01257
Angstrom unit	Inch	3.937×10^{-9}
Angstrom unit	Meter	1×10^{-10}
Angstrom unit	Micron or (Mu)	1×10^{-4}
Are	Acre (US)	.02471
Ares	sq. yards	119.60
ares	acres	0.02471
ares	sq meters	100.0
Astronomical Unit	Kilometers	1.495×10^8
atmospheres	Ton/sq. inch	.007348
atmospheres	cms of mercury	76.0
atmospheres	ft of water (at 4° C)	33.90
atmospheres	in. of mercury (at 0° C)	29.92
atmospheres	kgs/sq cm	1.0333
atmospheres	kgs/sq meter	10,332.
atmospheres	pounds/sq in.	14.70
atmospheres	tons/sq ft	1.058

B

To Convert	Into	Multiply by
barrel	cubic meter	0.15899
Barrels (U.S., dry)	cu. inches	7056.
Barrels (U.S., dry)	quarts (dry)	105.0
Barrels (U.S., liquid)	gallons	31.5
barrels (oil)	gallons (oil)	42.0
bars	atmospheres	0.9869
bars	dynes/sq cm	10^6
bars	kgs/sq meter	1.020×10^4
bars	pounds/sq ft	2,089.
bars	pounds/sq in.	14.50
Baryl	Dyne/sq. cm.	1.000
Bolt (US Cloth)	Meters	36.576
BTU	Liter—Atmosphere	10.409
BTU	ergs	1.0550×10^{10}
BTU	foot-lbs	778.3
BTU	gram-calories	252.0
BTU	horsepower-hrs	3.931×10^{-4}
BTU	joules	1,054.8
BTU	kilogram-calories	0.2520
BTU	kilogram-meters	107.5
BTU	kilowatt-hrs	2.928×10^{-4}

CONVERSION FACTORS (*continued*)

To Convert	Into	Multiply by
BTU/hr	foot-pounds/sec	0.2162
BTU/hr	gram-cal/sec	0.0700
BTU/hr	horsepower-hrs	3.929×10^{-4}
BTU/hr	watts	0.2931
BTU/lbm	kJ/kg	2.326
BTU/min	foot-lbs/sec	12.96
BTU/min	horsepower	0.02356
BTU/min	kilowatts	0.01757
BTU/min	watts	17.57
BTU/sq ft/min	watts/sq in.	0.1221
Bucket (Br. dry)	Cubic Cm.	1.818×10^4
bushels	cu ft	1.2445
bushels	cu in.	2,150.4
bushels	cu meters	0.03524
bushels	liters	35.24
bushels	pecks	4.0
bushels	pints (dry)	64.0
bushels	quarts (dry)	32.0

C

To Convert	Into	Multiply by
Calories, gram (mean)	B.T.U. (mean)	3.9685×10^{-3}
Candle/sq. cm	Lamberts	3.142
Candle/sq. inch	Lamberts	.4870
centares (centiares)	sq meters	1.0
Centigrade	Fahrenheit	$(C° \times 9/5) + 32$
centigrams	grams	0.01
Centiliter	Ounce fluid (US)	.3382
Centiliter	Cubic inch	.6103
Centiliter	drams	2.705
centiliters	liters	0.01
centimeters	feet	3.281×10^{-2}
centimeters	inches	0.3937
centimeters	kilometers	10^{-5}
centimeters	meters	0.01
centimeters	miles	6.214×10^{-6}
centimeters	millimeters	10.0
centimeters	mils	393.7
centimeters	yards	1.094×10^{-2}
centimeter-dynes	cm-grams	1.020×10^{-3}
centimeter-dynes	meter-kgs	1.020×10^{-8}
centimeter-dynes	pound-feet	7.376×10^{-8}
centimeter-grams	cm-dynes	980.7
centimeter-grams	meter-kgs	10^{-5}
centimeter-grams	pound-feet	7.233×10^{-5}

To Convert	Into	Multiply by
centimeters of mercury	atmospheres	0.01316
centimeters of mercury	feet of water	0.4461
centimeters of mercury	kgs/sq meter	136.0
centimeters of mercury	pounds/sq ft	27.85
centimeters of mercury	pounds/sq in.	0.1934
centimeters/sec	feet/min	1.1969
centimeters/sec	feet/sec	0.03281
centimeters/sec	kilometers/hr	0.036
centimeters/sec	knots	0.1943
centimeters/sec	meters/min	0.6
centimeters/sec	miles/hr	0.02237
centimeters/sec	miles/min	3.728×10^{-4}
centimeters/sec/sec	feet/sec/sec	0.03281
centimeters/sec/sec	kms/hr/sec	0.036
centimeters/sec/sec	meters/sec/sec	0.01
centimeters/sec/sec	miles/hr/sec	0.02237
Chain	Inches	792.00
Chain	meters	20.12
Chains (surveyors' or Gunter's)	yards	22.00
circular mils	sq cms	5.067×10^{-6}
circular mils	sq mils	0.7854
Circumference	Radians	6.283
circular mils	sq inches	7.854×10^{-7}
Cords	cord feet	8
Cord feet	cu. feet	16
Coulomb	Statcoulombs	2.998×10^{9}
coulombs	faradays	1.036×10^{-5}
coulombs/sq cm	coulombs/sq in.	64.52
coulombs/sq cm	coulombs/sq meter	10^{4}
coulombs/sq in.	coulombs/sq cm	0.1550
coulombs/sq in.	coulombs/sq meter	1,550.
coulombs/sq meter	coulombs/sq cm	10^{-4}
coulombs/sq meter	coulombs/sq in.	6.452×10^{-4}
cubic centimeters	cu feet	3.531×10^{-5}
cubic centimeters	cu inches	0.06102
cubic centimeters	cu meters	10^{-6}
cubic centimeters	cu yards	1.308×10^{-6}
cubic centimeters	gallons (U.S. liq.)	2.642×10^{-4}

To Convert	Into	Multiply by
cubic centimeters	liters	0.001
cubic centimeters	pints (U.S. liq.)	2.113×10^{-3}
cubic centimeters	quarts (U.S. liq.)	1.057×10^{-3}
cubic feet	bushels (dry)	0.8036
cubic feet	cu cms	28,320.0
cubic feet	cu inches	1,728.0
cubic feet	cu meters	0.02832
cubic feet	cu yards	0.03704
cubic feet	gallons (U.S. liq.)	7.48052
cubic feet	liters	28.32
cubic feet	pints (U.S. liq.)	59.84
cubic feet	quarts (U.S. liq.)	29.92
cubic feet/min	cu cms/sec	472.0
cubic feet/min	gallons/sec	0.1247
cubic feet/min	liters/sec	0.4720
cubic feet/min	pounds of water/min	62.43
cubic feet/sec	million gals/day	0.646317
cubic feet/sec	gallons/min	448.831
cubic inches	cu cms	16.39
cubic inches	cu feet	5.787×10^{-4}
cubic inches	cu meters	1.639×10^{-5}
cubic inches	cu yards	2.143×10^{-5}
cubic inches	gallons	4.329×10^{-3}
cubic inches	liters	0.01639
cubic inches	mil-feet	1.061×10^{5}
cubic inches	pints (U.S. liq.)	0.03463
cubic inches	quarts (U.S. liq.)	0.01732
cubic meters	bushels (dry)	28.38
cubic meters	cu cms	10^{6}
cubic meters	cu feet	35.31
cubic meters	cu inches	61,023.0
cubic meters	cu yards	1.308
cubic meters	gallons (U.S. liq.)	264.2
cubic meters	liters	1,000.0
cubic meters	pints (U.S. liq.)	2,113.0
cubic meters	quarts (U.S. liq.)	1,057.
cubic yards	cu cms	7.646×10^{5}
cubic yards	cu feet	27.0
cubic yards	cu inches	46,656.0
cubic yards	cu meters	0.7646
cubic yards	gallons (U.S. liq.)	202.0
cubic yards	liters	764.6
cubic yards	pints (U.S. liq.)	1,615.9
cubic yards	quarts (U.S. liq.)	807.9

To Convert	Into	Multiply by
cubic yards/min	cubic ft/sec	0.45
cubic yards/min	gallons/sec	3.367
cubic yards/min	liters/sec	12.74

D

To Convert	Into	Multiply by
Dalton	Gram	1.650×10^{-24}
days	seconds	86,400.0
decigrams	grams	0.1
deciliters	liters	0.1
decimeters	meters	0.1
degrees (angle)	quadrants	0.01111
degrees (angle)	radians	0.01745
degrees (angle)	seconds	3,600.0
degrees/sec	radians/sec	0.01745
degrees/sec	revolutions/min	0.1667
degrees/sec	revolutions/sec	2.778×10^{-3}
dekagrams	grams	10.0
dekaliters	liters	10.0
dekameters	meters	10.0
Drams (apothecaries' or troy)	ounces (avoirdupois)	0.1371429
Drams (apothecaries' or troy)	ounces (troy)	0.125
Drams (U.S., fluid or apothecaries)	cubic cm.	3.6967
drams	grams	1.7718
drams	grains	27.3437
drams	ounces	0.0625
Dyne/cm	Erg/sq. millimeter	.01
Dyne/sq. cm.	Atmospheres	9.869×10^{-7}
Dyne/sq. cm.	Inch of Mercury at 0° C	2.953×10^{-5}
Dyne/sq. cm.	Inch of Water at 4° C	4.015×10^{-4}
dynes	grams	1.020×10^{-3}
dynes	joules/cm	10^{-7}
dynes	joules/meter (newtons)	10^{-5}
dynes	kilograms	1.020×10^{-6}
dynes	poundals	7.233×10^{-5}
dynes	pounds	2.248×10^{-6}
dynes/sq cm	bars	10^{-6}

E

To Convert	Into	Multiply by
Ell	Cm.	114.30
Ell	Inches	45

CONVERSION FACTORS (*continued*)

To Convert	Into	Multiply by
Em, Pica	Inch	.167
Em, Pica	Cm.	.4233
Erg/sec	Dyne—cm/sec	1.000
ergs	Btu	9.480×10^{-11}
ergs	dyne-centimeters	1.0
ergs	foot-pounds	7.367×10^{-8}
ergs	gram-calories	0.2389×10^{-7}
ergs	gram-cms	1.020×10^{-3}
ergs	horsepower-hrs	3.7250×10^{-14}
ergs	joules	10^{-7}
ergs	kg-calories	2.389×10^{-11}
ergs	kg-meters	1.020×10^{-8}
ergs	kilowatt-hrs	0.2778×10^{-13}
ergs	watt-hours	0.2778×10^{-10}
ergs/sec	Btu/min	$5,688 \times 10^{-9}$
ergs/sec	ft-lbs/min	4.427×10^{-6}
ergs/sec	ft-lbs/sec	7.3756×10^{-8}
ergs/sec	horsepower	1.341×10^{-10}
ergs/sec	kg-calories/min	1.433×10^{-9}
ergs/sec	kilowatts	10^{-10}

F

To Convert	Into	Multiply by
farads .	microfarads	10^6
Faraday/sec	Ampere (absolute)	9.6500×10^4
faradays	ampere-hours	26.80
faradays	coulombs	9.649×10^4
Fathom	Meter	1.828804
fathoms	feet	6.0
feet	centimeters	30.48
feet	kilometers	3.048×10^{-4}
feet	meters	0.3048
feet	miles (naut.)	1.645×10^{-4}
feet	miles (stat.)	1.894×10^{-4}
feet	millimeters	304.8
feet	mils	1.2×10^4
feet of water	atmospheres	0.02950
feet of water	in. of mercury	0.8826
feet of water	kgs/sq cm	0.03048
feet of water	kgs/sq meter	304.8
feet of water	pounds/sq ft	62.43
feet of water	pounds/sq in.	0.4335
feet/min	cms/sec	0.5080
feet/min	feet/sec	0.01667
feet/min	kms/hr	0.01829
feet/min	meters/min	0.3048
feet/min	meters/sec	0.00508

CONVERSION FACTORS (*continued*)

To Convert	Into	Multiply by
feet/min	miles/hr	0.01136
feet/sec	cms/sec	30.48
feet/sec	kms/hr	1.097
feet/sec	knots	0.5921
feet/sec	meters/min	18.29
feet/sec	miles/hr	0.6818
feet/sec	miles/min	0.01136
feet/sec/sec	cms/sec/sec	30.48
feet/sec/sec	kms/hr/sec	1.097
feet/sec/sec	meters/sec/sec	0.3048
feet/sec/sec	miles/hr/sec	0.6818
feet/100 feet	per cent grade	1.0
Foot—candle	Lumen/sq. meter	10.764
foot-pounds	Btu	1.286×10^{-3}
foot-pounds	ergs	1.356×10^{7}
foot-pounds	gram-calories	0.3238
foot-pounds	hp-hrs	5.050×10^{-7}
foot-pounds	joules	1.356
foot-pounds	kg-calories	3.24×10^{-4}
foot-pounds	kg-meters	0.1383
foot-pounds	kilowatt-hrs	3.766×10^{-7}
foot-pounds/min	Btu/min	1.286×10^{-3}
foot-pounds/min	foot-pounds/sec	0.01667
foot-pounds/min	horsepower	3.030×10^{-5}
foot-pounds/min	kg-calories/min	3.24×10^{-4}
foot-pounds/min	kilowatts	2.260×10^{-5}
foot-pounds/sec	Btu/hr	4.6263
foot-pounds/sec	Btu/min	0.07717
foot-pounds/sec	horsepower	1.818×10^{-3}
foot-pounds/sec	kg-calories/min	0.01945
foot-pounds/sec	kilowatts	1.356×10^{-3}
Furlongs	miles (U.S.)	0.125
furlongs	rods	40.0
furlongs	feet	660.0

G

To Convert	Into	Multiply by
gallons	cu cms	3,785.0
gallons	cu feet	0.1337
gallons	cu inches	231.0
gallons	cu meters	3.785×10^{-3}
gallons	cu yards	4.951×10^{-3}
gallons	liters	3.785
gallons (liq. Br. Imp.)	gallons (U.S. liq.)	1.20095
gallons (U.S.)	gallons (Imp.)	0.83267

To Convert	Into	Multiply by
gallons of water	pounds of water	8.3453
gallons/min	cu ft/sec	2.228×10^{-3}
gallons/min	liters/sec	0.06308
gallons/min	cu ft/hr	8.0208
gausses	lines/sq in.	6.452
gausses	webers/sq cm	10^{-8}
gausses	webers/sq in.	6.452×10^{-8}
gausses	webers/sq meter	10^{-4}
gilberts	ampere-turns	0.7958
gilberts/cm	amp-turns/cm	0.7958
gilberts/cm	amp-turns/in	2.021
gilberts/cm	amp-turns/meter	79.58
Gills (British)	cubic cm.	142.07
gills	liters	0.1183
gills	pints (liq.)	0.25
Grade	Radian	.01571
grain	mg	64.8
Grains	drams (avoirdupois)	0.03657143
grains (troy)	grains (avdp)	1.0
grains (troy)	grams	0.06480
grains (troy)	ounces (avdp)	2.0833×10^{-3}
grains (troy)	pennyweight (troy)	0.04167
grains/U.S. gal	parts/million	17.118
grains/U.S. gal	pounds/million gal	142.86
grains/Imp. gal	parts/million	14.286
grams	dynes	980.7
grams	grains	15.43
grams	joules/cm	9.807×10^{-5}
grams	joules/meter (newtons)	9.807×10^{-3}
gr/scf	mg/scm	2,280.0
grams	kilograms	0.001
grams	milligrams	1,000.
grams	ounces (avdp)	0.03527
grams	ounces (troy)	0.03215
grams	poundals	0.07093
grams	pounds	2.205×10^{-3}
grams/cm	pounds/inch	5.600×10^{-3}
grams/cu cm	pounds/cu ft	62.43
grams/cu cm	pounds/cu in	0.03613
grams/cu cm	pounds/mil-foot	3.405×10^{-7}
grams/liter	grains/gal	58.417
grams/liter	pounds/1,000 gal	8.345
grams/liter	pounds/cu ft	0.062427

425

To Convert	Into	Multiply by
grams/liter	parts/million	1,000.0
grams/sq cm	pounds/sq ft	2.0481
gram-calories	Btu	3.9683×10^{-3}
gram-calories	ergs	4.1868×10^{7}
gram-calories	foot-pounds	3.0880
gram-calories	horsepower-hrs	1.5596×10^{-6}
gram-calories	kilowatt-hrs	1.1630×10^{-6}
gram-calories	watt-hrs	1.1630×10^{-3}
gram-calories/sec	Btu/hr	14.286
gram-centimeters	Btu	9.297×10^{-8}
gram-centimeters	ergs	980.7
gram-centimeters	joules	9.807×10^{-5}
gram-centimeters	kg-cal	2.343×10^{-8}
gram-centimeters	kg-meters	10^{-5}

H

To Convert	Into	Multiply by
Hand	Cm.	10.16
hectares	acres	2.471
hectares	sq feet	1.076×10^{5}
hectograms	grams	100.0
hectoliters	liters	100.0
hectometers	meters	100.0
hectowatts	watts	100.0
henries	millihenries	1,000.0
Hogsheads (British)	cubic ft.	10.114
Hogsheads (U.S.)	cubic ft.	8.42184
Hogsheads (U.S.)	gallons (U.S.)	63
horsepower	Btu/min	42.44
horsepower	foot-lbs/min	33,000.
horsepower	foot-lbs/sec	550.0
horsepower (metric) (542.5 ft lb/sec)	horsepower (550 ft lb/sec)	0.9863
horsepower (550 ft lb/sec)	horsepower (metric) (542.5 ft lb/sec)	1.014
horsepower	kg-calories/min	10.68
horsepower	kilowatts	0.7457
horsepower	watts	745.7
horsepower (boiler)	Btu/hr	33.479
horsepower (boiler)	kilowatts	9.803
horsepower-hrs	Btu	2,547.
horsepower-hrs	ergs	2.6845×10^{13}
horsepower-hrs	foot-lbs	1.98×10^{6}
horsepower-hrs	gram-calories	641,190.
horsepower-hrs	joules	2.684×10^{6}

CONVERSION FACTORS (*continued*)

To Convert	Into	Multiply by
horsepower-hrs	kg-calories	641.1
horsepower-hrs	kg-meters	2.737×10^5
horsepower-hrs	kilowatt-hrs	0.7457
hours	days	4.167×10^{-2}
hours	weeks	5.952×10^{-3}
Hundredweights (long)	pounds	112
Hundredweights (long)	tons (long)	0.05
Hundredweights (short)	ounces (avoirdupois)	1600
Hundredweights (short)	pounds	100
Hundredweights (short)	tons (metric)	0.0453592
Hundredweights (short)	tons (long)	0.0446429

I

To Convert	Into	Multiply by
inches	centimeters	2.540
inches	meters	2.540×10^{-2}
inches	miles	1.578×10^{-5}
inches	millimeters	25.40
inches	mils	1,000.0
inches	yards	2.778×10^{-2}
inches of mercury	atmospheres	0.03342
inches of mercury	feet of water	1.133
inches of mercury	kgs/sq cm	0.03453
inches of mercury	kgs/sq meter	345.3
inches of mercury	pounds/sq ft	70.73
inches of mercury	pounds/sq in.	0.4912
inches of water (at 4°C)	atmospheres	2.458×10^{-3}
inches of water (at 4°C)	inches of mercury	0.07355
inches of water (at 4°C)	kgs/sq cm	2.540×10^{-3}
inches of water (at 4°C)	ounces/sq in	0.5781
inches of water (at 4°C)	pounds/sq ft	5.204
inches of water (at 4°C)	pounds/sq in.	0.03613
International Ampere	Ampere (absolute)	.9998
International Volt	Volts (absolute)	1.0003

427

CONVERSION FACTORS (*continued*)

To Convert	Into	Multiply by
J		
joules	Btu	9.480×10^{-4}
joules	ergs	10^7
joules	foot-pounds	0.7376
joules	kg-calories	2.389×10^{-4}
joules	kg-meters	0.1020
joules	watt-hrs	2.778×10^{-4}
joules/cm	grams	1.020×10^4
joules/cm	dynes	10^7
joules/cm	joules/meter (newtons)	100.0
joules/cm	poundals	723.3
joules/cm	pounds	22.48
K		
kilograms	dynes	980,665.
kilograms	grams	1,000.0
kilograms	joules/cm	0.09807
kilograms	joules/meter (newtons)	9.807
kilograms	poundals	70.93
kilograms	pounds	2.205
kilograms	tons (long)	9.842×10^{-4}
kilograms	tons (short)	1.102×10^{-3}
kilograms/cu meter	grams/cu cm	0.001
kilograms/cu meter	pounds/cu ft	0.06243
kilograms/cu meter	pounds/cu in.	3.613×10^{-5}
kilograms/cu meter	pounds/mil-foot	3.405×10^{-10}
kilograms/meter	pounds/ft	0.6720
kilogram/sq. cm.	Dynes	980,665
kilograms/sq cm	atmospheres	0.9678
kilograms/sq cm	feet of water	32.81
kilograms/sq cm	inches of mercury	28.96
kilograms/sq cm	pounds/sq ft	2,408.
kilograms/sq cm	pounds/sq in.	14.22
kilograms/sq meter	atmospheres	9.678×10^{-5}
kilograms/sq meter	bars	98.07×10^{-6}
kilograms/sq meter	feet of water	3.281×10^{-3}
kilograms/sq meter	inches of mercury	2.896×10^{-3}
kilograms/sq meter	pounds/sq ft	0.2048
kilograms/sq meter	pounds/sq in.	1.422×10^{-3}
kilograms/sq mm	kgs/sq meter	10^6
kilogram-calories	Btu	3.968
kilogram-calories	foot-pounds	3,088.

To Convert	Into	Multiply by
kilogram-calories	hp-hrs	1.560×10^{-3}
kilogram-calories	joules	4,186.
kilogram-calories	kg-meters	426.9
kilogram-calories	kilojoules	4.186
kilogram-calories	kilowatt-hrs	1.163×10^{-3}
kilogram meters	Btu	9.294×10^{-3}
kilogram meters	ergs	9.804×10^{7}
kilogram meters	foot-pounds	7.233
kilogram meters	joules	9.804
kilogram meters	kg-calories	2.342×10^{-3}
kilogram meters	kilowatt-hrs	2.723×10^{-6}
kilolines	maxwells	1,000.0
kiloliters	liters	1,000.0
kilometers	centimeters	10^{5}
kilometers	feet	3,281.
kilometers	inches	3.937×10^{4}
kilometers	meters	1,000.0
kilometers	miles	0.6214
kilometers	millimeters	10^{6}
kilometers	yards	1,094.
kilometers/hr	cms/sec	27.78
kilometers/hr	feet/min	54.68
kilometers/hr	feet/sec	0.9113
kilometers/hr	knots	0.5396
kilometers/hr	meters/min	16.67
kilometers/hr	miles/hr	0.6214
kilometers/hr/sec	cms/sec/sec	27.78
kilometers/hr/sec	ft/sec/sec	0.9113
kilometers/hr/sec	meters/sec/sec	0.2778
kilometers/hr/sec	miles/hr/sec	0.6214
kilowatts	Btu/min	56.92
kilowatts	foot-lbs/min	4.426×10^{4}
kilowatts	foot-lbs/sec	737.6
kilowatts	horsepower	1.341
kilowatts	kg-calories/min	14.34
kilowatts	watts	1,000.0
kilowatt-hrs	Btu	3,413.
kilowatt-hrs	ergs	3.600×10^{13}
kilowatt-hrs	foot-lbs	2.655×10^{6}
kilowatt-hrs	gram-calories	859,850.
kilowatt-hrs	horsepower-hrs	1.341
kilowatt-hrs	joules	3.6×10^{6}
kilowatt-hrs	kg-calories	860.5

429

To Convert	Into	Multiply by
kilowatt-hrs	kg-meters	3.671×10^5
kilowatt-hrs	pounds of water evaporated from and at 212° F	3.53
kilowatt-hrs	pounds of water raised from 62° to 212° F.	22.75
knots	feet/hr	6,080.
knots	kilometers/hr	1.8532
knots	nautical miles/hr	1.0
knots	statute miles/hr	1.151
knots	yards/hr	2,027.
knots	feet/sec	1.689

L

league	miles (approx.)	3.0
Light year	Miles	5.9×10^{12}
Light year	Kilometers	9.46091×10^{12}
lines/sq cm	gausses	1.0
lines/sq in.	gausses	0.1550
lines/sq in.	webers/sq cm	1.550×10^{-9}
lines/sq in.	webers/sq in.	10^{-8}
lines/sq in.	webers/sq meter	1.550×10^{-5}
links (engineer's)	inches	12.0
links (surveyor's)	inches	7.92
liters	bushels (U.S. dry)	0.02833
liters	cu cm	1,000.0
liters	cu feet	0.03531
liters	cu inches	61.02
liters	cu meters	0.001
liters	cu yards	1.308×10^{-3}
liters	gallons (U.S. liq.)	0.2642
liters	pints (U.S. liq.)	2.113
liters	quarts (U.S. liq.)	1.057
liters/min	cu ft/sec	5.886×10^{-4}
liters/min	gals/sec	4.403×10^{-3}
lumens/sq ft	foot-candles	1.0
Lumen	Spherical candle power	.07958
Lumen	Watt	.001496
Lumen/sq. ft.	Lumen/sq. meter	10.76
lux	foot-candles	0.0929

M

maxwells	kilolines	0.001
maxwells	webers	10^{-8}

CONVERSION FACTORS (*continued*)

To Convert	Into	Multiply by
megalines	maxwells	10^6
megohms	microhms	10^{12}
megohms	ohms	10^6
meters	centimeters	100.0
meters	feet	3.281
meters	inches	39.37
meters	kilometers	0.001
meters	miles (naut.)	5.396×10^{-4}
meters	miles (stat.)	6.214×10^{-4}
meters	millimeters	1,000.0
meters	yards	1.094
meters	varas	1.179
meters/min	cms/sec	1.667
meters/min	feet/min	3.281
meters/min	feet/sec	0.05468
meters/min	kms/hr	0.06
meters/min	knots	0.03238
meters/min	miles/hr	0.03728
meters/sec	feet/min	196.8
meters/sec	feet/sec	3.281
meters/sec	kilometers/hr	3.6
meters/sec	kilometers/min	0.06
meters/sec	miles/hr	2.237
meters/sec	miles/min	0.03728
meters/sec/sec	cms/sec/sec	100.0
meters/sec/sec	ft/sec/sec	3.281
meters/sec/sec	kms/hr/sec	3.6
meters/sec/sec	miles/hr/sec	2.237
meter-kilograms	cm-dynes	9.807×10^7
meter-kilograms	cm-grams	10^5
meter-kilograms	pound-feet	7.233
mg	grains	0.01543
mg/scm	gr/scf	0.0004385
microfarad	farads	10^{-6}
micrograms	grams	10^{-6}
microhms	megohms	10^{-12}
microhms	ohms	10^{-6}
microliters	liters	10^{-6}
Microns	meters	1×10^{-6}
miles (naut.)	miles (statute)	1.1516
miles (naut.)	yards	2,027.
miles (statute)	centimeters	1.609×10^5
miles (statute)	feet	5,280.

To Convert	Into	Multiply by
miles (statute)	inches	6.336×10^4
miles (statute)	kilometers	1.609
miles (statute)	meters	1,609.
miles (statute)	miles (naut.)	0.8684
miles (statute)	yards	1,760.
miles/hr	cms/sec	44.70
miles/hr	feet/min	88.
miles/hr	feet/sec	1.467
miles/hr	kms/hr	1.609
miles/hr	kms/min	0.02682
miles/hr	knots	0.8684
miles/hr	meters/min	26.82
miles/hr	miles/min	0.1667
miles/hr/sec	cms/sec/sec	44.70
miles/hr/sec	feet/sec/sec	1.467
miles/hr/sec	kms/hr/sec	1.609
miles/hr/sec	meters/sec/sec	0.4470
miles/min	cms/sec	2,682.
miles/min	feet/sec	88.
miles/min	kms/min	1.609
miles/min	knots/min	0.8684
miles/min	miles/hr	60.0
mil-feet	cu inches	9.425×10^{-6}
milliers	kilograms	1,000.
Millimicrons	meters	1×10^{-9}
Milligrams	grains	0.01543236
milligrams	grams	0.001
milligrams/liter	parts/million	1.0
millihenries	henries	0.001
milliliters	liters	0.001
millimeters	centimeters	0.1
millimeters	feet	3.281×10^{-3}
millimeters	inches	0.03937
millimeters	kilometers	10^{-6}
millimeters	meters	0.001
millimeters	miles	6.214×10^{-7}
millimeters	mils	39.37
millimeters	yards	1.094×10^{-3}
million gals/day	cu ft/sec	1.54723
mils	centimeters	2.540×10^{-3}
mils	feet	8.333×10^{-5}
mils	inches	0.001
mils	kilometers	2.540×10^{-8}
mils	yards	2.778×10^{-5}

432

CONVERSION FACTORS (*continued*)

To Convert	Into	Multiply by
miner's inches	cu ft/min	1.5
Minims (British)	cubic cm.	0.059192
Minims (U.S., fluid)	cubic cm.	0.061612
minutes (angles)	degrees	0.01667
miles (naut.)	feet	6,080.27
miles (naut.)	kilometers	1.853
miles (naut.)	meters	1,853.
minutes (angles)	quadrants	1.852×10^{-4}
minutes (angles)	radians	2.909×10^{-4}
minutes (angles)	seconds	60.0
myriagrams	kilograms	10.0
myriameters	kilometers	10.0
myriawatts	kilowatts	10.0

N

nepers	decibels	8.686
Newton	Dynes	1×10^5

O

OHM (International)	OHM (absolute)	1.0005
ohms	megohms	10^{-6}
ohms	microhms	10^6
ounces	drams	16.0
ounces	grains	437.5
ounces	grams	28.349527
ounces	pounds	0.0625
ounces	ounces (troy)	0.9115
ounces	tons (long)	2.790×10^{-5}
ounces	tons (metric)	2.835×10^{-5}
ounces (fluid)	cu inches	1.805
ounces (fluid)	liters	0.02957
ounces (troy)	grains	480.0
ounces (troy)	grams	31.103481
ounces (troy)	ounces (avdp.)	1.09714
ounces (troy)	pennyweights (troy)	20.0
ounces (troy)	pounds (troy)	0.08333
ounce/sq. inch	Dynes/sq. cm.	4309
ounces/sq in.	pounds/sq in.	0.0625

P

Parsec	Miles	19×10^{12}
Parsec	Kilometers	3.084×10^{13}
parts/million	grains/U.S. gal	0.0584
parts/million	grains/Imp. gal	0.07016
parts/million	pounds/million gal	8.345

To Convert	Into	Multiply by
Pecks (British)	cubic inches	554.6
Pecks (British)	liters	9.091901
Pecks (U.S.)	bushels	0.25
Pecks (U.S.)	cubic inches	537.605
Pecks (U.S.)	liters	8.809582
Pecks (U.S.)	quarts (dry)	8
pennyweights (troy)	grains	24.0
pennyweights (troy)	ounces (troy)	0.05
pennyweights (troy)	grams	1.55517
pennyweights (troy)	pounds (troy)	4.1667×10^{-3}
pints (dry)	cu inches	33.60
pints (liq.)	cu cms.	473.2
pints (liq.)	cu feet	0.01671
pints (liq.)	cu inches	28.87
pints (liq.)	cu meters	4.732×10^{-4}
pints (liq.)	cu yards	6.189×10^{-4}
pints (liq.)	gallons	0.125
pints (liq.)	liters	0.4732
pints (liq.)	quarts (liq.)	0.5
Planck's quantum	Erg—second	6.624×10^{-27}
Poise	Gram/cm. sec.	1.00
Pounds (avoirdupois)	ounces (troy)	14.5833
poundals	dynes	13,826.
poundals	grams	14.10
poundals	joules/cm	1.383×10^{-3}
pounds	joules/meter (newtons)	4.448
poundals	kilograms	0.4536
poundals	pounds	0.03108
pounds	drams	256.
pounds	dynes	44.4823×10^4
pounds	grains	7,000.
pounds	grams	453.5924
pounds	joules/cm	0.04448
pounds	joules/meter (netwons)	4.448
pounds	kilograms	0.4536
pounds	ounces	16.0
pounds	ounces (troy)	14.5833
pounds	poundals	32.17
pounds	pounds (troy)	1.21528
pounds	tons (short)	0.0005
pounds (troy)	grains	5,760.
pounds (troy)	grams	373.24177
pounds (troy)	ounces (avdp.)	13.1657

434

CONVERSION FACTORS (*continued*)

To Convert	Into	Multiply by
pounds (troy)	ounces (troy)	12.0
pounds (troy)	pennyweights (troy)	240.0
pounds (troy)	pounds (avdp.)	0.822857
pounds (troy)	tons (long)	3.6735×10^{-4}
pounds (troy)	tons (metric)	3.7324×10^{-4}
pounds (troy)	tons (short)	4.1143×10^{-4}
pounds of water	cu feet	0.01602
pounds of water	cu inches	27.68
pounds of water	gallons	0.1198
pounds of water/min	cu ft/sec	2.670×10^{-4}
pound-feet	cm-dynes	1.356×10^{7}
pound-feet	cm-grams	13,825.
pound-feet	meter-kgs	0.1383
pounds/cu ft	grams/cu cm	0.01602
pounds/cu ft	kgs/cu meter	16.02
pounds/cu ft	pounds/cu in.	5.787×10^{-4}
pounds/cu ft	pounds/mil-foot	5.456×10^{-9}
pounds/cu in.	gms/cu cm	27.68
pounds/cu in.	kgs/cu meter	2.768×10^{4}
pounds/cu in.	pounds/cu ft	1,728.
pounds/cu in.	pounds/mil-foot	9.425×10^{-6}
pounds/ft	kgs/meter	1.488
pounds/in.	gms/cm	178.6
pounds/mil-foot	gms/cu cm	2.306×10^{6}
pounds/sq ft	atmospheres	4.725×10^{-4}
pounds/sq ft	feet of water	0.01602
pounds/sq ft	inches of mercury	0.01414
pounds/sq ft	kgs/sq meter	4.882
pounds/sq ft	pounds/sq in.	6.944×10^{-3}
pounds/sq in.	atmospheres	0.06804
pounds/sq in.	feet of water	2.307
pounds/sq in.	inches of mercury	2.036
pounds/sq in.	kgs/sq meter	703.1
pounds/sq in.	pounds/sq ft	144.0
ppm by volume for CO	mg/scm	1.0–1.1†
ppm by volume for HCl	mg/scm	1.4–1.5†
ppm by volume for CO_2	mg/scm	1.8–2.0†
ppm by volume for NO_2	mg/scm	1.8–2.0†
ppm by volume for SO_2	mg/scm	2.6–2.7†

† Varies with oxygen content and with degrees of background gases.

To Convert	Into	Multiply by
Q		
quadrants (angle)	degrees	90.0
quadrants (angle)	minutes	5,400.0
quadrants (angle)	radians	1.571
quadrants (angle)	seconds	3.24×10^5
quarts (dry)	cu inches	67.20
quarts (liq.)	cu cms	946.4
quarts (liq.)	cu feet	0.03342
quarts (liq.)	cu inches	57.75
quarts (liq.)	cu meters	9.464×10^{-4}
quarts (liq.)	cu yards	1.238×10^{-3}
quarts (liq.)	gallons	0.25
quarts (liq.)	liters	0.9463
R		
radians	degrees	57.30
radians	minutes	3,438.
radians	quadrants	0.6366
radians	seconds	2.063×10^5
radians/sec	degrees/sec	57.30
radians/sec	revolutions/min	9.549
radians/sec	revolutions/sec	0.1592
radians/sec/sec	revs/min/min	573.0
radians/sec/sec	revs/min/sec	9.549
radians/sec/sec	revs/sec/sec	0.1592
revolutions	degrees	360.0
revolutions	quadrants	4.0
revolutions	radians	6.283
revolutions/min	degrees/sec	6.0
revolutions/min	radians/sec	0.1047
revolutions/min	revs/sec	0.01667
revolutions/min/min	radians/sec/sec	1.745×10^{-3}
revolutions/min/min	revs/min/sec	0.01667
revolutions/min/min	revs/sec/sec	2.778×10^{-4}
revolutions/sec	degrees/sec	360.0
revolutions/sec	radians/sec	6.283
revolutions/sec	revs/min	60.0
revolutions/sec/sec	radians/sec/sec	6.283
revolutions/sec/sec	revs/min/min	3,600.0
revolutions/sec/sec	revs/min/sec	60.0
Rod	Chain (Gunters)	.25
Rod	Meters	5.029
Rods (Surveyors' meas.)	yards	5.5
rods	feet	16.5

To Convert	Into	Multiply by
	S	
Scruples	grains	20
seconds (angle)	degrees	2.778×10^{-4}
seconds (angle)	minutes	0.01667
seconds (angle)	quadrants	3.087×10^{-6}
seconds (angle)	radians	4.848×10^{-6}
Slug	Kilogram	14.59
Slug	Pounds	32.17
Sphere	Steradians	12.57
square centimeters	circular mils	1.973×10^5
square centimeters	sq feet	1.076×10^{-3}
square centimeters	sq inches	0.1550
square centimeters	sq meters	0.0001
square centimeters	sq miles	3.861×10^{-11}
square centimeters	sq millimeters	100.0
square centimeters	sq yards	1.196×10^{-4}
square feet	acres	2.296×10^{-5}
square feet	circular mils	1.833×10^8
square feet	sq cms	929.0
square feet	sq inches	144.0
square feet	sq meters	0.09290
square feet	sq miles	3.587×10^{-8}
square feet	sq millimeters	9.290×10^4
square feet	sq yards	0.1111
square inches	circular mils	1.273×10^6
square inches	sq cms	6.452
square inches	sq feet	6.944×10^{-3}
square inches	sq millimeters	645.2
square inches	sq mils	10^6
square inches	sq yards	7.716×10^{-4}
square kilometers	acres	247.1
square kilometers	sq cms	10^{10}
square kilometers	sq ft	10.76×10^6
square kilometers	sq inches	1.550×10^9
square kilometers	sq meters	10^6
square kilometers	sq miles	0.3861
square kilometers	sq yards	1.196×10^6
square meters	acres	2.471×10^{-4}
square meters	sq cms	10^4
square meters	sq feet	10.76
square meters	sq inches	1,550.
square meters	sq miles	3.861×10^{-7}
square meters	sq millimeters	10^6
square meters	sq yards	1.196

To Convert	Into	Multiply by
square miles	acres	640.0
square miles	sq feet	27.88×10^6
square miles	sq kms	2.590
square miles	sq meters	2.590×10^6
square miles	sq yards	3.098×10^6
square millimeters	circular mils	1,973.
square millimeters	sq cms	0.01
square millimeters	sq feet	1.076×10^{-5}
square millimeters	sq inches	1.550×10^{-3}
square mils	circular mils	1.273
square mils	sq cms	6.452×10^{-6}
square mils	sq inches	10^{-6}
square yards	acres	2.066×10^{-4}
square yards	sq cms	8,361.
square yards	sq feet	9.0
square yards	sq inches	1,296.
square yards	sq meters	0.8361
square yards	sq miles	3.228×10^{-7}
square yards	sq millimeters	8.361×10^5

T

To Convert	Into	Multiply by
temperature ($^\circ$ C) +273	absolute temperature ($^\circ$ K)	1.0
temperature ($^\circ$ C) +17.78	temperature ($^\circ$ F)	1.8
temperature ($^\circ$ F) +460	absolute temperature ($^\circ$ R)	1.0
temperature ($^\circ$ F) −32	temperature ($^\circ$ C)	5/9
tons (long)	kilograms	1,016.
tons (long)	pounds	2,240.
tons (long)	tons (short)	1.120
tons (metric)	kilograms	1,000.
tons (metric)	pounds	2,205
tons (short)	kilograms	907.1848
tons (short)	ounces	32,000.
tons (short)	ounces (troy)	29,166.66
tons (short)	pounds	2,000.
tons (short)	pounds (troy)	2,430.56
tons (short)	tons (long)	0.89287
tons (short)	tons (metric)	0.9078
tons (short)/sq ft	kgs/sq meter	9,765.
tons (short)/sq ft	pounds/sq in.	2,000.
tons of water/24 hrs	pounds of water/hr	83.333
tons of water/24 hrs	gallons/min	0.16643
tons of water/24 hrs	cu ft/hr	1.3349

CONVERSION FACTORS (*continued*)

To Convert	Into	Multiply by
	V	
Volt/inch	Volt/cm.	.39370
Volt (absolute)	Statvolts	.003336
	W	
watts	Btu/hr	3.4129
watts	Btu/min	0.05688
watts	ergs/sec	107.
watts	foot-lbs/min	44.27
watts	foot-lbs/sec	0.7378
watts	horsepower	1.341×10^{-3}
watts	horsepower (metric)	1.360×10^{-3}
watts	kg-calories/min	0.01433
watts	kilowatts	0.001
Watts (Abs.)	B.T.U. (mean)/min.	0.056884
Watts (Abs.)	joules/sec.	1
watt-hours	Btu	3.413
watt-hours	ergs	3.60×10^{10}
watt-hours	foot-pounds	2,656.
watt-hours	gram-calories	859.85
watt-hours	horsepower-hrs	1.341×10^{-3}
watt-hours	kilogram-calories	0.8605
watt-hours	kilogram-meters	367.2
watt-hours	kilowatt-hrs	0.001
Watt (International)	Watt (absolute)	1.0002
webers	maxwells	10^8
webers	kilolines	10^5
webers/sq in.	gausses	1.550×10^7
webers/sq in.	lines/sq in.	10^8
webers/sq in.	webers/sq cm	0.1550
webers/sq in.	webers/sq meter	1,550.
webers/sq meter	gausses	10^4
webers/sq meter	lines/sq in.	6.452×10^4
webers/sq meter	webers/sq cm	10^{-4}
webers/sq meter	webers/sq in.	6.452×10^{-4}
	Y	
yards	centimeters	91.44
yards	kilometers	9.144×10^{-4}
yards	meters	0.9144
yards	miles (naut.)	4.934×10^{-4}
yards	miles (stat.)	5.682×10^{-4}
yards	millimeters	914.4

439

References

1. Study by Dr. Arnold Shecter, reported in the *New York Times*, 18 December 87.

2. Vogg, H., and L. Stieglitz. "Thermal Behavior of PCDD in Fly Ash from Municipal Incinerators." Presented at the Fifth International Symposium on Chlorinated Dioxins and Related Compounds, Bayreuth, FRG, 16–19 September 1985.

3. Commoner, B. Testimony before the Hazardous Waste and Toxic Substances Subcommittee of the U.S. Senate, 6 August 1987.

4. Environment Canada. "The National Incinerator Testing and Evaluation Program: Two-Stage Combustion (Prince Edward Island)," September 1985.

5. Quirk, T.P. "Economic Aspects of Incineration vs. Incineration-drying." *Water Pollution Control Federation Journal*, November 1964, pp. 1355–1367.

6. Engdahl, R.B. "Solid Waste Processing," Report Sw-4c, prepared for the Bureau of Solid Waste Management by the staff of the Battelle Memorial Institute, 1969.

7. Committee on Refuse Disposal, American Public Works Association, *Municipal Refuse Disposal*, Public Administration Service, Chicago, 1966.

8. Taylor, F.B. "Poisonous Agents." In *Environmental Engineers' Handbook*, ed. B.G. Lipták, vol. 1. Radnor: Chilton, 1974, p. 435.

9. Smith, R.G. "Noxious Metals." In *Environmental Engineers' Handbook*, ed. B.G. Lipták, vol. 2. Radnor: Chilton, 1974, p. 196.

10. U.S. EPA, Office of Solid Waste and Emergency Response. "Municipal Waste Combustion Study: Report to Congress." EPA/530-SW-87-021a, June 1987.

11. Stanley, J.S., et al. "PCDDs and PCDFs in Human Adipose Tissue from the EPA FY82 NHATS Repository." Presented at the Fifth International Symposium on Chlorinated Dioxins and Related Com-

pounds, Bayreuth, FRG, 16–19 September 1985.

12. Peterson, I. "Toxic Ash and Costs." *New York Times,* Metropolitan News, 15 November 1987.

13. Santhanam, C.J. "Incinerators: Types, Features and Operation." In *Environmental Engineers' Handbook,* ed. B.G. Lipták, vol. 3. Radnor: Chilton, 1974.

14. Goder, R. "On Site Incineration of Wastes." In *Environmental Engineers' Handbook,* ed. B.G. Lipták, vol. 3. Radnor: Chilton, 1974.

15. Liu, D.H.F. "Solid Waste Characterization." In *Environmental Engineers' Handbook,* ed. B.G. Lipták, vol. 3. Radnor: Chilton, 1974.

16. Scher, J.A. "Solid Waste Characterization Techniques." *Chemical Engineering Progress* 67 (no. 3, 1971): 81–84.

17. Rabosky, J.G. "The Municipal Incineration Process." In *Environmental Engineers' Handbook,* ed. B.G. Lipták, vol. 3. Radnor: Chilton, 1974.

18. Velzy, C.O., and R.S. Hecklinger. "Incineration," sec. 7.4. In *Marks' Standard Handbook for Mechanical Engineers,* ed. T. Baumeister and E.A. Avallone, 9th ed. New York: McGraw-Hill, 1987.

19. Essenhigh, R.H. "Incinerators—The Incineration Process." In *Environmental Engineers' Handbook,* ed. B.G. Lipták, vol. 2. Radnor: Chilton, 1974.

20. Shah, I.S. "Scrubbers," secs. 5.12–5.21. In *Environmental Engineers' Handbook,* ed. B.G. Lipták, vol. 2. Radnor: Chilton, 1974.

21. Jorgensen, R. "Fans." In *Marks' Standard Handbook for Mechanical Engineers,* ed. T. Baumeister, 8th ed. New York: McGraw-Hill, pp. 14.53.

22. Lipták, B.G. *Optimization of Unit Operations.* Radnor: Chilton, 1987.

23. Lipták, B.G., ed. *Instrument Engineers' Handbook: Process Control.* Radnor: Chilton, 1985.

24. Babcock & Wilcox Company. *Steam,* 37th ed., 1955.

25. Lipták, B.G., ed. *Instrument Engineers' Handbook: Process Measurement.* Radnor: Chilton, 1982.

26. Sebastian, F.P. In *Environmental Engineers' Handbook,* ed. B.G. Lipták, vol. 3. Radnor: Chilton, 1974.

27. Bennett, R.L. In *Environmental Engineers' Handbook,* ed. B.G. Lipták, vol. 3. Radnor: Chilton, 1974.

28. Rabosky, J.G. In *Environmental Engineers' Handbook,* ed. B.G. Lipták, vol. 3. Radnor: Chilton, 1974.

29. Tuttle, K.L. "Combustion Generated Particulate Emissions." National Waste Processing Conference, Denver, 1986 (ASME, 1986).

30. Law, I.J. "Duluth Codisposal Facility Update." National Waste Processing Conference, Denver, 1986 (ASME, 1986).

31. Engdahl, R.B. "Energy Recovery from Raw Versus Refined Municipal Wastes." National Waste Processing Conference, Denver, 1986 (ASME, 1986).

32. Wilson, D.G., ed. *Handbook of Solid Waste Management.* Van Nostrand-Reinhold, 1977, p. 6.

33. Battelle Institute. "Refuse-Fired Energy Systems in Europe." Report SW771, November 1979, U.S. EPA Office of Water and Waste Management.

34. Bucholz, E., W. Jockel, and S. Temelli. "Change of Emissions of NO_x and CO after Optimization of the Refuse Incineration Installation at Wüppertal-Remscheid," Mull und Abfall, January 1985.

35. Rinaldi, G.M., et al. "An Evaluation of Emission Factors for Waste to Energy Systems." U.S. EPA, Report 600/7-80-135, July 1980.

36. Stettler, A. "Results of Investigations of the Five Gases of the Refuse Incineration Installation at Zurich-Josephstrasse Regarding Chlorinated Dioxins and Furans." Proceedings of The 44th Technical Waste Conference, University of Stuttgart, Germany, 1983.

37. Stelian, J., and H.L. Greene. "Operating Experience and Performance of Two Wet Ash Handling Systems." National Waste Processing Conference, Denver, 1986 (ASME, 1986).

38. Hasselriis, F. "Minimizing Trace Organic Emissions from Combustion of Municipal Solid Waste by the Use of Carbon Monoxide Monitors." National Waste Processing Conference, Denver, 1986 (ASME, 1986).

39. Larson, E.D., M.H. Ross, and R.H. Williams. "Beyond the Era of Materials." Scientific American, June 1986.

40. Czuczwa, J. and R. Hites. "Airborne Dioxins and Dibenzofurans: Sources and Fates." National Waste Processing Conference, Denver, 1986 (ASME, 1986).

41. World Health Organization Summary Report. "Working Group on Risks to Public Health of Dioxins from Incineration of Sewage Sludge and Municipal Solid Waste." Naples, March 1986.

42. Niessen, W.R. "The Role and Importance of Polychlorinated Dibenzo-p-Dioxins (CDDs) and Polychlorinated Dibenzo Furans (CDFs) from Resource Recovery Facilities." National Waste Processing Conference, Denver, 1986 (ASME, 1986).

43. Kemp, C.C. "Panel Session on Dioxin." National Waste Processing Conference, Denver, 1986 (ASME, 1986).

44. Weiand, H., et al. "Air Emissions Testing at the Würzburg, West Germany, Waste-to-Energy Facility." National Waste Processing Conference, Denver, 1986 (ASME, 1986).

45. Konheim, C.S. "Panel Session on Dioxin." 1986 National Waste Processing Conference, Denver, 1986 (ASME, 1986).

46. Domalski, E.S., A.E. Ledford, Jr., S.S. Bruce, and K.L. Churney. "The Chlorine Content of Municipal Solid Waste from Baltimore County, Maryland, and Brooklyn, New York." National Waste Processing Conference, Denver, 1986 (ASME, 1986).

47. Hay, D.J., A. Finkelstein, and R. Klicius. "The National Incinerator Testing and Evaluation Program: An Assessment of Pilot Scale Emission Control." National Waste Processing Conference, Denver, 1986 (ASME, 1986).

48. Finkelstein, A., D.J. Hay, and R. Klicius. "The National Incinerator Testing and Evaluation Program (NITEP)." National Waste Processing Conference, Denver, 1986 (ASME, 1986).

49. Ferretti, J.W.T. "Types of Air Pollution Control Equipment" and "Air Contaminant Threshold Limits." In Environmental Engineers' Handbook, ed. B.G. Lipták, vol. 2. Radnor: Chilton, 1974.

50. Santhanam, C.J. "Bag Filters and Centrifugal Collectors," secs. 5.5–5.11. In Environmental Engineers' Handbook, ed. B.G. Lipták, vol. 2. Radnor: Chilton, 1974.

51. Hatcher, W.J. Jr. "Electrostatic Precipitators," secs. 5.22–5.24. In Environmental Engineers' Handbook, ed. B.G. Lipták, vol. 2. Radnor: Chilton, 1974.

52. Rushton, J.D. "Boilers and Heaters," sec. 7.2. In Environmental Engineers' Handbook, ed. B.G. Lipták, vol. 2. Radnor: Chilton, 1974.

53. Feindler, K.S. "Long-Term Results of Operating Ta Luft Acid Gas Scrubbing Systems." National Waste Processing Conference, Denver, 1986 (ASME, 1986).

54. Sommers, E.J., et al. "Mass-Fired Energy Conversion Efficiency." National Waste Processing Conference, Denver, 1986 (ASME, 1986).

55. Schanche, G.W., and K.E. Griggs. "Features and Operating Experiences of Heat Recovery Incinerators." National Waste Processing Conference, Denver, 1986 (ASME, 1986).

56. Franco, H.A. "The Hamilton Swaru Retrofit." National Waste Processing Conference, Denver, 1986 (ASME, 1986).

57. Wheless, E., and M. Selna. "Commerce Refuse-to-Energy Facility: An Alternate to Landfilling." National Waste Processing Conference, Denver, 1986 (ASME, 1986).

58. Hurst, B.E., and C.M. White. "Thermal DeNO$_x$." 1986 National Waste Processing Conference, Denver, 1986 (ASME, 1986).

59. Cadle, R.D. "Sulfur Oxides." In *Environmental Engineers' Handbook,* ed. B.G. Lipták, vol. 2. Radnor: Chilton, 1974.

60. Singer, F.S. "Theories on the Global Consequences of Pollution." In *Environmental Engineers' Handbook,* ed. B.G. Lipták, vol. 2. Radnor: Chilton, 1974.

61. Canter, L.W. "Emission Sources." In *Environmental Engineers' Handbook,* ed. B.G. Lipták, vol. 2. Radnor: Chilton, 1974.

62. Hermann, J. "Experience over the Last Two Decades with Testing the Performance of Refuse-Fired Boilers Used as Calorimeters." National Waste Processing Conference, Denver, 1986 (ASME, 1986).

63. Matteson, M.J. "Sulfur Oxide Control." In *Environmental Engineers' Handbook,* ed. B.G. Lipták, vol. 2. Radnor: Chilton, 1974.

64. Carrera, P. "The Sorain-Cecchini System for Material Resource Recovery." National Waste Processing Conference, Denver, 1986 (ASME, 1986).

65. Churney, K.L., et al. "Assessing the Credibility of the Calorific Value of Municipal Solid Waste." National Waste Processing Conference, Denver, 1986 (ASME, 1986).

66. Rapier, P.M., and K.A. Roe. "Environmental Control," sec. 18. In *Marks' Standard Handbook for Mechanical Engineers,* ed. T. Baumeister and E.A. Avallone, 9th ed. McGraw-Hill, 1987.

67. Lipták, B.G., ed. *Instrument Engineers' Handbook: Process Measurement.* Radnor: Chilton, 1982, sec. 2.13, p. 115.

68. Matteson, M.J. In *Environmental Engineers' Handbook,* ed. B.G. Lipták, vol. 2. Radnor: Chilton, 1974, secs. 6.6, 6.7.

69. *Environmental Assessment of NH$_3$ Injection,* vol. 1. *Technical Results.* PB86-159852, February 1986.

70. Hepp, M.P., R.M. Nethercutt, and J.F. Wood. "Considerations in the Design of High Temperature and Pressure Refuse-Fired Power Boilers." National Waste Processing Conference, Denver, 1986 (ASME, 1986).

71. Delibert, J.C. "Steam Boilers," sec. 9.2. In *Marks' Standard Handbook for Mechanical Engineers,* ed. T. Baumeister and E.A. Avallone, 9th ed. New York: McGraw-Hill, 1987.

72. Goldman, M. "Use of Turbine Performance Curves in Predictions of Energy Production for a Cogeneration Resource Recovery Facility." National Waste Processing Conference, Denver, 1986 (ASME, 1986).

73. Daniel, P.L., J.L. Barna, and J.D. Blue. "Furnace-Wall Corrosion in Refuse-Fired Boilers." National Waste Processing Conference, Denver, 1986 (ASME, 1986).

74. Barr, T.D., K.C. O'Brien, and D.F. Moats. "Working Out the Kinks: An Update on Columbus' Refuse and Coal-Fired Municipal Electric Plant." National Waste Processing Conference, Denver, 1986 (ASME, 1986).

75. Frounfelker, R. "Heat Recovery Incineration for the City and Borough of Sitka, Alaska." National Waste Processing Conference, Denver, 1986 (ASME, 1986).

76. Stabenow, G. Comments given at the National Waste Processing Conference, Denver, 1986.

77. Sommerlad, R.E. "The Boiler-Calorimeter Method for Determining Capacity and Performance." National Waste Processing Conference, Denver, 1986 (ASME, 1986).

78. Hermann, J.D. "Experience over the Last Two Decades with Testing the Performance of Refuse-Fired Boilers Used as Calorimeters." National Waste Processing Conference, Denver, 1986 (ASME, 1986).

79. Baumann, E., and W. Greller. "Vergleichende Untersuchungen zur Bestimmung des Heizwertes von Mull." Mitt. der VGB 90 (June 1964).

80. Hecklinger, R.S. "Experience in Sampling Unprocessed Municipal Solid Waste." National Waste Processing Conference, Denver, 1986 (ASME, 1986).

81. Snell, J.R. "Size Reduction and Compaction Equipment." In Environmental Engineers' Handbook, ed. B.G. Lipták, vol. 3. Radnor: Chilton, 1974.

82. Santhanam, C.J. "Flotation Techniques." In Environmental Engineers' Handbook, ed. B.G. Lipták, vol. 3. Radnor: Chilton, 1974.

83. Paul, G. "Reclamation and Salvaging Processes," secs. 3.1–3.5, 3.8. In Environmental Engineers' Handbook, ed. B.G. Lipták, vol. 3. Radnor: Chilton, 1974.

84. Mack, W.A. "Plastic Waste Recycling." In Environmental Engineers' Handbook, ed. B.G. Lipták, vol. 3. Radnor: Chilton, 1974.

85. Lipták, B.G. "The Complete Solid Waste Reclamation Plant." In Environmental Engineers' Handbook, ed. B.G. Lipták, vol. 3. Radnor: Chilton, 1974.

86. Buchanan, R.D. "Interrelationships between the Various Forms of Pollution." In Environmental Engineers' Handbook, ed. B.G. Lipták, vol. 3. Radnor: Chilton, 1974.

87. Bergvall, G., and J. Hult. "Technology, Economics and Environmental Effects of Solid Waste Treatment." Final Report #3033, DRAV Project 85:11, Sweden, July 1985.

88. Glaub, J.C., and G.J. Trezek. "Processing and Economics of Mixed Paper Waste as an Energy Source." National Waste Processing Conference, Denver, 1986 (ASME, 1986).

89. Savage, G.M., and L.F. Diaz. "Key Issues Concerning Waste Processing Design." National Waste Processing Conference, Denver, 1986 (ASME, 1986).

90. Kenny, G., E.J. Sommer, and J.A. Kearley. "Operating Experience and Economics of a Simplified MSW Materials Recovery and Fuel-Enhancement Facility." National Waste Processing Conference, Denver, 1986 (ASME, 1986).

91. Hill, R.M. "Three Types of Low-Speed Shredder Designs." National Waste Processing Conference, Denver, 1986 (ASME, 1986).

92. Cotter, D.A. "San Francisco's Integrated Recycling Programs." National Waste Processing Conference, Denver, 1986 (ASME, 1986).

93. Nollet, A.R., D.S. Hedlund, and F. Hasselriis. "Test of Rotary Drum Air-Classifier at Albany, New York, Solid Waste Energy Recovery System." Na-

tional Waste Processing Conference, Denver, 1986 (ASME, 1986).

94. Cashin, F.J., and F. Carrera. "The Sorain-Cecchini System for Material Resource Recovery." National Waste Processing Conference, Denver, 1986 (ASME, 1986).

95. Kaminski, D. "Performance of the RDF Delivery and Boiler-Feed System at the Lawrence, Massachusetts, Facility." National Waste Processing Conference, Denver, 1986 (ASME, 1986).

96. Chesner, W.H., R.J. Collins, and T. Fiesinger. "The Characterization of Incinerator Residue in the City of New York." National Waste Processing Conference, Denver, 1986 (ASME, 1986).

97. Bergvall, G., and J. Hult. "Technology, Economics and Environmental Effects of Solid Waste Treatment." Report 3033, Publication 85:11. *Naturvardsverket,* July 1985.

98. Parker, F.L. "Fertilizer Manufacture." In *Environmental Engineers' Handbook,* ed. B.G. Lipták, vol. 3. Radnor: Chilton, 1974.

99. Snell, J.R. "Composting Processes." In *Environmental Engineers' Handbook,* ed. B.G. Lipták, vol. 3. Radnor: Chilton, 1974.

100. Sebastian, F.P. "Multiple Hearth Incineration." In *Environmental Engineers' Handbook,* ed. B.G. Lipták, vol. 3. Radnor: Chilton, 1974.

101. Rice, F. "Where Will We Put All That Garbage?" *Fortune,* 11 April 1988, p. 96.

102. Moats, D.F., J. Mathews, and K.C. O'Brien. "A Performance Update for the Columbus Project." National Waste Processing Conference, Philadelphia, 1988. New York: ASME, 1988.

103. Spencer, D.B., L.A. Sowa, and W.T. Sloop. "Akron Recycle Energy System: An RDF Success Story." National Waste Processing Conference, Philadelphia, 1988. New York: ASME, 1988.

104. Russell, S.H. "Solid Waste Preprocessing: The Return of an Alternative to Mass Burn," National Waste Processing Conference, Philadelphia, 1988. New York: ASME, 1988.

105. Sommerland, R.E., W.R. Seeker, A. Finkelstein, and J.D. Kilgroe. "Environmental Characterization of Refuse-Derived-Fuel Incinerator Technology." National Waste Processing Conference, Philadelphia, 1988. New York: ASME, 1988.

106. Getter, R.D. "Emission Estimates for Modern Resource Recovery Facilities." National Waste Processing Conference, Philadelphia, 1988. New York: ASME, 1988.

107. Licata, A.J., R.W. Herbert, and U. Kaiser. "Design Concepts to Minimize Superheater Corrosion in Municipal Waste Combustors." National Waste Processing Conference, Philadelphia, 1988. New York: ASME, 1988.

108. Smith, D.J. "Can Recycling Save Waste-to-Energy?" *Power Engineering,* September 1988, p. 21.

109. Personal visit with G. Dubbeldam, plant manager of Recycling Zoetermeer BV in Rotterdam, the Netherlands.

110. National Solid Waste Management Association. *New York Times,* 23 October 1988, p. F4.

111. New York State Department of Environmental Conservation, *New York Times,* 23 October 1988, p. F4.

112. Franklin Associates. *New York Times,* 23 October 1988, p. F4.

113. Echelberger, W.F. "Sanitary Landfills." In *Environmental Engineers' Handbook,* ed. B.G. Lipták, vol. 3. Radnor: Chilton, 1974.

114. Buchanan, R.D. "Sanitary Landfills." In *Environmental Engineers' Hand-*

book, ed. B.G. Lipták, vol. 3. Radnor: Chilton, 1974.

115. Snell, J.R. "Sanitary Landfills." In *Environmental Engineers' Handbook*, ed. B.G. Lipták, vol. 3. Radnor: Chilton, 1974.

116. Wilcomb, M.J., and H.L. Hickman. "Sanitary Landfill Design, Construction and Evaluation." Washington, D.C.: U.S. GPO.

117. "Guide to the RCRA Permitting Process," prepared by the EPA, Region 5, Chicago.

118. *New York Times*, 4 December 1988, p. E5.

119. *New York Times*, "Do Disposable Diapers Ever Go Away?" 10 December 1988, p. 33.

120. Snell, John R. "Ocean Dumping." In *Environmental Engineers' Handbook*, ed. B.G. Lipták, vol. 3. Radnor: Chilton, 1974.

121. Lund, M. "The Great Sea Sickness." *Signature*, September 1971, pp. 36–40.

122. Gunnerson, C.G., R.P. Brown, and D.D. Smith. "Marine Disposal of Solid Wastes." *Journal of the Sanitary Engineering Division*, December 1970, pp. 1387–1397.

123. Hinesly, T.D., and B. Sosewitz. "Digested Sludge Disposal on Crop Land," May 1960, Part 1, pp. 822–30.

124. *New York Times*, 22 May 1988, p. L31.

125. Gough, W.C. "Environmental Interrelationships." In *Environmental Engineers' Handbook*, ed. B.G. Lipták, vol. 3. Radnor: Chilton, 1974.

126. "Scientist Warns of Chemicals in Sewage Sludge." *New York Times*, 25 December 1988, p. 51.

127. "Everyone's Loss." *Globe and Mail*, Toronto, 23 April 1988, p. D3.

128. Lipták, B.G. Introduction to *Environmental Engineers' Handbook*. Radnor: Chilton, 1974.

129. "New York City Confronts Prospect of Recycling Law." *New York Times*, 23 February 1989.

130. "Study Finds More New York Trash, Despite Crisis." *New York Times*, 26 March 1989.

131. Lilienthal, N., and M. de Kadt. "Solid Waste Management: The Garbage Challenge for New York City," 1989.

132. "State Preparing to Act on Full Landfills." *New York Times*, 26 February 1989, sec 12.

133. Waste Management, Inc. 1986 Annual Report.

134. "EPA Proposes Giving States Say on Pesticides." *New York Times*, 29 February 1988, p. B9.

135. Snell, J.R. "The Open or Burning Dumps." In *Environmental Engineers' Handbook*, ed. B.G. Lipták, vol. 3. Radnor: Chilton, 1974.

136. U.S. EPA. "Superfund: Looking Back, Looking Ahead." Office of Public Affairs, OPA-87-007, April 1987.

137. Greenhouse, S. "U.N. Conference Supports Curbs on Exporting of Hazardous Waste." *New York Times*, 23 March 1989.

138. "Suits Mounting on Toxic Waste." *New York Times*, 15 February 1988.

139. Shabecoff, P. "Industry to Give Vast New Data On Toxic Perils." *New York Times*, 14 February 1988.

140. Smothers, R. "S. Carolina Challenges U.S. on Waste." *New York Times*, 9 March 1989.

141. Frost and Sullivan, Inc. "Hazardous Waste Resource Recovery in the U.S." Report #1618, 1988.

142. Hamilton, R.A. "Industries Lower Toxic Waste 15%" *New York Times*, 20 December 1987.

143. Sims, C. "New Methods Of Handling Toxic Waste." *New York Times*, 23 December 1988.

144. Hanley, R. "New Jersey Wrestles with Sludge Dilemma." *New York Times*, 26 April 1989, p. B1.

145. Henmi, M., K. Okazawa, and K. Sota. "Energy Saving in Sewage Sludge Incineration with Indirect Heat Drier." National Waste Processing Conference, Denver, 1986 (ASME, 1986).

146. Sebastian, F. "Incinerator Economics," sec. 5.26. In *Environmental Engineers' Handbook*, ed. B.G. Lipták, vol. 1. Radnor: Chilton, 1974.

147. Snell, J.R. "Economics and Past Experience," sec. 8.4. In *Environmental Engineers' Handbook*, ed. B.G. Lipták, vol. 1. Radnor: Chilton, 1974.

148. Snell, J.R. "Flash Drying or Incineration," sec. 8.6. In *Environmental Engineers' Handbook*, ed. B.G. Lipták, vol. 1. Radnor: Chilton, 1974.

149. Snell, J.R. "Air Drying of Digested Sludge," sec. 8.8. In *Environmental Engineers' Handbook*, ed. B.G. Lipták, vol. 1. Radnor: Chilton, 1974.

150. Lipták, P. "Restoration of Strip Mining Sites," sec. 8.11. In *Environmental Engineers' Handbook*, ed. B.G. Lipták, vol. 1. Radnor: Chilton, 1974.

151. U.S. Department of the Interior, Federal Water Pollution Control Administration. *A Study of Sludge Handling and Disposal*. Publ. WP-20-4, May, 1968.

152. Quirk, T.P. "Economic Aspects of Incineration vs. Incineration-Drying." *Water Pollution Control Federation Journal*, November 1964, pp. 1355–1367.

153. Gunnerson, C.G., R.P. Brown, and D.D. Smith. "Marine Disposal of Solid Wastes." *Journal of the Sanitary Engineering Division, ASCE*, December 1970, pp. 1387–1397.

154. *New York Times*, 2 August 1988, p. C4.

155. *Wall Street Journal*, 26 July 1988, p. 35.

156. Kleiman, D. "A Simple Domestic Chore Becomes a Cause." *New York Times*, 26 July 1989, p. C1.

157. Brown, P.L. "Ingenuity Is Creating a Home for Waste." *New York Times*, 27 July 1989, p. C1.

158. Hinds deCourcy, M. "In Sorting Trash, Householders Get Little Help from Industry." *New York Times*, 29 July 1989, p. 50.

159. Terry, E. "Success Imperils Japan's Waste Line." *Globe and Mail*, Toronto, 15 April 1989, p. D5.

160. Stevens, W.K. "When the Trash Leaves the Curb: New Methods Improve Recycling." *New York Times*, 2 May 1989, p. C1.

161. "The Markets Page." *Recycling Times*, 25 April 1989, p. 3.

162. Adams, V.A. "To Solve the Waste Disposal Problem." *New York Times*, 30 July 1989, p. C14.

163. Schmidt, W.E. "Most Throwaway Plastics Face Ban in Minneapolis." *New York Times*, 1 April 1989, p. L6.

164. Holusha, J. "Plastic Trash." *New York Times*, 3 May 1989, p. D9.

165. Holusha, J. "7 Polystyrene Makers Form Recycling Project." *New York Times*, 13 June 1989.

166. *New York Times*, 30 August 1988, p. C4.

167. *New York Times*, 25 September 1988, p. F13.

168. Hinds deCourcy, M. "Health Risks of Used Batteries Prompt Action on Disposal." *New York Times*, 25 February 1989.

169. Verbal communication with J.F. Ras, director of "Maltha," Rotterdam, the Netherlands.

170. *New York Times*, 28 September 1988.

171. Bennett, R.L. "Incineration in Metal Salvage and Industrial Waste," sec. 2.25. In *Environmental Engineers' Handbook*, ed. B.G. Lipták, vol. 3.

172. Salimando, J. "Contracts and Deals Abound." *Recycling Times*. 1 August 1989, p. 12.

173. Crawford, S.L. "A Skeptic Speaks." *Recycling Times*. 1 August 1989, p. 2.

174. Meade, K. "Battery of Legislation Deals with Batteries." *Recycling Times*, 1 August 1989, p. 4.

175. Misner, M. "Aluminum Can Prices Drop." *Recycling Times*, 1 August 1989.

176. *New York Times*, 4 July 1988, p. L7.

177. *New York Times*, 23 October 1988.

178. U.S. Bureau of the Census, *Statistical Abstract of the United States: 1989*, 109th ed. Washington, D.C., 1988.

179. Rathje, W.L. "Rubbish!" *Atlantic Monthly*, December 1989.

180. Paul, B. "For Recyclers, the News Is Looking Bad." *Wall Street Journal*, 31 August 1989.

181. Gold, A.R. "For Recycling Cops" *New York Times*, 23 November 1990.

182. Davis, A., and S. Kinsella. "Recycled Paper." *Garbage*, May/June 1990.

183. Holusha, J. "The Push to Recycle Newsprint." *New York Times*, 7 November 1990.

184. Breen, B. "Landfills Are #1." *Garbage*, September/October 1990.

185. Schmidt, W.E. "Trying to Solve the Side Effects of Converting Trash to Energy." *New York Times*, 27 May 1990.

186. Holmes, H. "Cities Fight for Right to Recycle." *Garbage*, Sept. 1990.

187. Sullivan, W. "Sewage Dump Site Teems with Sea Life." *New York Times*, 10 October 1990.

188. Donnelly, J. "Degradable Plastics." *Garbage*, May/June 1990.

189. Martinelli, J. "Packaging." *Garbage*, May/June 1990.

190. "Garbage Index." *Garbage*, September/October 1990. Sources: *Garbage in the Cities* (Texas A&M University Press); *New York Environment Book* (Island Press).

191. "Quotes and Facts." *EcoSource*, November/December 1990.

192. Egan, T. "Mob Looks at Recycling and Sees Green." *New York Times*, 28 November 1990.

193. Foderaro, L.W. "Trying to Hold Down the Garbage Pile." *New York Times*, 30 November 1990.

194. Holusha, J. "Making the Town Dump Sanitary." *New York Times*, 13 September 1989.

195. Schneider, K. "Radiation Danger Found in Oilfields across Nation." *New York Times*, Dec. 3, 1990.

196. Study avail. William Eager, BASF Corp. Parsippany, NJ 07054.

197. Simons, M. "U.S. Paper Recycling Hurts Europe's System," *New York Times*, 11 December 1990.

198. Passell, P. "Economists Start to Fret Again about Population." *New York Times*, 18 December 1990.

199. Holusha, J. "In Solid Waste, It's the Breakdown That Counts," *New York Times*, 31 March 1991.

200. Schneider, K., "As Recycling becomes a Growth Industry," *New York Times*, 20 January 1991.

201. Bloom, T., "Recycled Trash is Put to Use in Construction," *New York Times*, 1 May 1991.

202. Standards of Performance for Incinerators, Federal Register Volume 54, Number 243, pages 52,207, 52,209, 52,251; 20 December 1989.

203. Holusha, J., "Pricing Garbage to Reduce Waste," *New York Times*, 29 May 1991.

Index

Page numbers in *italics* refer to figures.
Page numbers followed by "t" refer to tables.

449